OXOCARBONS

OXOCARBONS

Edited by

ROBERT WEST

Department of Chemistry
University of Wisconsin
Madison, Wisconsin

1980

ACADEMIC PRESS

A Subsidiary of Harcourt Brace Jovanovich, Publishers

New York London Toronto Sydney San Francisco

ACADEMIC PRESS, INC.
111 Fifth Avenue, New York, New York 10003

United Kingdom Edition published by
ACADEMIC PRESS, INC. (LONDON) LTD.
24/28 Oval Road, London NW1 7DX

Library of Congress Cataloging in Publication Data

Main entry under title:

Chemistry of the oxocarbons.

Includes bibliographical references and index.
1. Carbonyl compounds. 2. Enols. I. West, Robert, Date.
QD305.A6C48 547'.03 80-515
ISBN 0-12-744580-3

PRINTED IN THE UNITED STATES OF AMERICA

80 81 82 83 9 8 7 6 5 4 3 2 1

Contents

List of Contributors ix
Preface xi

1 HISTORY OF THE OXOCARBONS
Robert West

I. Introduction 1
II. Early History: Croconate and Rhodizonate 2
III. Squaric Acid and the Aromatic Oxocarbons 3
IV. Deltic Acid 5
V. Diamagnetic Anisotropy and Aromaticity in Croconate Ion 6
VI. Oxidation and Reduction Products 8
VII. Theoretical Calculations and Electronic Spectra 10
VIII. Pseudo-Oxocarbons 12
References 13

2 THIOXOCARBON DIANIONS AND THEIR DERIVATIVES
Gunther H. R. Seitz

I. Introduction 15
II. Thioxocarbon Dianions 16
III. Mixed C_4 Oxo-Thioxocarbon Dianions 19
IV. Dithiosquaric Acid Diamides 22
V. Thiodeltic Acid Derivatives 36
VI. Mixed C_5 Oxo-Thioxocarbon Dianions 38
References 40

3 PHYSICAL CHEMISTRY OF AQUEOUS OXOCARBONS
Lowell M. Schwartz, Robert I. Gelb, and Daniel A. Laufer

I. Introduction 43
II. Survey of Oxocarbon Acid pK Determinations 44
III. Thermodynamic Data 46

IV. Structures of Aqueous Oxocarbons 49
V. A New Type of Aqueous Complex 56
References 57

4 NEW BOND-DELOCALIZED (DICYANOMETHYLIDENE)CROCONATE DERIVATIVES: "CROCONATE VIOLET" AND "CROCONATE BLUE"
Alexander J. Fatiadi
I. Introduction 59
II. Reaction of Croconates with Malononitrile 62
III. Electrical Conductivity of Some Bond-Delocalized Salts 73
IV. Summary 76
References 76

5 EXCITED STATES OF OXOCARBON DIANIONS
Josef Michl and Robert West
I. Introduction 79
II. The Perimeter Model for $(4N+2)$-Electron $[n]$Annulenes 80
III. The Perimeter Model for Oxocarbon Dianions 88
IV. Comparison with Experiment 95
References 99

6 THE MYCOTOXIN "MONILIFORMIN" AND RELATED SUBSTANCES
H.-D. Scharf and H. Frauenrath
I. Moniliformin 101
II. Syntheses and Properties of Semisquaric Acid and Its Derivatives 109
III. The Series of Semioxocarbons $CH(CO)_m{}^{(-)}M^{(+)}$ 114
IV. Naturally Occurring Phenylog Substances 116
References 117

7 RAMAN SPECTRA AND JAHN–TELLER EFFECTS OF OXOCARBON DIANIONS
Mitsuo Ito, Koji Kaya, and Machiko Takahashi
I. Introduction 121
II. Raman Intensity and Electronic States 122
III. Electronic Absorption Spectra of Oxocarbon Ions 127
IV. Raman Spectra of $C_nO_n{}^{2-}$ 128
V. Jahn–Teller Effect Suggested by Raman Intensities 133
VI. Jahn–Teller Deformation of Ions 135
VII. Excitation Profile and Jahn–Teller Effect 138
VIII. Conclusion 139
References 140

8 THE STRUCTURAL PHASE TRANSITION AND DIELECTRIC PROPERTIES OF SQUARIC ACID
Jens Feder

I. Introduction 141
II. Structural Determinations 143
III. Raman and Infrared Spectra 147
IV. NMR Spectra 149
V. Birefringence, Dielectric, and Elastic Properties 152
VI. Theory 160
VII. Conclusion 165
References 166

9 SYNTHESES OF HIGHLY OXIDIZED CYCLOBUTANES VIA [2 +2] CYCLOADDITION REACTIONS OF KETENES
Daniel Belluš

I. Introduction 169
II. Moniliformin 173
III. Alkylmoniliformins 175
IV. Arylmoniliformins 179
V. Squaric Acid 180
References 183

10 THE CHEMISTRY OF SQUARAINES
Arthur H. Schmidt

I. Introduction 185
II. Structure and Nomenclature 186
III. Identification of Squaraines 190
IV. Methods of Preparation 190
V. Tables of Squaraines 197
VI. Reactions of Squaraines 197
VII. Final Remarks 229
References 230

Index 233

List of Contributors

Numbers in parentheses indicate the pages on which the authors' contributions begin.

Daniel Belluš (169), Central Research Laboratories, Ciba-Geigy AG, CH-4002, Basel, Switzerland

Alexander J. Fatiadi (59), National Measurement Laboratory, National Bureau of Standards, Washington, D.C. 20234

Jens Feder (141), Institute of Physics, University of Oslo, Oslo 3, Norway

H. Frauenrath (101), Institut für Organische Chemie der RWTH Aachen, D-5100 Aachen, Germany

Robert I. Gelb (43), Department of Chemistry, University of Massachusetts, Boston, Massachusetts 02125

Mitsuo Ito (121), Department of Chemistry, Faculty of Science, Tohoku University, Sendai 980, Japan

Koji Kaya (121), Department of Chemistry, Faculty of Science, Tohoku University, Sendai 980, Japan

Daniel A. Laufer (43), Department of Chemistry, University of Massachusetts, Boston, Massachusetts 02125

Josef Michl (79), Department of Chemistry, University of Utah, Salt Lake City, Utah 84112

H.-D. Scharf (101), Institut für Organische Chemie der RWTH Aachen, D-5100 Aachen, Germany

Arthur H. Schmidt (185), Abteilung für Organische Chemie und Biochemie der Fachhochschule Fresenius, D-6200 Wiesbaden, Germany

Lowell M. Schwartz (43), Department of Chemistry, University of Massachusetts, Boston, Massachusetts 02125

Gunther H. R. Seitz (15), Institute for Pharmaceutical Chemistry, Phillipps University, Lahn, Germany

Machiko Takahashi (121), Department of Chemistry, Faculty of Science, Tohoku University, Sendai 980, Japan

Robert West (1, 79), Department of Chemistry, University of Wisconsin, Madison, Wisconsin 53706

Preface

Oxocarbons, first synthesized 155 years ago, were little known until the rebirth of the field following the discovery of squaric acid in 1959. Only since then have chemists generally come to appreciate the remarkable structural, chemical, and electronic properties of these polycarbonylated organic species.

This volume represents a collection of chapters by experts on diverse aspects of the science of oxocarbons. Subjects treated include the spectroscopy and chemical physics of oxocarbons, as well as their reaction chemistry. Included also are several chapters on the substituted derivatives of cyclic polycarbonyls now called "pseudooxocarbons." The colors used on the cover of this volume were chosen to match those of the first known oxocarbon anions. The yellow of the lettering matches the color of croconate ion, and the red of the cover itself is similar to that of rhodizonate ion.

It has been a pleasure to work with the authors of these chapters of this, the first book to be published on oxocarbon chemistry.

Robert West

1

History of the Oxocarbons

Robert West

I. Introduction 1
II. Early History: Croconate and Rhodizonate 2
III. Squaric Acid and the Aromatic Oxocarbons 3
IV. Deltic Acid 5
V. Diamagnetic Anisotropy and Aromaticity in Croconate Ion 6
VI. Oxidation and Reduction Products 8
A. Neutral Species 8
B. Anions and Anion-Radicals 8
VII. Theoretical Calculations and Electronic Spectra 10
VIII. Pseudo-Oxocarbons 12
References 13

I. Introduction

The term "oxocarbon," first suggested in 1963 [*1*], designates compounds in which all or nearly all of the carbon atoms are bonded to carbonyl or enolic oxygens or their hydrated or deprotonated equivalents. About 20 years ago the cyclic oxocarbon anions were recognized as members of an aromatic series, stabilized by electron delocalization of π electrons around the ring [*2*]. Known monocyclic oxocarbon anions include deltate (**1**), squarate (**2**), croconate (**3**), rhodizonate (**4**), and the tetraanion of tetrahydroxy-*p*-benzoquinone (**5**). This introductory chapter is a brief historical review of oxocarbon chemistry. Its purpose is to set into context the chapters dealing with particular aspects of oxocarbon chemistry which make up the rest of the volume. Two reviews on oxocarbons were published about 10 years ago, one emphasizing physical prop-

OXOCARBONS

ISBN 0-12-744580-3

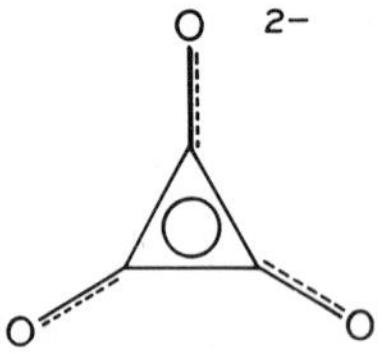

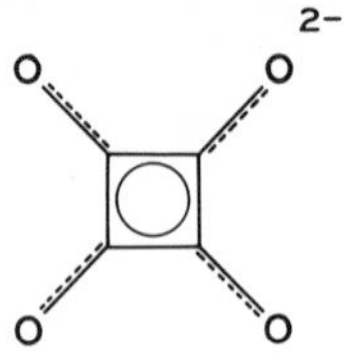

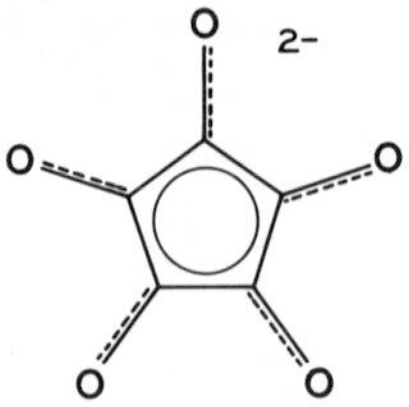

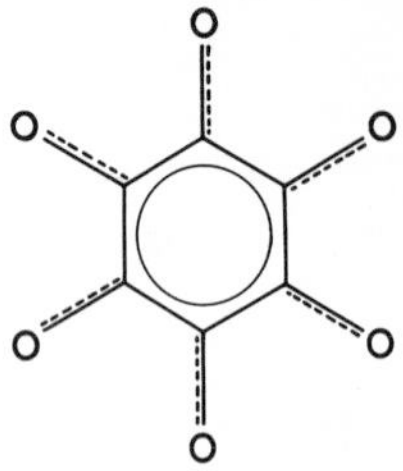

(1) (2) (3) (4, 2^-) (5, 4^-)

erties [3] and one dealing with synthesis [4]. A more recent review mainly covers developments within the past decade [5].

II. Early History: Croconate and Rhodizonate

Although the special properties of oxocarbons have come to light only recently, their history begins at the very dawn of chemistry and involves many of the great chemists of the early nineteenth century. The first workers to have in hand an oxocarbon were probably Berzelius, Wöhler, and Kindt, who in 1823 observed a black powdery residue formed in the reaction of potassium hydroxide with carbon in what was apparently a pioneer attempt at an industrial synthesis of potassium [6]. Dipotassium croconate and croconic acid (named from the Greek *krokos,* yellow, the color of the acid and its salts) were isolated from this residue by Gmelin in 1825 [7]. This early date is quite significant, for it is the same year in which benzene was obtained from illuminating gas oil by Michael Faraday.*

These very early experiments on oxocarbons are historically important in another respect. Croconic acid and rhodizonic acid are both known to be products of microbiological oxidation of myoinositol, a compound that is common in plants [8,9]. Gmelin's preparation of croconic acid was therefore a synthesis of a natural product—an "organic" compound in the original sense—from inorganic starting materials. It was perhaps the first such synthesis ever carried out, predating Wöhler's classic synthesis of urea by 3 years.

In 1834, Liebig discovered that a product similar to that from the potassium hydroxide–carbon reaction could be obtained by the reaction of potassium metal with carbon monoxide [10]. A few years later Heller isolated rhodizonic acid (named from the Greek *rhodizein,* rose red) from the products of this reaction and deduced that rhodizonate is the precursor of croconate ion [11].

The nature of these transformations has become clear mostly as a result of more recent studies. At low temperatures, alkali metals react with carbon

*Benzene has usually been assumed to be the first aromatic compound to be isolated, but it is now evident that it must share this distinction with croconate ion.

monoxide to give dialkali salts of dihydroxyacetylene $2M^+[OC{\equiv}CO]^{2-}$ [*12*]. This anion can also be regarded formally as an oxocarbon, representing the first member of the oxocarbon series $C_nO_n^{2-}$. However, the bond lengths in this dianion, 1.20 Å for C—O, suggest that there is little charge delocalization away from the oxygen [*13*].

Upon heating, the acetylene diolate salts cyclotrimerize to benzenehexol hexaanion $C_6O_6^{6-}$ (**6**). This is the first species that can be isolated when the reaction of metals with CO is carried out at higher temperatures, as in the early studies [*10,11*]. The anion **6**, which can be thought of as species in which the oxocarbon and conventional aromatic series intersect, undergoes oxidation rapidly in air to rhodizonate ion:

$$2\ M^+[OC{\equiv}CO]^{2-} \xrightarrow[\Delta]{} \underset{\mathbf{6}}{6\ M^+[C_6O_6]^{6-}} \xrightarrow{[O]_2} 2\ M^+[C_6O_6]^{2-}$$

However, in Gmelin's early experiment he isolated not rhodizonate but the five-membered ring croconate ion. The latter arises in a rather surprising oxidative ring-contraction reaction, the nature of which was not clarified until many years later. When air or oxygen is bubbled through an alkaline aqueous solution of rhodizonate, the following sequence of changes takes place [*13*]:

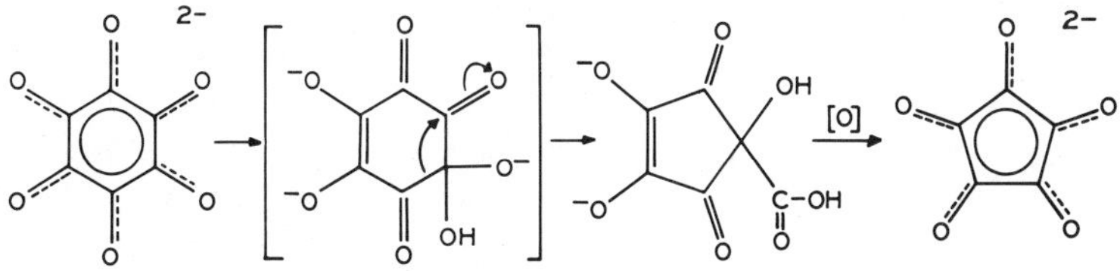

This ring contraction is an example of an α-oxo alcohol rearrangement, in turn related to the benzilic acid rearrangement. It provides such a convenient method for synthesis of croconates that, to this date, no other preparative method has been developed.

III. Squaric Acid and the Aromatic Oxocarbons

In the 125 years following the discovery of croconic and rhodizonic acids, these substances were investigated sporadically. Some improved synthesis of rhodizonate (and hence croconate) were developed, and the properties of rhodizonate as an oxidation–reduction indicator were exploited, but no detailed consideration was given to the unusual chemical bonding in these anions. The first suggestion of cyclic delocalization in an oxocarbon was apparently made in 1958 by Yamada [*14*], but, because this idea appeared only at the end of a lengthy article dealing with other matters, it was overlooked until later.

The modern era of oxocarbon chemistry can be dated from the famous accidental synthesis of squaric acid (**7**) by Cohen, Lacher, and Park a year later [*15*]:

(7)

Squaric acid was found to be a remarkably strong acid for an enol, having pK values close to those for sulfuric acid. (The unusual acidity of the cyclic oxocarbons is discussed in Chapter 3.) Cohen *et al.* interpreted the high acid strength as evidence that squarate dianion was greatly resonance-stabilized. The delocalized structure proposed for **2** led to the suggestion that the squarate ion was aromatic and that the oxocarbon anions might constitute a previously unrecognized aromatic series [*1*]. Evidence from vibrational spectroscopy soon confirmed that squarate and croconate are planar and symmetric (D_{4h} and D_{5h}, respectively), with delocalized π bonding around the ring [*16*]. X-ray crystallographic studies confirmed the delocalized, symmetric nature of **2, 3,** and **4.** The bond lengths for these species, shown in Table I [*17–19*] indicate that the carbon–carbon π bonding is substantial in the oxocarbon anions, although weaker than in benzene. All three of the oxocarbon salts in the table have layerlike structures in which the anions are close enough to raise the possibility of interionic charge-transfer interaction in the salts. Such interionic effects may be responsible for the unusual low-energy electronic absorption of some rhodizonate salts [*20*]. Semiconductivity probably resulting from intermolecular charge transfer in croconate derivatives is discussed in Chapter 4.

As mentioned above, there are several satisfactory syntheses for **3**, **4**, and **5**, but convenient routes to squaric acid have been lacking. A promising new method is the electrochemical reduction of carbon monoxide in polar aprotic solvents [*21*], which can be carried out to yield 20 gm of **7** in a single reaction (35% based on CO consumed). It is remarkable that **2** is formed in electrochemical reduction, considering that it has never been observed in chemical reductions of CO. The synthesis of squaric acid and related compounds is treated in Chapter 9, and in a recent review [*22*].

Table I. Bond Lengths (r) and π-Bond Orders (ρ) in Oxocarbon Anions

Species	r_{CC} (Å)	ρ_{CC}	r_{CO} (Å)	ρ_{CO}	Ref.
$K_2C_4O_4 \cdot H_2O$	1.469	0.38	1.259	0.78	*17*
$(NH_4)_2C_5O_5$	1.457	0.44	1.262	0.78	*18*
$Rb_2C_6O_6$	1.488	0.27	1.213	1.00	*19*

IV. Deltic Acid

The possible existence of numerous unknown members of the oxocarbon family was suggested in 1963 [*1*]. Of the species then unknown, the smallest possible oxocarbon, $C_3O_3^{2-}$ (**1**), seemed particularly significant. Attempts in various laboratories to obtain this species were unsuccessful until 1976. The key to the synthesis was formation of bis(trimethylsiloxy)cyclopropenone (**9**) by photochemical decarbonylation of bis(trimethylsiloxy)cyclobutenedione (**8**), which can easily be made from squaric acid. Unlike other fully substituted cyclopropenones, **9** undergoes solvolysis without undergoing ring opening to give either deltic acid (**10**) or the dilithium salt of deltate ion [*23*]:

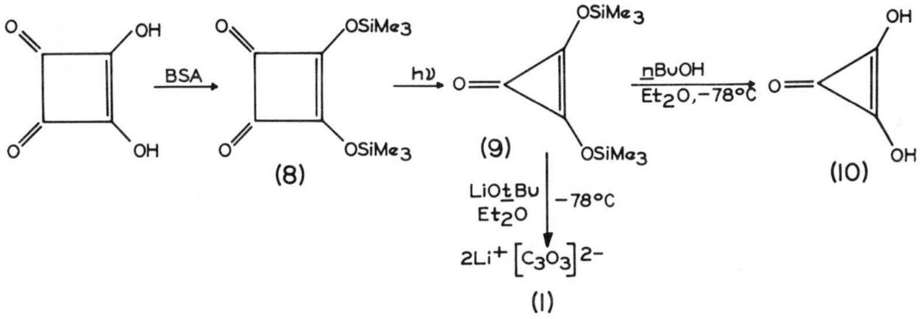

Deltic acid and the deltate salts are colorless compounds that slowly decompose in water. As predicted from theory [*24*], **1** and **10** are less stable than the other monocyclic oxocarbon acids and their anions, but nevertheless deltic acid survives heating to 100°C. Compound **10** is a somewhat weaker acid than the other monocyclic oxocarbons (Chapter 3). For the second dissociation, the greater p*K* value for deltic acid is reasonable in view of the larger charge repulsion in **1** than in **2–4**.

A second synthesis for deltic acid has been reported, starting with di-*tert*-butoxyacetylene [*25*]:

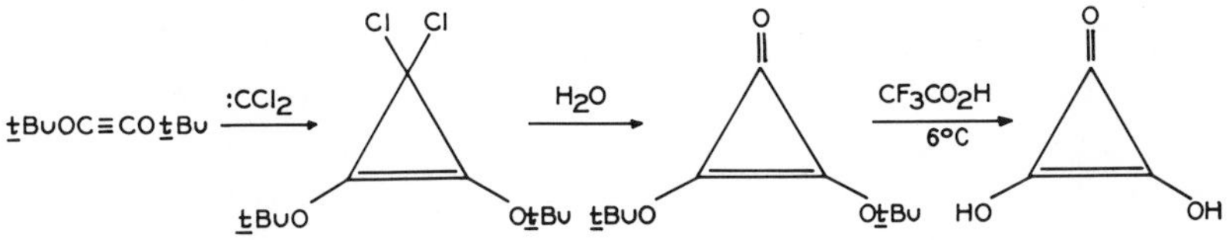

An X-ray crystal structure is not yet available for a deltate salt, but a complete virbrational analysis of the Raman and infrared spectra of deltate ion has been carried out [*26*]. Urey–Bradley force constants (Table II) allow a comparison of bonding in **1** and other aromatic species to be made. The carbon–carbon stretching force constants K_{CC} for squarate and croconate are between 3.5 and 4.0 mdyn/Å, distinctly greater than for C—C single bonds but lower than in ben-

Table II. Urey-Bradley Force Constants for Aromatic Species[a]

Species	K_{CC} (mdyn/Å)	K_{CO} (mdyn/Å)
$C_3O_3^{2-}$ (**1**)	5.57	5.50
$C_4O_4^{2-}$ (**2**)	3.95	5.60
$C_5O_5^{2-}$ (**3**)	3.50	6.72
C_6H_6	5.19	
$C_5H_5^-$	5.39	
$C_3H_3^+$	6.59	
$C_3Cl_3^+$	6.22	

[a] From Eggerding *et al.* [*26*].

zene. For deltate ion K_{CC} is about 5 mdyn/Å, nearly as large as that for benzene (~5.2). Deltate seems, therefore, to resemble the other three-membered ring aromatic species, $C_3Cl_3^+$, $C_3H_3^+$, and $C_3Ph_3^+$, which also have exceptionally short and strong C—C bonds.

V. Diamagnetic Anisotropy and Aromaticity in Croconate Ion

If the oxocarbon anions are in fact aromatic species, they should be expected to sustain a diamagnetic ring current. However, this is difficult to demonstrate by the usual NMR techniques, since compounds **1–4** contain no protons, and the ^{13}C-NMR chemical shifts are dominated by shielding effects due to the negative charges so that any small contributions from a ring current cannot be easily detected [*27*].

Some years ago the diamagnetic anisotropy of diammonium croconate $(NH_4)_2C_5O_5$ was investigated in our laboratories. This salt was chosen because its crystal structure had been determined earlier [*18*]. The method used was the "flip-angle" method of Krishnan, in which a large single crystal is oriented in a magnetic field and the force required to reorient the crystal to another magnetic axis is measured [*28*]. This direct method is now seldom used, but it was extensively employed in the 1920's and 1930's to obtain all of the early data that led to the Pauling and London theories of diamagnetic ring current.*

Although we were able to determine the value of the diamagnetic anisotropy ($\Delta\chi$) for **3**, at that time it was impossible to separate the total anisotropy into that part resulting from local contributions ($\Delta\chi_{loc}$) and any part due to a diamagnetic ring current ($\Delta\chi_{arom}$). However, more recent studies by Benson, Norris, Flygare,

*This method has two disadvantages: Rather large single crystals are required, and the crystal structure of the solid must be known. However, now that it is a relatively simple matter to determine crystal structures, the Krishnan method may again come into favor.

and Beak have led to a table of local atom and group anisotropies for planar compounds containing carbon, hydrogen, and oxygen [29]. The data in their table enable one to make at least an approximate calculation of the ring current anisotropy for croconate ion.

Our experiment established the diamagnetic anisotropy of croconate ion to be -51.5 ± 3 cgs units [30] (the ammonium ions have T_d symmetry, and hence $\Delta\chi = 0$).* We can now assume a localized structure for croconate. The logical one is **11**, containing five sp² carbon atoms, three carbonyl oxygens, and two alkoxide

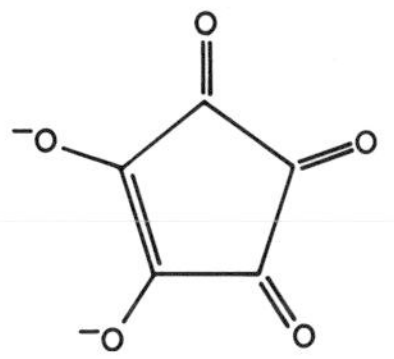

(15)

oxygens. The tables of Benson *et al.* [29] give local $\Delta\chi$ values of -4.4 for sp² carbon, -6.5 for carbonyl oxygen, and $+2.0$ for ether-type oxygen. The value for the alkoxide oxygens is somewhat uncertain because there are apparently no data on magnetic anisotropies of alkoxides. However, we believe that the value of $+2.0$ is reasonable. For C—O⁻, $\Delta\chi_{loc}$ should in any event be positive, as are the $\Delta\chi_{loc}$ values for C—Cl and for the isoelectronic C—F. Comparison of available data for halogen compounds and analogous hydrocarbons suggests $\Delta\chi_{loc} \sim +1$ for C—Cl and $+1.5$–3.0 for C—F [31].

Taking the Benson value of $+2.0$ for the alkoxide oxygens, we calculate for **3**, $\Delta\chi_{loc} = -37.5$. Subtracting from the measured value of -51.5 for **3** gives -14 as the probable value for $\Delta\chi_{arom}$, the diamagnetic anisotropy due to the ring current. The value of $\Delta\chi_{arom}$ for **3** is compared with that for other cyclic molecules in Table III. The ring current anisotropy in croconate appears to be a little less than that for furan, and about half that of benzene. In contrast, the nonaromatic molecules tropone and α-pyrone show nearly zero values of $\Delta\chi_{arom}$ [32]. The result, therefore, confirms other evidence for aromatic behavior in this oxocarbon anion.

*The torsion constant of the quartz fibers, 1 m in length, was calculated from measurements on crystals of 1,3,5-triphenylbenzene. As a further calibration the magnetic anisotropy of hexachlorobenzene was redetermined, with results in good agreement with those in the literature [30a]. In crystalline diammonium croconate the $C_5O_5^{2-}$ ions are perpendicular to the c (needle) axis. Typical azimuths at the flip angle, with c parallel and b perpendicular to the magnetic field were 146°, 147°, 146°, and 149°; similarly, for $c_{\parallel}$ and $a_{\perp}$, azimuths were 150°, 152°, 150°, and 150°. The anisotropy between the a and b axes was too small to be determined. The diamagnetic anisotropy $\Delta\chi$ $(c - b = c - a)$ was calculated as -52.5 ± 3 cgs units from the usual formula, $\Delta\chi = Mk(\alpha_c - \pi/4)/mH^2$ [28].

Table III. Diamagnetic Anisotropy and Ring Current Anisotropy in Planar Species

Species	$\Delta\chi_{exp}$	$\Delta\chi_{loc}$	$\Delta\chi_{arom}$
Benzene	−59.7	−26	−33
Furan	−38.7	−16	−23
Croconate	−52.5	−37	−15
Tropone	−36.0	−37	0
2-Pyrone	−24.8	−26	0
Cyclopenten-3-one	−16.8	−17	0

VI. Oxidation and Reduction Products

A. NEUTRAL SPECIES

The oxidation of croconic and rhodizonic acids leads to compounds that have long been known as leuconic acid ($C_5H_{10}O_{10}$) and triquinoyl octahydrate ($C_6H_{16}O_{14}$), respectively [*4*]. These substances are obtained by treating the parent acids or their alkali salts with oxidants, such as free halogens or nitric acid. The analogous four-membered ring compound $C_4(OH)_8$ was synthesized from squaric acid with bromine or nitric acid in 1963 [*33*]. All of the oxidation products can be reduced back to the parent oxocarbon acids with SO_2.

The structure of $C_4(OH)_8$ has been shown both by virbrational spectroscopy [*33*] and X-ray crystallography [*34*] to be octahydroxycyclobutane. The infrared spectra of octahydroxycyclobutane, leuconic acid, and triquinoyl are all similar; none of the compounds shows C=O absorption. It is highly probable that all three compounds have the fully hydroxylated structures shown in Scheme 1.

The dehydrated molecules corresponding to the perhydroxycycloalkanes shown in the scheme would be of great interest. These molecules, which would be oligomers of carbon monoxide, $(CO)_n$, are almost surely strongly endothermic, and all attempts to prepare them have been unsuccessful.

Reduction of the oxocarbons is well understood only for the six-membered ring compounds. Rhodizonate ion or rhodizonic acid can be reduced with SO_2 or other mild reagents, first to tetrahydroxy-*p*-benzoquinone and later to hexahydroxybenzene. The reduction takes place in successive two-electron steps and is essentially reversible.

B. ANIONS AND ANION-RADICALS

In addition to the oxocarbon anions themselves (**1**–**5**), the species $C_6O_6^{6-}$ and $C_6O_6^{4-}$ [*35*] are well established. Formation of anion-monoradicals and anion-triradicals by oxocarbons was predicted some years ago, but, because these

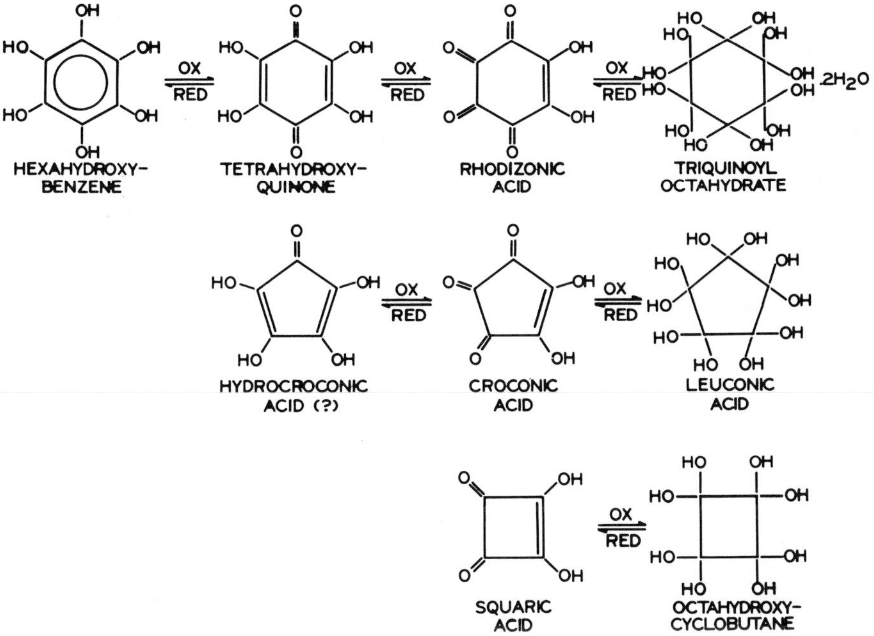

Scheme 1

species are unstable in water, they could not be investigated until salts of oxocarbons with the $Ph_3PNPPh_3^+$ cation were prepared and found to be soluble in aprotic solvents [*36*]. Electrolytic oxidation of CH_2Cl_2 solutions of these salts of **2**, **3**, and **4** led to ESR signals assigned to the anion-monoradicals (Table IV). The g values increase with increasing ring size, indicating (in agreement with other evidence and with theoretical calculations) that spin density is shifting gradually toward the oxygen atoms, which have larger spin–orbit coupling than carbon. Only for $C_5O_5^{\overline{\cdot}}$ were ^{13}C side bands observed, but the measured splitting constant of 4.1 G for this species is of special interest. From known Q values for π radicals [*37*], this splitting indicates a spin density $\Sigma\rho_c$ of 0.08 on the carbon

Table IV. ESR Spectra for Oxocarbon Anion-Radicals[a]

Species	g	$a_{^{13}C}$ (G)
$C_4O_4^{\overline{\cdot}}$	2.00584	
$C_5O_5^{\overline{\cdot}}$	2.00624	4.10
$C_6O_6^{\overline{\cdot}}$	2.00652	
$C_6O_6^{3\overline{\cdot}}$	2.00457	

[a] From Patton and West [*36*].

atoms, and hence $\Sigma\rho_o = 0.92$ on the oxygens. This is in excellent agreement with results from Hückel–McLachlan MO calculations, which predict $\Sigma\rho_c = 0.10$, $\Sigma\rho_o = 0.90$ [*36*].

A radical-trianion was obtained upon electrolytic reduction only for the six-membered ring. This finding is consistent with the fact that the tetraanion (**5**) is known for the six-membered ring but not for the smaller ones. The $C_6O_6^{3-\cdot}$ radical has a somewhat lower *g* value than the radical-monoanions, suggesting that in this species the unpaired electron is associated more with the carbon atoms and less with the oxygen.

The following scheme shows the possible anions and anion-radicals for the monocyclic oxocarbons. Known species appear in boldface type; the others are possible but have not yet been established:

$$
\begin{array}{ccccccccccc}
\mathbf{C_6O_6^{6-}} & \rightleftarrows & C_5O_6^{5-\cdot} & \rightleftarrows & \mathbf{C_6O_6^{4-}} & \rightleftarrows & \mathbf{C_6O_6^{3-\cdot}} & \rightleftarrows & \mathbf{C_6O_6^{2-}} & \rightleftarrows & \mathbf{C_6O_6^{1-\cdot}} \\
 & & C_5O_5^{5-\cdot} & \rightleftarrows & C_5O_5^{4-} & \rightleftarrows & C_5O_5^{3-\cdot} & \rightleftarrows & \mathbf{C_5O_5^{2-}} & \rightleftarrows & \mathbf{C_5O_5^{1-\cdot}} \\
 & & & & C_4O_4^{4-} & \rightleftarrows & C_4O_4^{3-\cdot} & \rightleftarrows & \mathbf{C_4O_4^{2-}} & \rightleftarrows & \mathbf{C_4O_4^{1-\cdot}} \\
 & & & & & & C_3O_3^{3-\cdot} & \rightleftarrows & \mathbf{C_3O_3^{2-}} & \rightleftarrows & C_3O_3^{1-\cdot}
\end{array}
$$

VII. Theoretical Calculations and Electronic Spectra

Because of their high degree of symmetry, the oxocarbon anions lend themselves especially well to theoretical studies. The π-electron energy levels for oxocarbons, from early Hückel MO calculations, are shown in Fig. 1. The patterns of the π orbitals can also be predicted qualitatively from symmetry considerations, as explained in Chapter 5. As shown in the figure, each of the oxocarbon dianions possesses a nondegenerate HOMO and a doubly degenerate LUMO. This arrangement is similar to that in two-electron annulenes such as cyclopropenium ion or cyclobutadiene dication.*

The earliest calculations on oxocarbons were of the simple Hückel type, but more recently theoretical studies have been carried out using more advanced methods, including extended Hückel [*38*], LCAO–CI [*39*], semiempirical SCF–CI [*40*], CNDO and CNDO–CI [*41*], MINDO/2 and CNDO–CI [*42*], and unrestricted Hartree–Fock [*43*]. Several of the more advanced techniques give qualitatively good agreement with observed energies for the first π–π^* transition. Of these papers the one by Leibovici [*42*] is particularly useful because it deals with the entire series of oxocarbons and permits comparisons to be made between theoretical results and experimentally observed quantities. Table V gives results of the MINDO/2 calculations by Leibovici. The calculated heats of formation, normalized for ring size, suggest equivalent stabilization for **2** and **3**,

*From this orbital arrangement the anion $C_6O_6^{4-}$ (**5**) is predicted to be a triplet-state species. Magnetic measurements, however, indicate that **5** is diamagnetic [*3*]; the degeneracy of the partly occupied orbitals is probably removed by Jahn–Teller distortion.

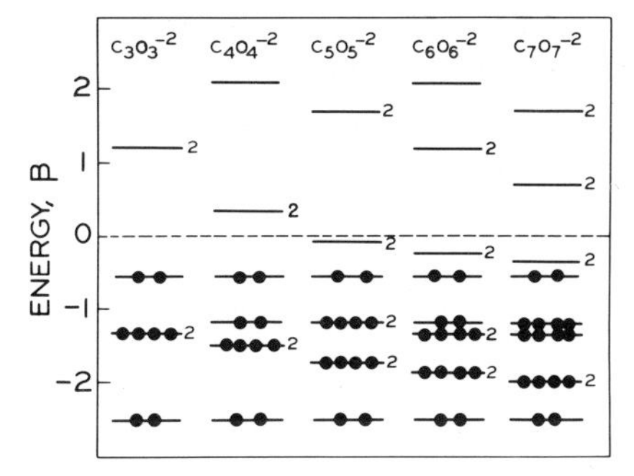

Fig. 1. π-Electron energy levels for oxocarbon dianions, from Hückel LCAO–MO calculations.

slightly less for **4**, and much less for **1**, in agreement with the observed stabilities of these anions. The calculated bond lengths accurately portray the observed trends, which are that the C—C bond length increases and C—O decreases with increasing ring size (see Table I). Unusually short C—C and long C—O bonds are predicted for **1** due to ring strain effects.

Calculations of electronic distribution in the anions indicate that there is a very strong charge separation between oxygen and carbons, which decreases with increasing ring size. The double-bond character is predicted from the π-bond orders (Table V) and associated Wiberg bond indices to increase for C—O and decrease for C—C bonds as the ring size increases. This corresponds very well with the trends in force constants derived from vibrational studies for **1**, **2**, and **3** [*16,26*] (see Table II). Identical Wiberg bond indices are calculated for C—C and C—O in deltate ion, in good agreement with the nearly identical values of the Urey–Bradley force constants. Even the strong transannular repulsion interaction between carbon atoms in 1,3 positions in squarate, evident from the vibrational studies, is consistent with the high values of Wiberg populations calculated for these nonbonded carbons [*42*].

For a quantitative treatment of the electronic spectra of oxocarbons, calculations, including configuration interaction, are necessary and, as several workers

Table V. Results of MINDO/2 Molecular Orbital Calculations for Oxocarbon Anions $C_nO_n^{2-a}$

n	d_{CC} (Å)	d_{CO} (Å)	p^{π}_{CC}	p^{π}_{CO}	$-\Delta H_f/n$ (kcal/mole)
3	1.396	1.262	0.570	0.470	26.3
4	1.440	1.249	0.427	0.607	40.3
5	1.458	1.243	0.369	0.687	41.2
6	1.471	1.240	0.340	0.736	37.6

^{a} From Leibovici [*42*].

have found, doubly excited as well as singly excited states must be included in the treatment. The VESCF–CI calculations by Sakamoto and I'Haya [*40*] correspond fairly well to the observed frequencies for the first π–π^* transition. The electronic spectra of oxocarbons have been carefully studied and are the subject of Chapters 5 and 7.

VIII. Pseudo-Oxocarbons

In recent years a number of oxocarbon analogs have been prepared in which one or more of the oxygen atoms are replaced by other groups, such as S, Se, NR, or $C(CN)_2$. Because of their resemblance to the oxocarbons, such compounds are sometimes called pseudo-oxocarbons. The remarkable chemistry of the sulfur compounds (thioxocarbons) and that of the dicyanomethylidene derivatives are the subjects of Chapters 2 and 4, respectively. Since they are not covered elsewhere in this volume, some mention should be made of the interesting tetranitrogen derivatives of squaric acid, studied by Hünig and his students [*44*]. From the reaction of squaric acid with *o*-diaminobenzene they obtained the 1,3-diamide **12,** which can be condensed to form the green dication **13:**

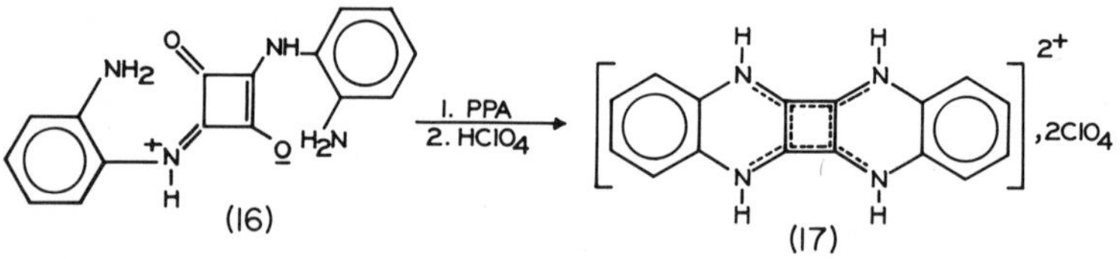

As shown in Scheme 2, **13** can be converted to several other interesting species, including the diamidines **14**, the radialene derivative **15**, the dianion **16**,

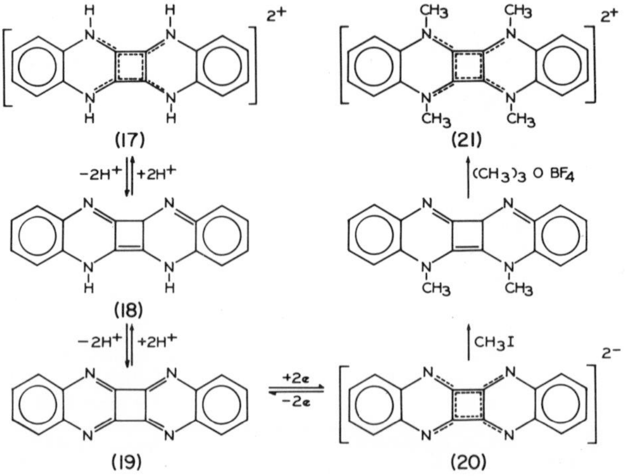

Scheme 2

and the tetramethyl dication **17**. These species show an elaborate oxidation–reduction chemistry; the dication **17** can be reduced to a radical cation, and **15** undergoes reversible oxidation–reduction to **16** through an intermediate electron-delocalized radical-anion.

More complex derivatives of the oxocarbons have also been prepared. Those of squaric acid are especially significant because of their unusual electronic properties. The chemistry of these cyanine-like derivatives, the "squaraines," is discussed in Chapter 10. Special mention should be made of the photoconductivity of compounds such as **18.** Photovoltaic cells constructed of **18** with appropriate contact materials show the highest efficiencies for conversion of sunlight to electrical energy ever recorded for an organic substance [*45*].

O^- HO Me₂N OH O $\overset{+}{N}Me_2$

(22)

References

1. R. West and D. L. Powell, *J. Am. Chem. Soc.* **85,** 2577 (1963).
2. R. West, H.-Y. Niu, D. L. Powell, and M. V. Evans, *J. Am. Chem. Soc.* **82,** 6204 (1960).
3. R. West and J. Niu, *in* "Non-Benzenoid Aromatics" (J. Snyder, ed.), Vol. 1, p. 311. Academic Press, New York, 1969.
4. R. West and J. Niu, *in* "The Chemistry of the Carbonyl Group" (J. Zabicky, ed.), Vol. 2, p. 241. Wiley (Interscience), New York, 1970.
5. R. West, *Isr. J. Chem.* in press (1980).
6. C. Brunner, *Schweigger's J.* **38,** 517 (1823).
7. L. Gmelin, *Ann. Phys. (Leipzig)* [2] **4,** 1 (1825).
8. A. J. Fatiadi, H. S. Isbell, and W. F. Sager, *J. Res. Natl. Bur. Stand., Sect. A* **67,** 153 (1963).
9. A. J. Kluyver, T. Hof, and A. G. J. Boezaardt, *Enzymologia* **7,** 257 (1939); *Chem. Abstr.* **34,** 6322 (1940).
10. J. Liebig, *Ann. Chem.* **11,** 182 (1834).
11. J. F. Heller, *Justus Liebig's Ann. Chem.* **24,** 1 (1837).
12. E. Weiss and W. Büchner, *Helv. Chim. Acta* **46,** 1121 (1963); *Z. Anorg. Allg. Chem.* **330,** 251 (1964); *Chem. Ber.* **98,** 126 (1965).
13. R. Nietzki, *Ber. Dtsch. Chem. Ges.* **20,** 1617, 2114 (1887).
14. K. Yamada, M. Mizuno, and Y. Hirata, *Bull. Chem. Soc. Jpn.* **31,** 543 (1958).
15. S. Cohen, J. R. Lacher, and J. D. Park, *J. Am. Chem. Soc.* **81,** 3480 (1959).
16. M. Ito and R. West, *J. Am. Chem. Soc.* **85,** 2580 (1963).
17. W. M. McIntyre and M. S. Werkema, *J. Chem. Phys.* **42,** 3563 (1964).
18. N. C. Baenziger and J. J. Hegenbarth, *J. Am. Chem. Soc.* **86,** 3250 (1964).
19. M. A. Neuman, *Diss. Abstr.* **26,** 6394 (1966).
20. J. Aihara, *Bull. Chem. Soc. Jpn.* **47,** 2899 (1974).
21. G. Silvestri, G. Salvatore, G. Filardo, M. Guazzini, and E. Raffaele, *Gazz. Chim. Ital.* **182,** 818 (1972); G. Silvestri, S. Gambino, G. Filardo, G. Spadaro, and L. Palmisano, *Electrochim. Acta* **23,** 413 (1978).

22. A. H. Schmidt and W. Reid, *Synthesis,* p. 869 (1978).
23. D. Eggerding and R. West, *J. Am. Chem. Soc.* **98,** 3641 (1976).
24. C. Leibovici, *J. Mol. Struct.* **13,** 185 (1972).
25. M. A. Percias and F. Serratoso, *Tetrahedron Lett.,* pp. 4430, 4437 (1977).
26. D. Eggerding, R. West, J. Perkins, D. Handy, and E. C. Tuazon, *J. Am. Chem. Soc.* **101,** 1710 (1979).
27. W. Staedli, R. Hollenstein, and W. von Phillipsborn, *Helv. Chim. Acta* **60,** 948 (1977).
28. K. Lonsdale and K. S. Krishnan, *Proc. R. Soc. London, Ser. A* **156,** 597 (1936).
29. R. C. Benson, C. L. Norris, W. H. Flygare, and P. Beak, *J. Am. Chem. Soc.* **95,** 5392 (1973); C. L. Norris, R. C. Benson, P. Beak, and W. H. Flygare, *ibid.,* p. 2766.
30. R. Ma, M.S. Research, University of Wisconsin, Madison, 1963.

30a. K. S. Krishnan and S. Banerjee, *Proc. R. Soc. London, Ser. A* **234,** 265 (1935).

31. A. A. Bothner-By and J. A. Pople, *Annu. Rev. Phys. Chem.* **16,** 43 (1965).
32. W. H. Flygare and R. C. Benson, *J. Mol. Phys.* **20,** 225 (1970).
33. R. West, H. Y. Niu, and M. Ito, *J. Am. Chem. Soc.* **85,** 2584 (1963).
34. C. M. Bock, *J. Am. Chem. Soc.* **90,** 2748 (1968).
35. R. West and H. Y. Niu, *J. Am. Chem. Soc.* **84,** 1324 (1962).
36. E. Patton and R. West, *J. Phys. Chem.* **77,** 2652 (1973).
37. M. Broze and Z. Luz, *J. Chem. Phys.* **51,** 749 (1969).
38. K. Sakamoto and Y. I'Haya, *Bull. Chem. Soc. Jpn.* **44,** 1201 (1971).
39. J. C. Bert, *J. Chim. Phys. Phys.-Chim. Biol.* **67,** 586 (1970).
40. K. Sakamoto and Y. I'Haya, *J. Am. Chem. Soc.* **92,** 2636 (1970); *Theor. Chim. Acta* **13,** 220 (1969).
41. P. Cremaschi and A. Gamba, *J. Mol. Struct.* **13,** 241 (1972).
42. C. Leibovici, *J. Mol. Struct.* **13,** 185 (1972).
43. D. M. Phillips and J. C. Shug, *J. Chem. Phys.,* **60,** 1597 (1974).
44. S. Hünig and U. Pütter, *Angew. Chem., Int. Ed. Engl.* **11,** 431, 433 (1972); **12,** 149 (1973).
45. V. Y. Merritt and H. J. Hovel, *Appl. Phys. Lett.* **29,** 414 (1978).

Thioxocarbon Dianions and Their Derivatives

Gunther H. R. Seitz

I. Introduction 15
II. Thioxocarbon Dianions 16
A. Synthesis and Reactions of the C_4 Thioxocarbon Dianion . . . 16
B. Spectroscopic Data 18
III. Mixed C_4 Oxo–Thioxocarbon Dianions 19
A. Syntheses and Reactions of Monothiosquarate and 1,2-Dithiosquarate Dianions 19
B. Reaction of Thiosquarate Esters with Amines 21
C. Syntheses and Reactions of 1,3-Dithio- and 1,2,3-Trithiosquarate Dianions 22
IV. Dithiosquaric Acid Diamides 22
A. Syntheses and Properties of 1,2-Dithiosquaric Acid Diamides . . 23
B. Syntheses and Properties of "1,3-Dithiosquaric Acid Diamides" 25
C. Reactions with Nucleophiles 28
D. Reactions with Electrophiles 33
V. Thiodeltic Acid Derivatives 36
VI. Mixed C_5 Oxo–Thioxocarbon Dianions 38
A. Synthesis and Properties 38
B. Chemical Behavior 39
References 40

I. Introduction

In the 20 years since the oxocarbon anions $C_nO_n^{2-}$ were recognized as members of a new class of stabilized carbocyclic nonbenzenoid aromatic compounds [*1*], knowledge of their chemical and physical properties has been developing rapidly. Their unique electronic structures, high degree of symmetry, and esthet-

OXOCARBONS

ISBN 0-12-744580-3

ically beautiful geometry generated a fresh impetus to study the effect of replacing the oxygens by various other functional groups, such as nitrogen [*2*], sulfur [*3–7*], selenium [*8*], phosphorus [*9*], and the dicyanomethylene group [*10–14*]. Our interest has been focused on studies pertinent to the aromatic oxocarbon dianions in which the original carbonyl oxygen atoms in $C_nO_n^{2-}$ are either partially or completely replaced by sulfur. In the meantime a considerable volume of information about these fascinating sulfur-substituted analogs of the oxocarbons has accumulated in numerous publications scattered throughout the chemical literature [*3–7*]. Their well-known chemical behavior and their potential for further applications make a review of the chemistry of the sulfur derivatives of cyclic oxocarbon anions particularly opportune, especially because no comprehensive publication on these compounds has yet appeared.

II. Thioxocarbon Dianions

Analogous to the oxocarbon dianions **1**, the thioxocarbon dianions $C_nS_n^{2-}$ can be characterized by the common formula **2**, in which n is any positive integer. In the case of only partial replacement, as in formula **3** (see Scheme 1), the dianions have to be termed mixed oxo–thioxocarbons.

A. SYNTHESIS AND REACTIONS OF THE C_4 THIOXOCARBON DIANION

It is well known that addition–elimination reactions of various nucleophiles proceed smoothly with cyclobutenediones bearing leaving groups on the vinyl carbons [*1*]. This observation provided a convenient route for the synthesis of the first thioxocarbon dianion (**4**), the sulfur analog of the squarate dianion [*5,6*]. The 1,2-dithiosquaric acid diamide **5** [*15,16*] serves as a suitable precursor of **4.** Upon treatment with a freshly prepared solution of excess potassium hydrosulfide in dry ethanol, both the amine functions in **5** are replaced by sulfur to give **4** in nearly-quantitative yield [*5*]. Likewise, the 1,3-dithiosquaric acid diamide **6** [*17*] or the tosyl-substituted bis(amidines) of squaric acid (**7** or **8**) [*18,19*] serve as easily available starting materials for the synthesis of the C_4 thioxocarbon dianion by a similar reaction sequence, as shown in Scheme 2.

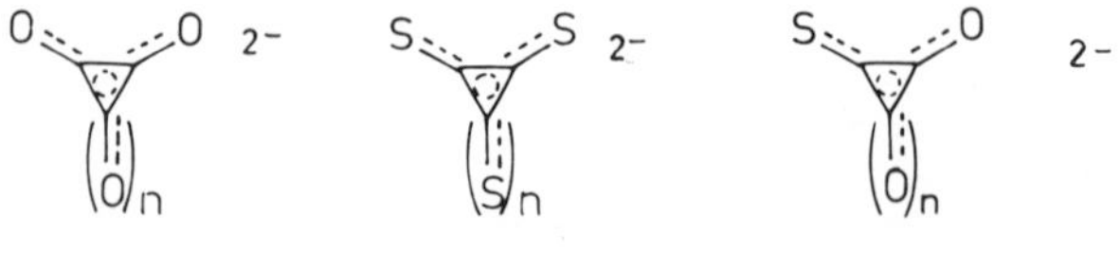

O O 2− S S 2− S O 2−

O n S n O n

(1) (2) (3)

Scheme 1

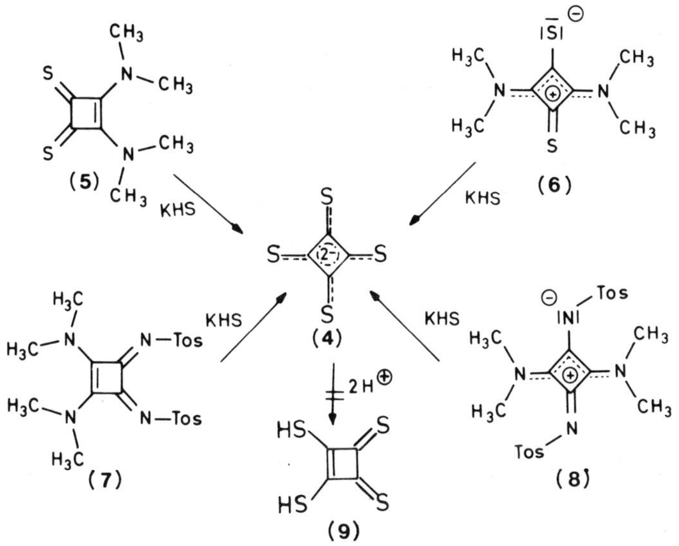

Scheme 2

These syntheses give **4** as the orange-yellow hydrate $K_2[C_4S_4]\cdot H_2O$, which can then be recrystallized from ethanol/water. Upon heating of **4** in air above 120°C, the water of crystallization is removed and a dark violet, strongly hygroscopic modification is formed, which is readily reconverted to the hydrated salt when exposed to the air. Compound **4** dissolves readily in water to give stable solutions.

Free tetrathiosquaric acid (**9**) cannot be obtained from the alkali metal salts of **4**, either by ion exchange on the aqueous solution with H^+-form cation-exchange resins, or by treating with the equivalent amount of sulfuric acid. In both cases an insoluble yellow powder is formed, probably a polymeric product of **9**.

On account of its high degree of symmetry and readily polarizable S atoms, $C_4S_4^{2-}$ appears to be especially suitable for the production of organometallic catena complexes, which are potential one-dimensional electrical conductors. Thus, the dianion of tetrathiosquaric acid (**4**) formed numerous compounds having bischelate structure with transition metals (Cr, Mo, W, Mn, Au) and various diamagnetic compounds with a chain structure (Ni, Pd, Pt), which exhibit interesting properties pertinent to electrical conductivity [*20*]. In addition, the search for more effective "organic metals" led to the investigation of some organic charge-transfer complexes prepared from 1,2-dithiolylium derivatives and tetrathiosquarate salts [*21*]. Redox reactions of **4** are of considerable interest (Scheme 3) because they must ultimately lead either to the neutral tetrathioxocyclobutane **10** via the radical-anion **11** or to the antiaromatic cyclobutadiene tet-

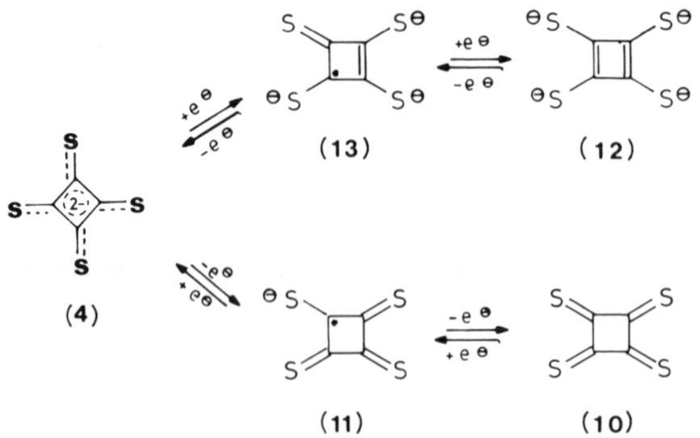

Scheme 3

rathiolate **12** via the radical-trianion **13.** While oxidation of **4** in many cases led to polymeric products, **4** in contrast to the $C_4O_4^{2-}$ dianion could be smoothly reduced polarographically in two one-electron steps (half-wave potentials −1.53 and −1.79 V) in aqueous solution to the tetraanion $C_4S_4^{4-}$ (**12**) [*20*]. Hence, further work on the redox products of the $C_4S_4^{2-}$ dianion would be especially useful.

B. SPECTROSCOPIC DATA

Spectroscopic data have contributed much to knowledge of the structure of oxocarbon anions [*1*]. Also, in the case of the C_4 thioxocarbon dianion **4**, there is no doubt that it has a planar structure with D_{4h} symmetry, consistent with complete π-electron delocalization and aromaticity [*5*]. The characteristic orange color of **4** is due to an intense UV absorption at 430 nm. The infrared spectrum is very simple and is dominated by some intense sharp bands around 1240 cm^{-1}, which can be assigned to C⋯C⋯S vibrational frequencies. The ^{13}C-NMR spectrum in D_2O shows one signal at $\delta = 229.2$ ppm, indicating that all carbon atoms in **4** are equivalent. An X-ray crystal structure determination confirms these findings [*5*]. Bond lengths for the symmetric and planar anion are $\langle C—C \rangle = 1.448(6)$ Å and $\langle C—S \rangle = 1.663(9)$ Å; the corresponding bond orders of about 1.25 for the C—C bond and 1.5 for the C—S bond are calculated from the accepted bond distance–bond order relationship [*22,23*]. An interesting finding is that bond orders in the C_4 oxocarbon and in the sulfur analog dianion are only slightly different, despite the fact that sulfur, in contrast to the 2p element oxygen, belongs to the third row of the periodic table with an increased radius of covalency (C, 0.74 Å; S, 1.04 Å), decreased tendency to form 2p–3p double bonds, larger polarizability, etc.

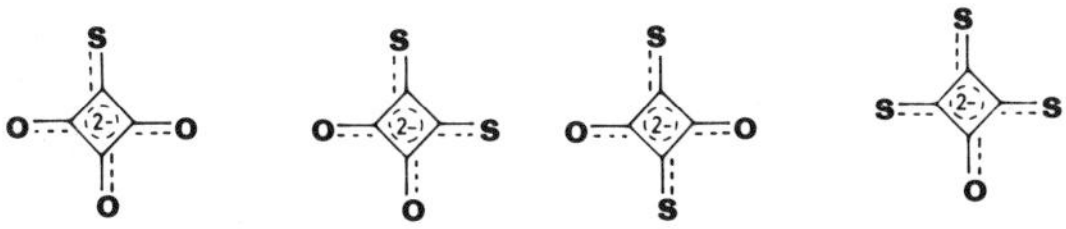

(14) (15) (16) (17)

Scheme 4

III. Mixed C_4 Oxo–Thioxocarbon Dianions

A formal, stepwise replacement of the oxygen atoms in the squarate dianion leads to the mixed C_4 oxo–thioxocarbon dianions formulated as **14**, **15**, **16**, and **17** in Scheme 4.

A. SYNTHESES AND REACTIONS OF MONOTHIOSQUARATE AND 1,2-DITHIOSQUARATE DIANIONS

The monothiosquarate ion **14** (MTS^{2-}) was first prepared by West and Eggerding from diethyl squarate (**18**) [*3*], as shown in Scheme 5. With 1 equivalent of potassium or tetramethylammonium hydrosulfide, displacement of only one ethoxy group occurs, yielding the 3-ethoxycyclobutenedione 4-thiolate anion **19**. Hydrolysis of **19** with hydroxide is nearly quantitative, giving MTS^{2-} (**14**), which has been isolated and characterized as the hygroscopic tetramethylammonium salt $(Me_4N)_2MTS$ (**14a**) and as the ternary zinc salt $(Me_4N)_2Zn(MTS)_2$ (**14b**). Another route to MTS^{2-} starts from the half-ester 1-hydroxy-2-methoxycyclobutenedione (**20**), which is obtained instead of the dimethyl ester when squaric acid is heated with methanol. Sulfhydrolysis of **20**, however, yields a mixture of the desired MTS^{2-} and the 1,2-dithiosquarate ion **15** (1,2-DTS^{2-}), as indicated by ^{13}C-NMR spectroscopy [*6*]. All attempts to separate these two products were unsuccessful.

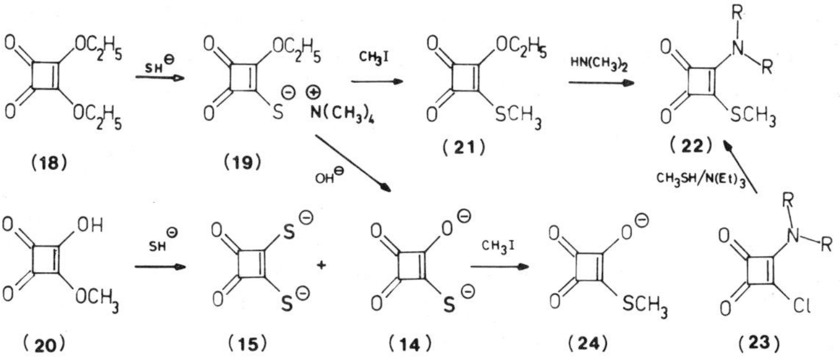

(18) (19) (21) (22) (20) (15) (14) (24) (23)

Scheme 5

The 3-ethoxycyclobutenedione 4-thiolate ion **19** is easily alkylated, forming 3-ethoxy-4-methylthiocyclobutenedione (**21**) [*4*]. Reaction of **21** with 1 equivalent of dimethylamine gives only displacement of the ethoxy group, yielding 3-dimethylamino-4-methylthiocyclobutenedione (**22**, R = CH_3), which is identical with the product obtained from the reaction of methylmercaptan and 4-chloro-3-dimethylaminocyclobutenedione (**23**) in the presence of triethylamine [*24*]. The MTS^{2-} ion reacts only at sulfur with excess iodomethane, to form the *S*-methylated ion **24** [*4*]. An interesting finding is the synthesis of a sulfur-bridged binuclear oxocarbon acid (**26**) by reaction of MTS^{2-} with 0.5 equivalent of copper(II) chloride [*25*]. The obtained yellow anion (**25**) can be smoothly transformed to its conjugate acid by treatment of a concentrated solution of **25** with cold 12 *N* aqueous hydrochloric acid, as shown in Scheme 6. The resulting 2,2′-thiobis(3,4-dioxocyclobuten-1-ol) (**26**) crystallizes as a dihydrate and decomposes above 125°C. Like squaric acid, **26** is an extremely strong acid with a p*K* less than zero, as indicated by spectrophotometric measurements [*25*].

Useful starting materials for the synthesis of the 1,2-dithiosquarate ion **15** are either diethyl squarate (**18**), which can smoothly be converted to **15** by treatment with 2 equivalents of potassium hydrosulfide [*3,4*], or the dithiosquaramide **27** (but = *n*-C_4H_9), which is cleanly hydrolyzed with potassium hydroxide [*6*] (Scheme 7). Disodium 1,2-DTS^{2-} (**15a**) has also been prepared in low yield from the bis(cyclohexylamide) of squaric acid (**28**) by the amide cleavage method of Shahak and Sasson [*26*]. Reaction of 1,2-DTS^{2-} (**15**) with alkyl iodides in DMF–water yields *S,S′*-dialkyl 1,2-dithiosquarates (**29**) [*4*], which have also been prepared by reactions of alkylmercaptans and 3,4-dichlorocyclobutenedione (**30**) [*4,27*]. Treatment of **15** with equal amounts of **30**, the "acid chloride" of squaric acid [*15,28*], yields the symmetric red-orange dithiin **31** [*6*]. The unsymmetric alternative structures (**31b** and **c**) can be ruled out by ^{13}C-NMR spectroscopy. Two signals at δ = 187.2 and 184.1 ppm are consistent only with structure **31a**, which can also be obtained through an independent route by treatment of **30** with the dimer of *p*-methoxyphenylthionophosphine sulfide [*29*].

Interesting anionic complexes of 1,2-DTS^{2-} as chelating ligand with transition metal cations—for instance, the potassium salts of the bis(dithiosquarato)nickelate(II) (**32**) and bis(dithiosquarato)zincate(II) (**33**)— have been prepared and

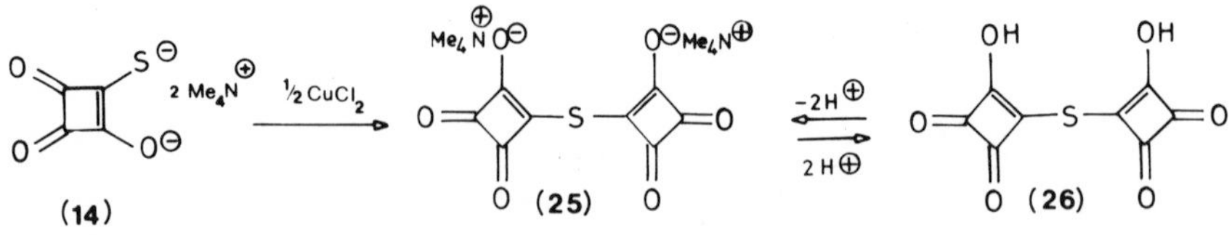

(14) 2 $Me_4N^{\oplus}$ ½ $CuCl_2$ → (25) $\underset{2H^{\oplus}}{\overset{-2H^{\oplus}}{\rightleftharpoons}}$ (26)

Scheme 6

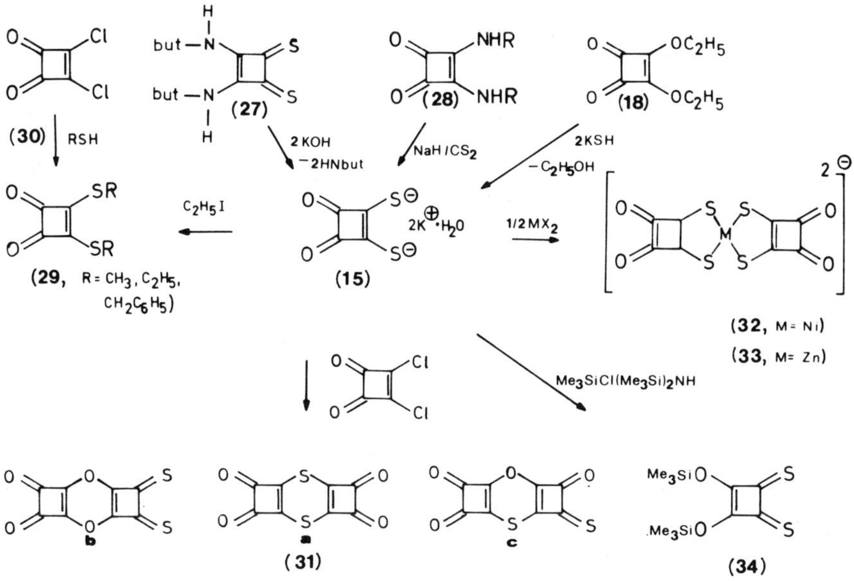

Scheme 7

characterized by X-ray crystal structure determination to confirm the intriguing mode of bonding as well as the formulation of the ligand as the 1,2-DTS^{2-} [*3,30*]. Protonation of 1,2-DTS^{2-} gave a yellow solid with an S—H band in the infrared spectrum, but rapid decomposition with loss of hydrogen sulfide discouraged further characterization [*4*]. Silylation of **15** yielded an extremely moisture sensitive red-orange compound (**34**), which readily dissolved in ether and carbon tetrachloride. Spectroscopic data provide evidence that **34** was the *O*-silylated derivative of 1,2-DTS^{2-} [*4*].

B. REACTION OF THIOSQUARATE ESTERS WITH AMINES

It is well known that reactions of dichlorocyclobutenedione (**30**) and diethyl squarate (**18**) resemble those of carboxylic acid chlorides and esters and that carboxylic thioesters react like esters. In contrast, chemical properties of alkylthiocyclobutenediones cannot be predicted from those of the alkoxycyclobutenediones. As shown in the reaction with diethylamine [*4*] (Scheme 8), the reactivity of **29** is quite different from that of **18**. Not the expected squaramide (**35**), but a fumaramide derivative (**36**) is obtained when **29** is treated with an excess of diethylamine. An obvious mechanism of this reaction via **37** is shown in Scheme 8.

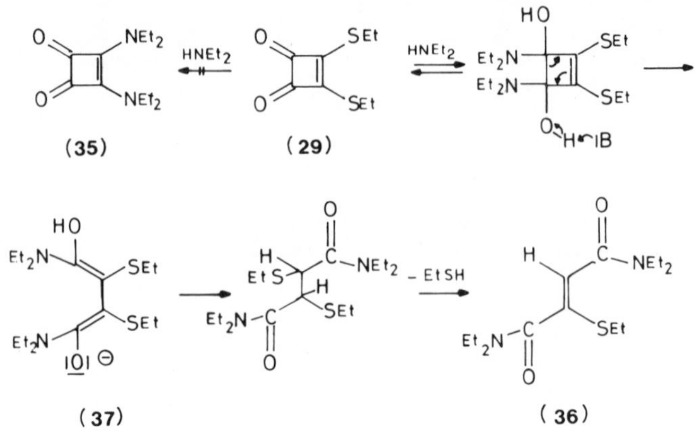

Scheme 8

C. SYNTHESES AND REACTIONS OF 1,3-DITHIO- AND 1,2,3-TRITHIOSQUARATE DIANIONS

In the same fashion as tetrathiosquarate ion **4**, 1,3-dithiosquarate ion **16**, (1,3-DTS^{2-}) and 1,2,3-trithiosquarate ion **17** (TTS^{2-}) are smoothly obtained by sulfhydrolysis of the corresponding amides **38** and **39**, respectively [*6*].

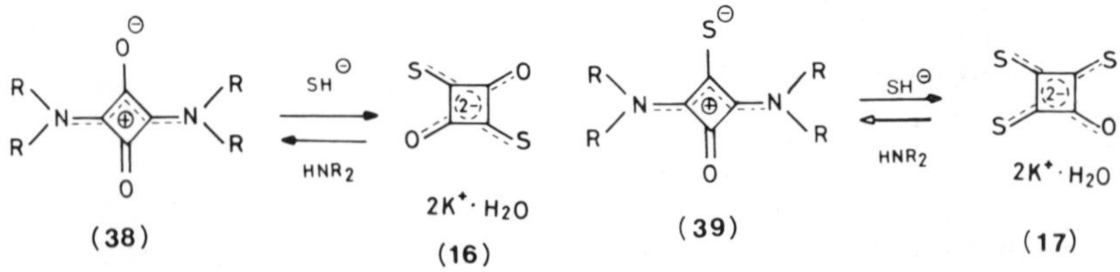

Scheme 9

Upon treatment with excess secondary amine, **16** and **17** are reconverted to the corresponding 1,3-squaric acid diamides [*6*], as shown in Scheme 9. The spectroscopic data of all the mixed oxo–thioxocarbon dianions (**14–17**) are consistent with a delocalized structure, but π-electron delocalization is less intensive than in the symmetric $C_4O_4^{2-}$ or $C_4S_4^{2-}$ dianions [*4,6*].

IV. DITHIOSQUARIC ACID DIAMIDES

Dithiosquaric acid diamides with the sulfur atoms in 1,2 and 1,3 positions have proved useful as starting materials for the synthesis of various pseudo-oxocarbons [*5,6*].

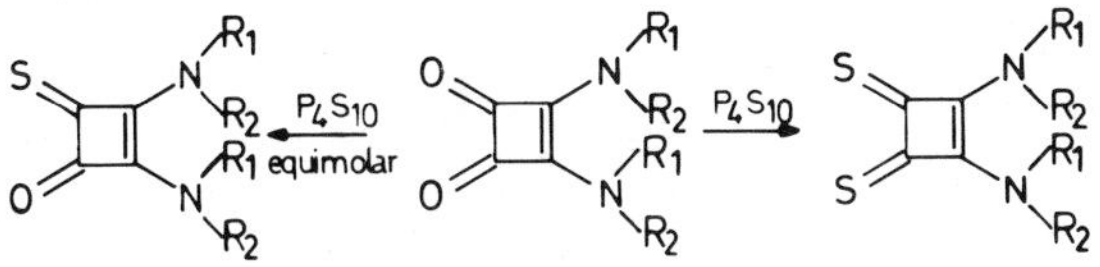

Scheme 10

A. SYNTHESES AND PROPERTIES OF 1,2-DITHIOSQUARIC ACID DIAMIDES

The 1,2-dithiosquaric acid diamides **40** (1,2-DTSD) are easily obtained in a stepwise sequence via **42** (Scheme 10) by reaction of the corresponding diamides **41** [*15*] with convenient sulfur transfer reagents such as phosphorus(V) sulfide [*31*] (method A), the dimer of *p*-methoxyphenylthionophosphine [*32,33*], or the recently discovered alkoxycarbonyl isothiocyanate [*17*] (method B). The yields of and some data for 1,2-DTSD are summarized in Table I. All of these compounds are high melting and stable, and their properties and spectra are in agreement with the open α-dithione structure **40A** (Scheme 11). The initially

Table I. Yields of and Data for 1,2-DTSD's (**40**) from the Corresponding Diamides with P_4S_{10} (Method A) or Alkoxycarbonyl Isothiocyanate (Method B)

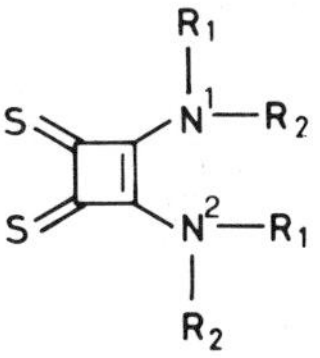

	N^1		N^2			Yield (%)	
Compound	R^1	R^2	R^1	R^2	mp (°C)	Method A	Method B
40a	CH_3	CH_3	CH_3	CH_3	235 (dec.)	70	54
40b	$(CH_2)_2$—O—$(CH_2)_2$		$(CH_2)_2$—O—$(CH_2)_2$		240 (dec.)	35	70
40c	—$(CH_2)_5$—		—$(CH_2)_5$—		243 (dec.)	72	78
40d	—$(CH_2)_4$—		—$(CH_2)_4$—		249 (dec.)	68	75
40e	H	$(CH_2)_3CH_3$	H	$(CH_2)_3CH_3$	225 (dec.)	76	—
40f	H	$C(CH_3)_3$	H	$C(CH_3)_3$	345 (dec.)	78	—
40g	H	$CH_2C_6H_5$	H	$CH_2C_6H_5$	228 (dec.)	81	—
40h	—$(CH_2)_5$—		—$(CH_2)_4$—		229 (dec.)	46	—
40i	$(CH_2)_2$—O—$(CH_2)_2$		CH_3	CH_3	198 (dec.)	52	—

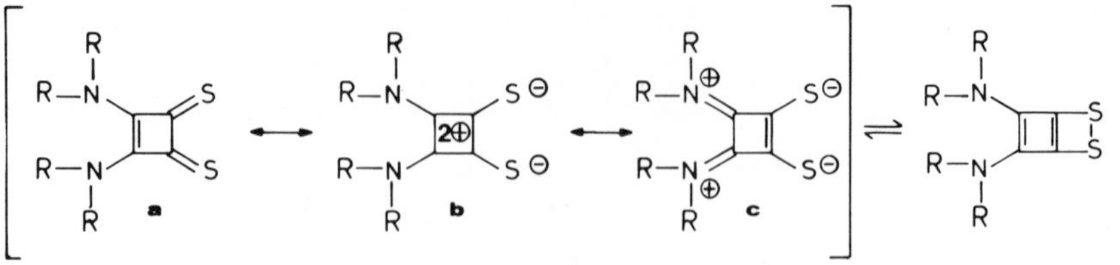

(40 A) (40 B)

Scheme 11

surprising stability of the intriguing α-dithione system [*34*] is in good accord with LCAO–MO calculations [*35*], which predict that conjugative electron release by the substituents stabilizes the α-dithione structure relative to the valence tautomeric 1,2-dithiete form (**40B**). In the case of 1,2-DTSD (**40**), the unfavorable 1,2-dithiete structure is additionally destabilized by the considerable angle strain of the bicyclic system, but primarily by the antiaromatic character of the resulting cyclobutadiene. As anticipated, spectroscopic evidence excludes the bicyclic valence tautomeric 1,2-dithiete structure (**40B**) for all these compounds [*36*]. In a first approximation the molecules may be described as hybrids of three resonance forms (**a**–**c**) and are probably best represented as zwitterionic (**40Ac**) with a high contribution to the actual electronic structure of **40A**. This corresponds to the high melting points and slight solubility in nonpolar aprotic solvents and is moreover strongly supported by spectroscopic data. The unusually high stretching vibration frequencies of the $>\overset{\oplus}{N}\cdots C\cdots C\cdots \overset{\oplus}{N}<$ system around 1680 and 1580 cm^{-1} indicate a relatively high π-bond order of the semicyclic C—N bond. In addition, strong preference for the polar resonance structure **40Ac** follows unambiguously from an X-ray structure determination of **40d** [*37*]. The atomic distances of the C—N bond [C—N = 1.313(4) Å] as well as of the C—S bond [C—S = 1.645(3) Å] fall between the known values for single and double bonds [C—N = 1.47 Å; C═N (from oximes) = 1.27 Å; and C—S = 1.81 Å; C═S = 1.60 Å, respectively] and demonstrate a degree of delocalization similar to that found in the closely related thioamides [*38*].

All the 1,2-DTSD's have a bright yellow color, whereas their oxygen analogs are colorless. The electronic spectra exhibit two intense absorptions around 290 and 400 nm, with a shoulder around 370 nm. The intense long-wavelength bands can probably be assigned to the $\pi\rightarrow\pi^*$ transition. An expected absorption of low intensity at about 450 nm due to the $n\rightarrow\pi^*$ transition cannot be detected and is perhaps hidden under the long-tailing, broad absorption around 400 nm. The mass spectra of all 1,2-DTSD's [*36*] are dominated by intense parent ion peaks, which are also the base peaks. Characteristic features of the fragmentation are the elimination of carbon monosulfide and ring contraction besides a scission of two parallel C—C bonds of the four-membered ring, as outlined in Scheme 12.

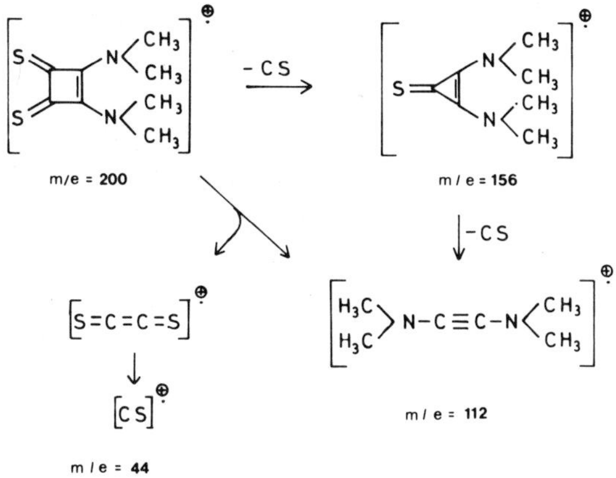

m/e = 200

m/e = 156

m/e = 112

m/e = 44

Scheme 12

B. SYNTHESES AND PROPERTIES OF "1,3-DITHIOSQUARIC ACID DIAMIDES"

In contrast to the facile synthesis of 1,2-DTSD's, reaction of "1,3-squaric acid diamides" with convenient sulfur transfer reagents in most cases yields mixtures of mono- and dithiosquaric acid diamides that are difficult to separate. Fortunately, this is not the case with the hitherto unknown dimethylamide **43**, which proved to be the most appropriate starting material for the synthesis of a great variety of secondary "1,3-dithiosquaric acid diamides" (**44**, 1,3-DTSD) [*36*]. Compound **43** can be prepared from squaric acid (**45**) [*36*], as outlined in Scheme 13. Upon treatment of 1,3-DTSD (**44a**) with excess secondary amines, replacement of the dimethylamine functions occurs (Scheme 13), as expected for an all-trans vinamidinium system [*2*], to yield the 1,3-DTSD's (**44b–d**) listed in

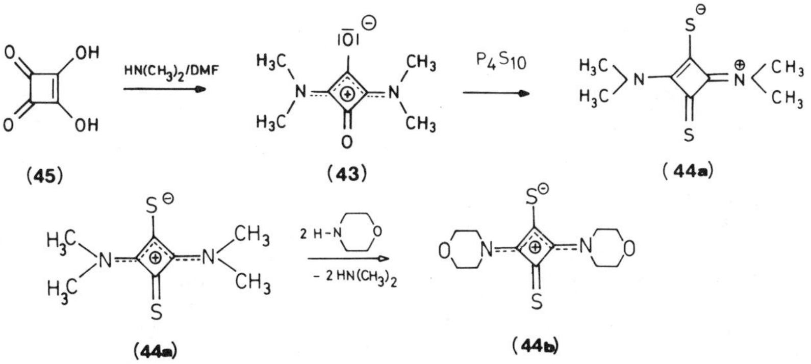

$HN(CH_3)_2$/DMF P_4S_{10}

(**45**) (**43**) (**44a**)

2 H-N(morpholine) $- 2\ HN(CH_3)_2$

(**44a**) (**44b**)

Scheme 13

Table II. Yields of and Data for 1,3-DTSD's Synthesized via 1,3 Displacement (A) from **44a** or via Rearrangement (B) from **40e**

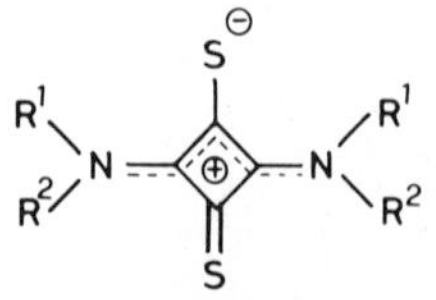

(44)

Compound	R^1	R^2	Yield (%)	mp (°C)
44b	$(CH_2)_2$—O—$(CH_2)_2$		70 (A)	304
44c	—$(CH_2)_5$—		97 (A)	290
44d	—$(CH_2)_4$—		78 (A)	295
44e	H	CH_3	67 (B)	218 (dec.)
44f	H	$(CH_2)_3CH_3$	70 (B)	165 (dec.)
44g	H	$CH_2C_6H_5$	85 (B)	260 (dec.)

Table II. Attempts to convert **44a** to primary diamides by the same method were not successful [*36*]. Moreover, various 1,3-DTSD's (**44**) could be synthesized in a most surprising way. Upon treatment of 1,2-DTSD (**40a** or **d**) with excess primary or secondary amine, the expected nucleophilic displacement of the amine is accompanied by rearrangement. The starting material is readily converted to the isomeric, thermodynamically more stable 1,3-DTSD (**44a** or **f**), summarized in Table II. The rearrangement is also achieved when 1,2-DTSD's are treated with traces of methylmercaptan as a very efficient catalyst in acetonitrile as solvent, yielding the corresponding 1,3-DTSD (**44**) quantitatively [*7,36*].

An obvious route of this interesting rearrangement is outlined in Scheme 14. In

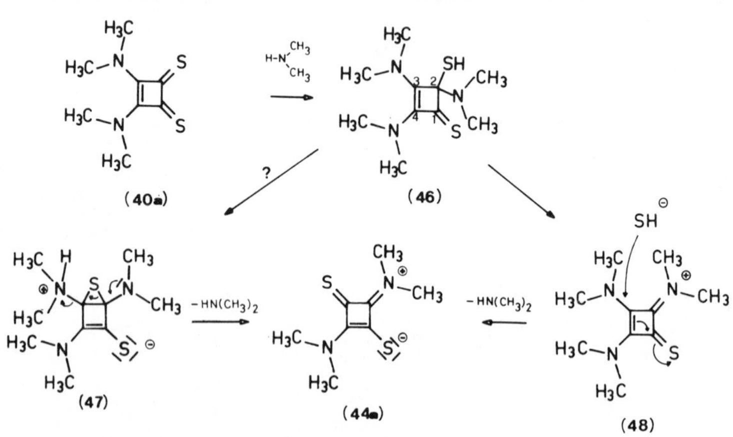

(40a) (46) (47) (44a) (48)

Scheme 14

the first step of the reaction, the nucleophilic reactant attacks the thiocarbonyl carbon atom of **40a**, yielding the intermediate *N,S*-acetal **46**. The migration of the sulfur atom from position 2 to 3 may proceed by an inter- or intramolecular mechanism via **47** or **48**, respectively, as intermediate. In both cases elimination of dimethylamine follows, and the 1,3 isomer (**44a**) of the starting material is the only reaction product in nearly quantitative yield.

Two crossover experiments, outlined in Scheme 15, produced evidence for an intermolecular mechanism for this arrangement [*7*]. Upon heating of equal molar amounts of two differently substituted 1,2-DTSD's (**40c** and **40d**) in the same acetonitrile solution in the presence of traces of methylmercaptan, three reaction products were detected—two symmetric and one mixed 1,3-DTSD—proving intermolecular migration at least of the amine functions. These findings were confirmed by a second experiment in which the starting material was the mixed 1,2-DTSD **40h**, which was treated under the same conditions as before, yielding the same three products. The formation of the two symmetric crossover products again indicates an intermolecular migration mechanism. Criteria for discriminating between 1,2- and 1,3-DTSD depend on the different constitution of these two structures [*36*]. Thus, the 1,3-DTSD's have an intramolecular salt structure and exhibit higher melting points. Because of the higher degree of symmetry, their IR spectra are simpler and dominated by an intense, broad band around 1600 cm^{-1}, which can be assigned to stretching vibrations of the $[\ {>}N\cdots C\cdots C\cdots C\cdots N{<}\ ]^+$ system. In contrast to the bright yellow 1,2-DTSD's, the 1,3-DTSD's are

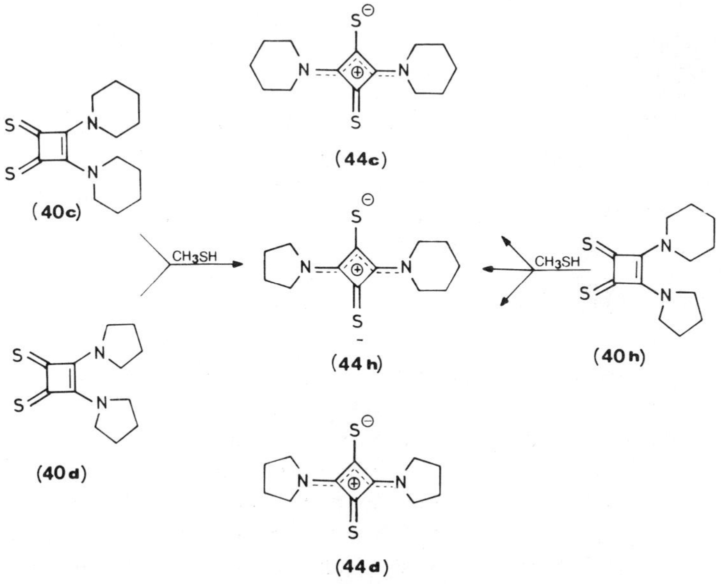

Scheme 15

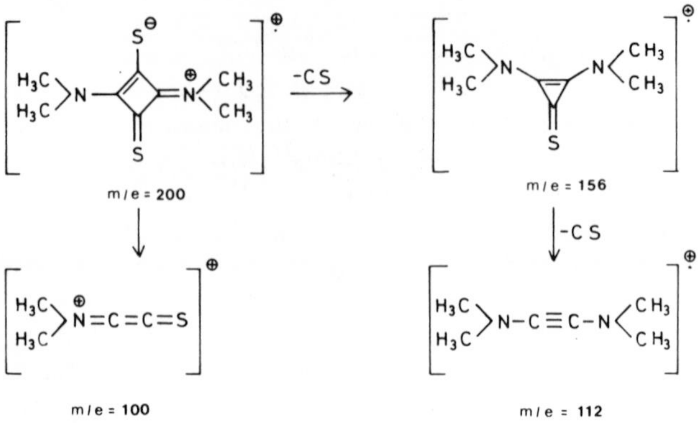

Scheme 16

orange-yellow. Their electronic spectra exhibit strong absorption maxima at about 430 nm and a band of lower intensity in the region of longer wavelengths at 490 nm, which can be assigned to the $\pi \rightarrow \pi^*$ and $n \rightarrow \pi^*$ transitions, respectively. The mass spectra show a very intense parent ion peak, which is also the base peak, indicating the high stability of these compounds. The main fragmentation pattern can be represented by Scheme 16. The parent ion may break down via two paths, either by the successive loss of two molecules of carbon monosulfide or by fragmentation in two identical ions with a pronounced peak at $m/e = \frac{1}{2}M^+$.

C. REACTIONS WITH NUCLEOPHILES

Because of their amphoteric character, dithiosquaric acid diamides can react with electrophiles and nucleophiles as well. Various nucleophiles prefer reaction on the vinylic carbon atoms of the four-membered ring and displace the substituents via an addition–elimination mechanism. These general remarks can be confirmed by some typical examples.

1. Sulfur-Containing Squarate Anions from Hydrolysis or Sulfhydrolysis of 1,2- or 1,3-DTSD

In view of their potential applicability as contact herbicides, new synthetic routes to variations of the recently discovered mycotoxin "moniliformin" [*39*], the potassium or sodium salt of semisquaric acid, have been extensively investigated [*40*]. Hydrolysis or sulfhydrolysis of 1,2- or 1,3-DTSD has provided important pathways to novel variations of this natural product, in which the vinylic hydrogen of the mycotoxin is replaced by a dimethylamino group [*41*]. Heating of 1,2-DTSD (**40a**) in water yields a new anion (**49**) with two vicinal sulfur atoms. Surprisingly, the corresponding 1,3-DTSD (**44a**) is not converted to the expected product under the same conditions; instead, the monothiosquaric

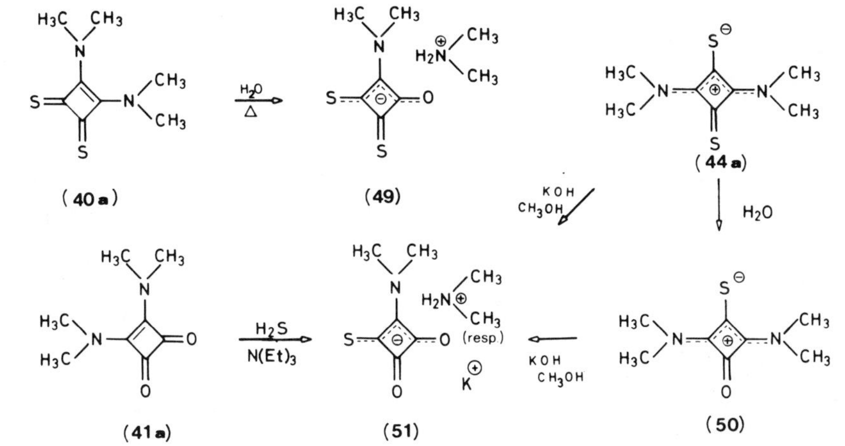

Scheme 17

acid diamide **50** is obtained, which upon treatment with potassium hydroxide in methanol gives the anion (**51**) with only one sulfur atom. Treatment of 1,2-squaric acid bis(dimethylamide) (**41a**) with $H_2S/N(Et)_3$ yields the same anion by an independent synthetic route as its dimethylammonium salt (see Scheme 17). The anion (**52**) with three adjacent sulfur atoms is obtained by sulfhydrolysis of either **40a** or **44a** (Scheme 18), whereas the synthesis of **53** with two sulfur atoms

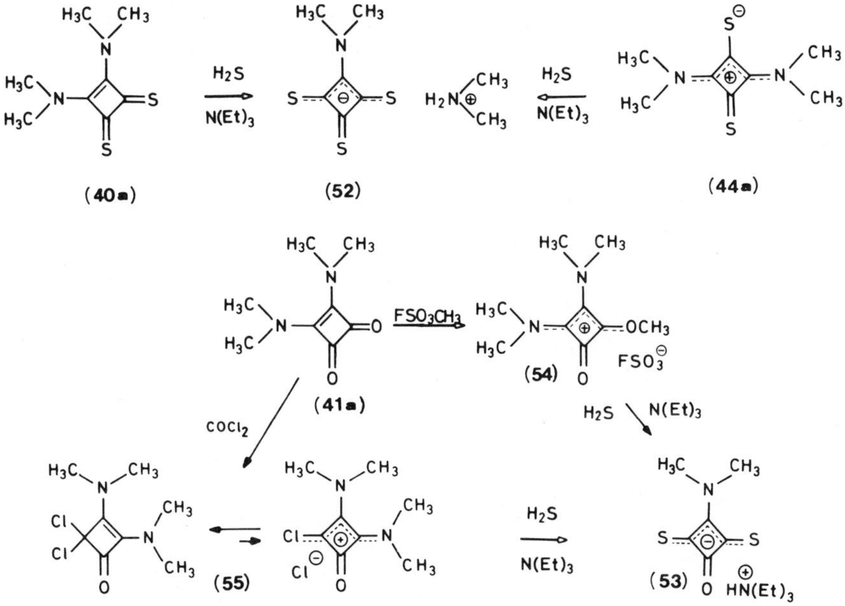

Scheme 18

in positions 1 and 3 can easily be achieved by sulfhydrolysis of the cations **54** or **55** [*13,41*]. All these derivatives of moniliformin have proved inactive in various pharmacological tests [*36*].

2. Reaction of 1,2-DTSD with Hydrazines

The reaction of 1,2-DTSD (**40e**) with the strongly nucleophilic hydrazines surprisingly leads to sulfur-free products [*36,42*], as shown in Scheme 19. Apparently all functional groups of the 1,2-DTSD are replaced by the excess nucleophile. However, the resulting compound is not the expected bishydrazidine (**56**), but its oxidation product (**57**), the tetrahydrazone of the hitherto unknown cyclobutatetraone. Under the same conditions the corresponding tetrahydrazones (**58** or **59**) are obtained upon treatment of **40e** with excess phenyl- or methylhydrazine, respectively. Thus far, all the experiments to reduce **59** to the corresponding bis(hydrazidine) have been unsuccessful. Formaldehyde, for instance, does not reduce (**59**); instead, ring closure to the novel tricyclic rings system **60** is observed. Further work pertinent to the redox properties of these tetrahydrazones (**57**–**60**) would be useful.

3. Reaction of 1,2-DTSD with Malononitrile or Cyanamide

Despite the fact that dicyanomethylene analogs of the deltate [*43*], squarate [*10–13*], and croconate dianions [*14*] have recently been reviewed [*44*], sulfur-containing derivatives of these most interesting variations of the oxocarbon dianions so far remained unexplored. The first example of this new class of pseudo-oxocarbons was obtained in the thiosquarate series [*45*]. In a two-step reaction sequence (see Scheme 20), the anion of malononitrile first reacts with 1,2-DTSD (**40a**) with replacement of only one dimethylamino group. The resulting salt (**61**) can be characterized by alkylation with methyl iodide to the neutral **62**. In the second step, **61** is treated with potassium hydrogen sulfide in dry ethanol to form

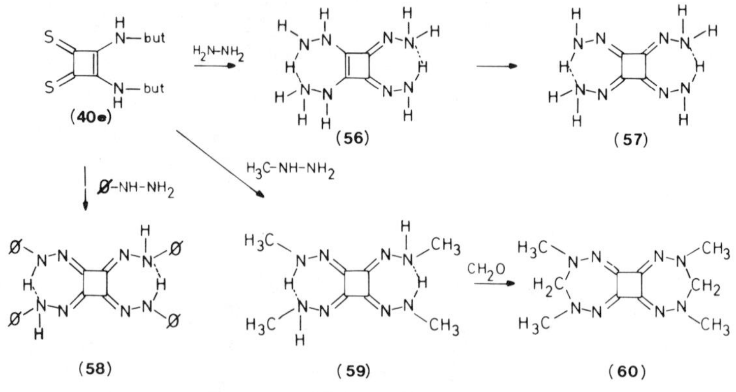

(40e) H_2N–NH_2 (56) (57)
Ø–NH–NH_2 H_3C–NH–NH_2 CH_2O
(58) (59) (60)

Scheme 19

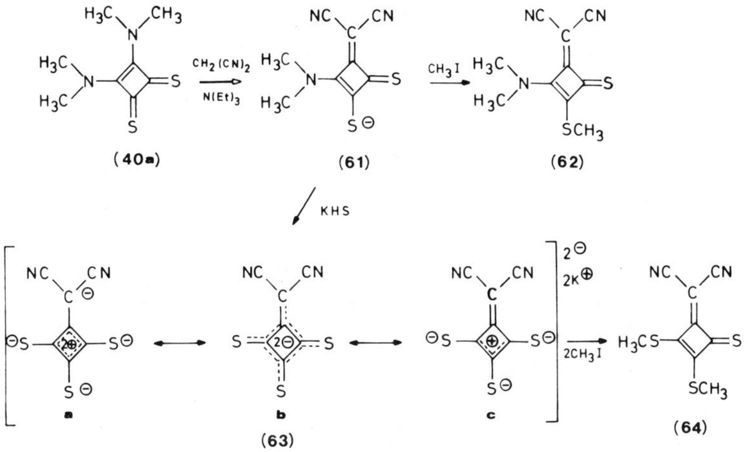

Scheme 20

the ochre-colored dipotassium salt **63**, which can be alkylated to **64**. The dianion of **63** is assigned a symmetric bond-delocalized structure comparable with that of the trithiosquarate dianion (**17**). The resonance forms (**63a–c**) are three of several possible forms contributing to the ground state of **63**. It receives considerable stabilization from the electron-withdrawing exocyclic substituent. However, π-electron delocalization among ring thiocarbonyl groups and the dicyanomethylene group is not as intensive as in the oxygen analog dianion **17**. This is indicated by the ^{13}C-NMR data shown in Scheme 21 and the relatively high frequency absorption of the semicyclic C—C bond at 1594 cm^{-1} in the IR spectrum of **63**.

A high yield of a similar novel pseudo-oxocarbon dianion is smoothly obtained from the reaction of 1,2-DTSD (**40a**) with 1 equivalent of the anion of cyanamide as nucleophilic reagent (Scheme 22). The main product of the first step of the replacement reaction is the intermediate anion **65**, which is accompanied by traces of two by-products, such as 1,3-DTSD (**44a**) and the neutral species **66**. In the latter, surprisingly one sulfur atom is substituted. Anion **65** can be cleanly transformed to a novel type of sulfur-containing oxocyanocarbon (**67**) by sulf-

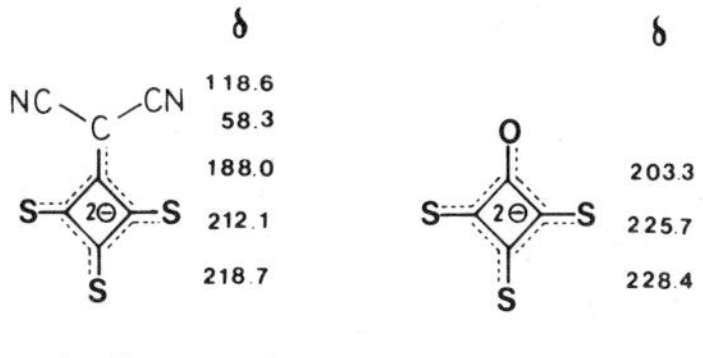

Scheme 21

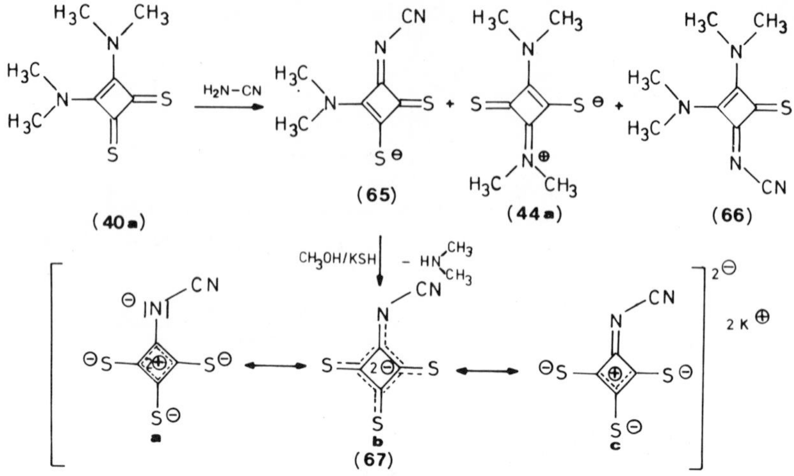

Scheme 22

hydrolysis, as shown in Scheme 22. Close examination of this orange-red dianion by spectroscopic methods revealed that it can also be regarded as a member of the aromatic pseudo-oxocarbons with a bond-delocalized electronic system. Structure **67a–c** are three of several resonance forms contributing to the ground state of **67**.

4. Reaction of 1,2-DTSD with Methylmercaptan

The results of most reactions of 1,2-DTSD with nucleophiles, which proceeded by loss of both the dimethylamino groups, suggested that the reaction with methylmercaptan might provide an attractive synthetic approach to the hitherto unknown tetrathiosquaric ester **68**, a potent precursor of the aromatic cyclobutadiene dication **69**. Surprisingly, reaction of 1,2-DTSD (**40a**) with excess methylmercaptan at elevated temperature did not lead to the desired product; instead, the 1,3 isomer (**44a**) of the starting material was obtained in quantitative yield via intermolecular rearrangement [7], as already mentioned. At room temperature, however, we succeeded in isolating an intermediate. Structure elucidation indicated that 2 equivalents of the nucleophile had been added to the same vinylic carbon atom. Via substitution of one dimethylamino group, the salt **70** is obtained, which is in reversible equilibrium with the starting material. Upon heating of a solution of the salt **70** in chloroform, 1,3-DTSD (**44a**) is obtained irreversibly. The anion **70** proved to be suitable starting material at least for a precursor of the desired tetrathiosquaric ester.

In a two-step reaction sequence, outlined in Scheme 23, the anion **70** could be converted to the cation **71** via the neutral **72**. Sulfhydrolysis of **71** in pyridine yielded a dark red crystalline product, which proved to be the thioketal (**73**) of

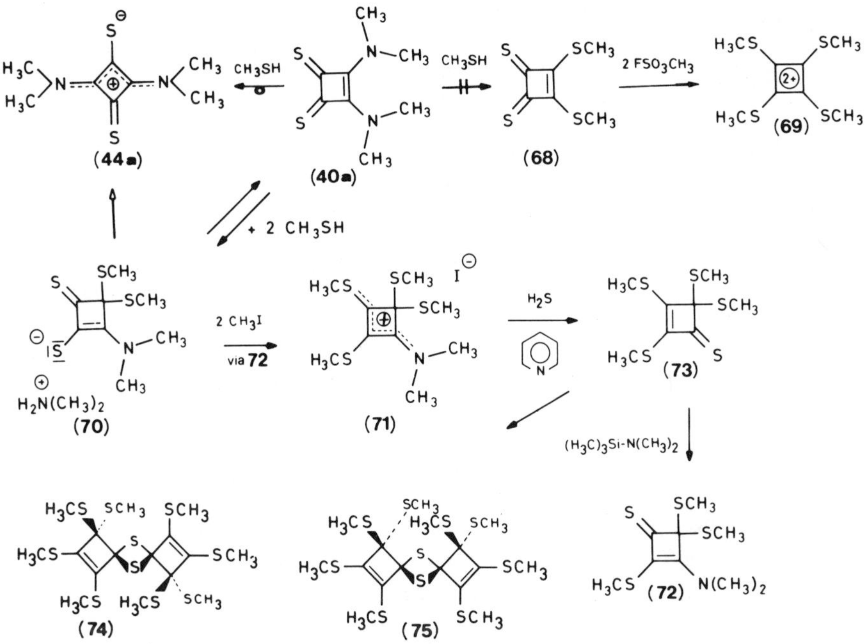

Scheme 23

tetrathiosquaric ester. The structure of this interesting precursor of **68** was established by its reaction with $(CH_3)_3SiN(CH_3)_2$ to regenerate **72** and furthermore by its spectroscopic data. Thus, the IR spectrum exhibits a strong absorption band at 1272 cm^{-1}, which probably can be assigned to the C=S stretching vibration. The UV spectrum shows three absorptions at λ_{max} (log ϵ) 274 (3.9), 395 (3.9, $\pi \rightarrow \pi^*$ transition), and 507 nm (2.2, $n \rightarrow \pi^*$ transition), which are typical of a vinylog dithio ester system. At least the presence of a thiocarbonyl group in **73** was unambiguously indicated by a signal at δ = 219 ppm in the ^{13}C-NMR spectrum. Compound **73** is sensitive to light both in solution and in the crystalline state, suggesting that the tetrathiosquaric ester **68** is not a very stable compound. When exposed to daylight the red solution of **73** in dichloromethane turned bright yellow; evaporation of the solvent gave a yellowish product, which turned out to be a dimer of **73** with one of the alternative structures **74** or **75** [7].

D. REACTIONS WITH ELECTROPHILES

1. Synthesis and Properties of a Stable Cyclobutadiene Dication

As already pointed out, electrophiles can also react with DTSD and attack predominantly at the sulfur atoms. Thus, "magic methyl" (FSO_3CH_3) alkylates

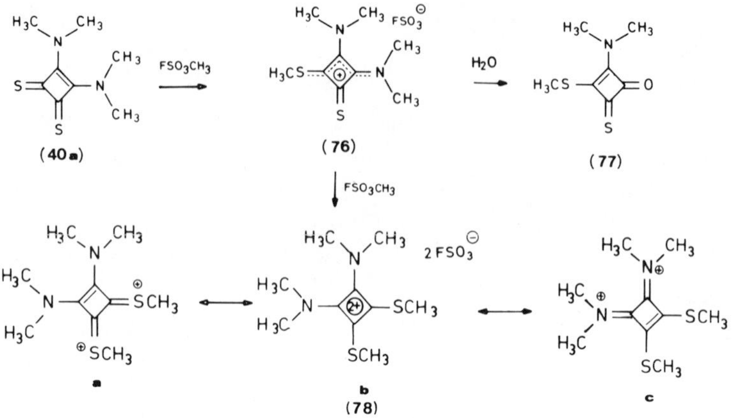

Scheme 24

1,2-DTSD (**40a**) in a two-step reaction via the cation **76** which is smoothly hydrolyzed to **77** at both the sulfur atoms, yielding the first isolable cyclobutadiene dication (**78**) [*27,46*] (Scheme 24). This interesting ''Hückel aromatic'' system containing $4n + 2$ ($n = 0$) π electrons obtains a strong stabilizing influence from the substituents [*47–49*]. On the one hand, delocalization of the positive charge occurs from the sulfur atoms into the ring; on the other hand, the amino groups strongly back-donate electron density to the carbons of the four-membered ring, as illustrated by the hybrid structures **78a–c**; the contribution by **78c** is expected to be most important. This is strongly supported by the spectroscopic data, which give insight into the electronic structure of the dication **78.** The C—N stretching frequency in the IR spectrum at 1758 cm^{-1} indicates an unusually high π-bond order for this semicyclic bond. Thus, the dication **78** undergoes relatively infrequent rotation about the C═N bond at room temperature, giving rise to nonequivalent methyl groups in the ^{1}H-NMR spectrum. Variable-temperature study showed coalescence of the *N*-methyls at 106°C, corresponding to a free activation enthalpy $\Delta G\ddagger = 20.6$ kcal/mole [*50*]. This rotational barrier seems to be relatively small compared to calculated values of other amino-substituted dications (**79** and **80**) [*47*] (Scheme 25). Steric hindrance of the adjacent dimethylamino groups certainly lowers the rotational barrier and thus can account for this discrepancy between theory and experiment.

The dication **78** is stable at room temperature in the absence of moisture and is, as anticipated, very sensitive to nucleophiles [*27,46*]. Some of the interesting chemical behavior [*51*] is outlined in Scheme 25. Hydrolysis yields the dithiosquaric acid *S,S'*-dimethyl ester **29.** Upon treatment with hydrogen sulfide, a deep red solution was obtained, but the expected tetrathiosquaric ester **68** could not be isolated. Reaction of **78** with dimethylamine leads to the ring-opened product **81**

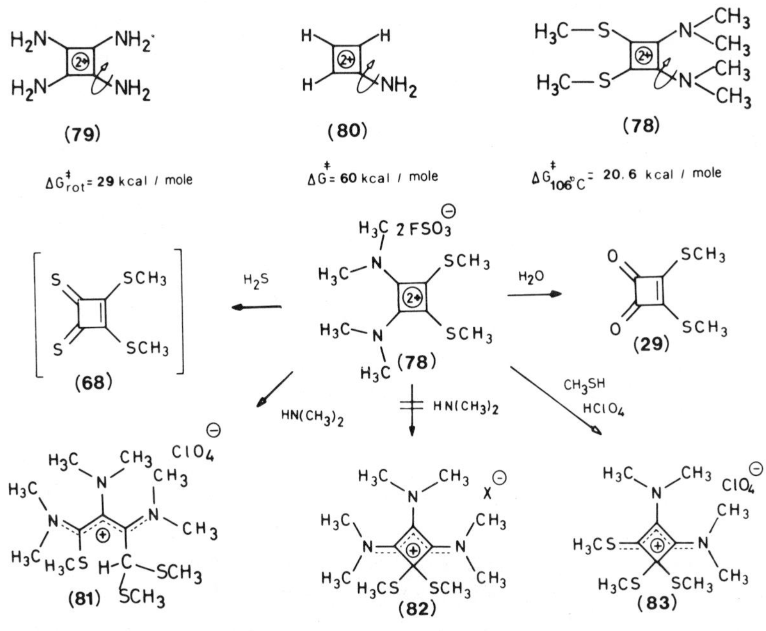

Scheme 25

and not to the vinamidinium cation **82**. With methylmercaptan the dication **78** surprisingly reacts to give the cation **83**.

In the same fashion as described above thioamides of semisquaric acid (**84** R = CH_3) are alkylated by "magic methyl" [*52*]. The resulting cation **85** is readily hydrolyzed to yield the corresponding thioester **86,** which has proved to be a potent reagent in [4+2] cycloadditions [*53*], as shown in Scheme 26. Reaction

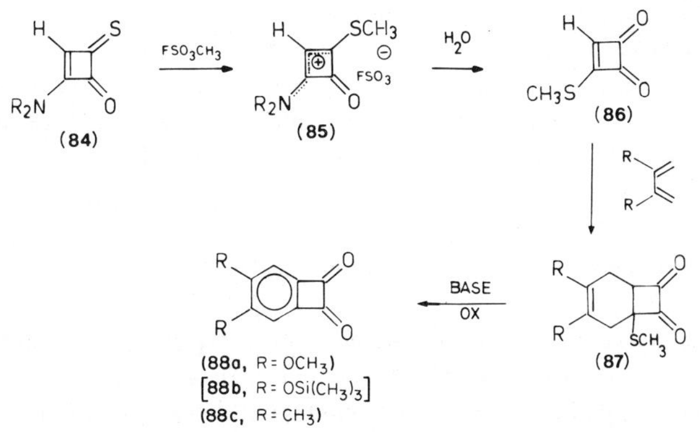

Scheme 26

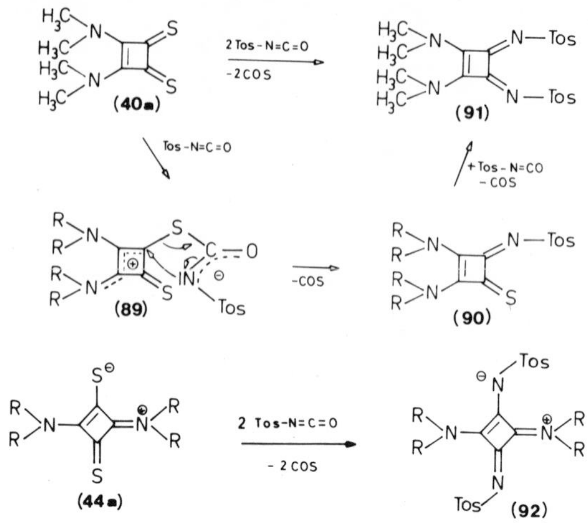

Scheme 27

with substituted butadienes leads to the cycloaddition products **87**, which on treatment with base eliminate methylmercaptan. Promoted by aromatization, subsequent oxidation yields the most interesting benzenoid vinylog of squaric acid esters [**88a,** R = OCH_3 [*54*]; **88b,** R = $OSi(CH_3)_3$] or other benzocyclobutenediones (**88c,** R = CH_3).

2. Reaction of 1,2- and 1,3-DTSD with Tosyl Isocyanate

Although well known in the thioamide series [*38*], acylation reactions of DTSD have not yet been reported. In contrast, the reaction with tosyl isocyanate has been extensively investigated. As shown in Scheme 27, the reaction of DTSD with this electrophilic reagent probably proceeds via initial attack at the sulfur atom, yielding a dipolar intermediate (**89**), which in the course of an S_Ni reaction or after cyclization to a 1,3-thiazetidin-2-one spontaneously eliminates carbon oxysulfide [*55,56*]. Treatment of the resulting amidine with excess tosyl isocyanate leads to the interesting bis(amidine) of squaric acid (**91**) in very good yield [*19*]. The corresponding 1,3 isomer (**92**) is prepared in the same fashion with 1,3-DTSD (**44a**) as starting material.

V. Thiodeltic Acid Derivatives

Since the successful synthesis of the fascinating deltate dianion **93** [*57,58*] (see Scheme 28), all our efforts to synthesize the corresponding sulfur analog (**94**), the three-carbon member of the thioxocarbon dianions [*59*], have proved unsuccessful. Only some interesting derivatives of the thiodeltate dianion **94**

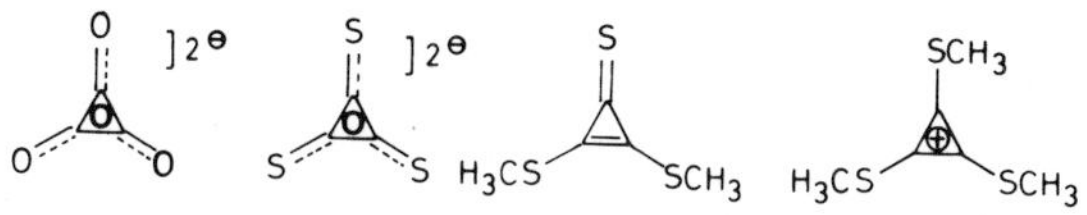

(93) (94) (95) (96)

Scheme 28

have recently been reported, such as the thiodeltic acid dimethyl ester **95** [*60*] and the aromatic trimethylthiocyclopropenyl cation **96** [*60–62*]. However, new sulfur-containing pseudo-oxocarbons in the aryl-substituted deltic acid series are easily obtained from the corresponding vinylog thioamides (**97**) [*63,64*]. These can be prepared in a two-step reaction from aryltrichlorocyclopropene (**98**) via the intermediate vinamidinium cation **99** (Scheme 29). Reaction with hydrogen sulfide in pyridine gives the thioamide **97** in high yield [*64,65*]. Upon treatment with potassium sulfide the aryl-substituted pseudo-oxocarbon **100** is obtained and can be isolated as the yellow, not very stable potassium salt. Compound **100** is best characterized by successive alkylation with methyl iodide, which leads to the aromatic cation **102** via the yellowish-brown dithioester **101** and by ^{13}C-NMR spectroscopy [*59*]. In addition, **102** can be obtained by an independent route from **98** upon treatment with silver fluoroborate and excess methylmercaptan.

In the same fashion as described above, reaction of **97** with potassium hydroxide or hydrogen selenide leads to the mixed pseudo-oxocarbons **103** and **104,** respectively (Scheme 30), which are stable enough to be characterized by ^{13}C-NMR spectroscopy in D_2O solution [*63*]. The corresponding malononitrile pseudo-oxocarbon anion (**106**) with one sulfur atom can be attained by the reaction sequence **99**→**106** [*59*]. The anion **106** is a stable compound, which can easily be characterized by spectroscopic data and alkylation to the triafulvene derivative **107.**

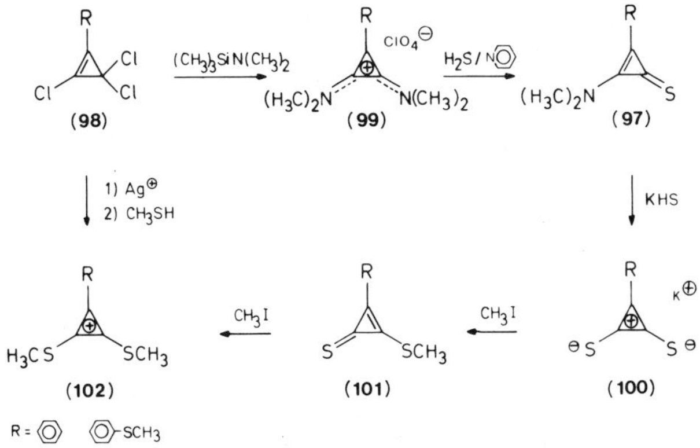

(98) $(CH_3)_3SiN(CH_3)_2$ (99) $ClO_4^{\ominus}$ H_2S / Py (97)

1) $Ag^{\oplus}$ 2) CH_3SH KHS

(102) CH_3I (101) CH_3I (100)

R = C_6H_5, C_6H_4-SCH_3

Scheme 29

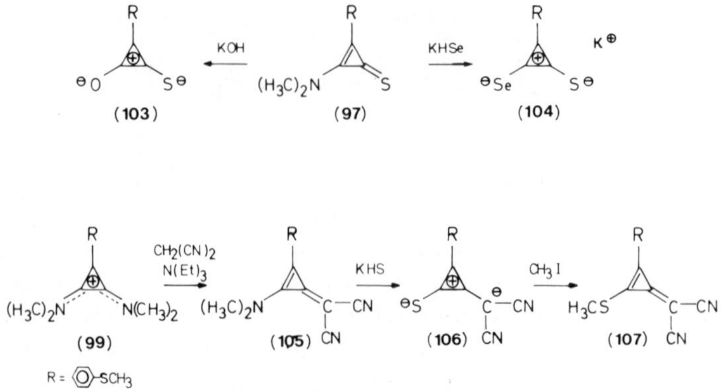

R R R
KOH KHSe K⊕
⊖O S⊖ (H3C)2N S ⊖Se S⊖
(103) (97) (104)

R R R R
CH2(CN)2
N(Et)3 KHS CH3I
(H3C)2N N(CH3)2 (H3C)2N C—CN ⊖S C—CN H3CS C—CN
(99) (105) CN (106) CN (107) CN
R = ⟨O⟩-SCH3

Scheme 30

VI. Mixed C_5 Oxo–Thioxocarbon Dianions

A. SYNTHESIS AND PROPERTIES

Although sulfur derivatives of the croconate dianion, the dinosaur among the oxocarbon dianions, were mentioned about 100 years ago [*66,67*], they remained uninvestigated until recently [*68,69*]. The synthesis of the dithiocroconate dianion **108**, the first C_5 pseudo-oxocarbon with two adjacent sulfur atoms, was easily achieved by reaction of either dimethyl croconate (**110**) [*68*] or croconic acid 1,2-bis(dimethylamide) (**109**) [*69*] with alcoholic alkali hydrogen sulfide (Scheme 31). The structure of this mixed C_5 oxo–thioxocarbon was established on the basis of spectroscopic data, especially of the ^{13}C-NMR spec-

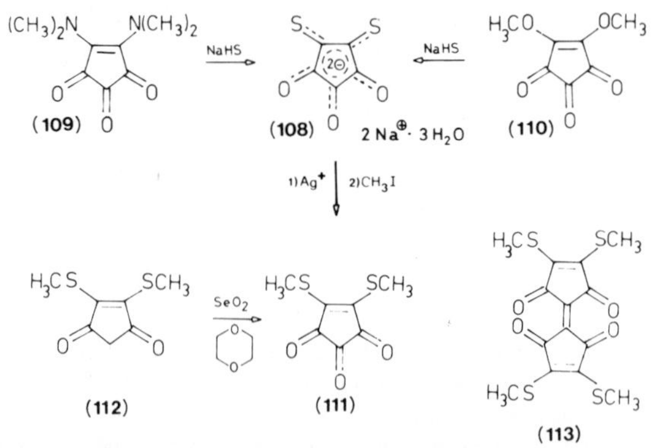

(CH3)2N N(CH3)2 S S H3CO OCH3
NaHS (2⊖) NaHS
O O O O O O
(109) O (108) O 2 Na⊕ · 3 H2O (110) O
1) Ag+ 2) CH3I
H3CS SCH3
H3CS SCH3 H3CS SCH3
SeO2
O O O O O O
O O
O
(112) (111) H3CS SCH3
(113)

Scheme 31

trum, which exhibited three signals at δ = 184.7 and 187. 5 ppm (C—O) and δ = 204.8 ppm (C—S). Visible–UV studies in aqueous solutions at different acidities demonstrate that the dithiocroconate spectrum exhibits recognizable alteration as the pH is lowered from 4 to 3, whereas the croconate spectrum is unchanged down to pH 2.7. This leads to the conclusion that dithiocroconate **108** is a stronger proton acceptor than is croconate and suggests that resonance stabilization in the croconate dianion must be substantially greater than in dithiocroconate [*68*] since simple thiols are more acidic than their oxygen counterparts in the absence of resonance effects. These findings are confirmed by investigation of the IR spectrum of $K_2C_5O_3S_2$ in KBr, which exhibits strong absorption bands at 1658(s), 1602(m), 1564 cm^{-1} (vs, broad), indicating an increased C—O bond order compared with that of the croconate dianion.

B. CHEMICAL BEHAVIOR

Some of the interesting chemical reactions of the dark red dithiocroconate dianion **108** are summarized in Scheme 31. The black silver salt, which is quantitatively precipitated from aqueous solution, dissolves in potassium iodide or sodium thiosulfate solutions to regenerate the dithiocroconate dianion. Treatment of dry silver dithiocroconate with excess methyl iodide yields the intensely red ''dithio ester'' of croconic acid (**111**). The observation of one singlet in the ^{1}H-NMR spectrum at δ = 3.07 ppm is consistent only with structure **111**, which can be additionally confirmed by an independent synthetic route. Thus, 4,5-bis(methylthio)-4-cyclopentene-1,3-dione (**112**) can be oxidized by selenium dioxide in dioxane, yielding a red reaction product identical with that from alkylation of the silver salt of **108** [*69*]. Under these conditions condensation of the starting material with the oxidation product leads to the intriguing by-product **113**, a possible precursor of a sulfur-substituted ''diaryl,'' consisting of two C_5 pseudo-oxocarbon rings [*69*]. Solutions containing the dithiocroconate dianion react to form intensely colored complexes of various hues with several transition metal ions, which have been investigated by spectral and ESR measurements [*68*]. These studies suggest that the 1,2-dithiocroconate is functioning as a bidentate ligand, with four coordination prefered for Ni(II) and Cu(II).

Acknowledgments

Our work on the thioxocarbon dianions and their derivatives described above was supported by grants from the Deutsche Forschungsgemeinschaft and from Fonds der Chemischen Industrie, Frankfurt, Germany, as well as by Chemische Werke Hüls, Germany, who supplied squaric acid derivatives. I would like to thank heartily many collaborators whose names appear in the reference section for their contributions to the work.

References

1. R. West and J. Niu, *in* "The Chemistry of the Carbonyl Group" (J. Zabicky, ed.), Vol. 2, p. 241. Wiley (Interscience), New York, 1970.
2. S. Hünig and H. Pütter, *Chem. Ber.* **110,** 2532 (1977); *Angew. Chem., Int. Ed. Engl.* **11,** 431 (1972); **12,** 149 (1973).
3. D. Coucouvanis, F. J. Hollander, R. West, and D. Eggerding, *J. Am. Chem. Soc.* **96,** 3006 (1974).
4. D. Eggerding and R. West, *J. Org. Chem.* **41,** 3904 (1976).
5. R. Allmann, T. Debaerdemaeker, K. Mann, R. Matusch, R. Schmiedel, and G. Seitz, *Chem. Ber.* **109,** 2208 (1976).
6. G. Seitz, K. Mann, R. Schmiedel, and R. Matusch, *Chem.-Ztg.* **99,** 90 (1975); *Chem. Ber.* **112,** 990 (1979).
7. R. Matusch, R. Schmiedel, and G. Seitz, *Justus Liebigs Ann. Chem.* p. 595 (1979).
8. A. H. Schmidt, W. Ried, and P. Pustoslemsek, *Chem.-Ztg.* **101,** 154 (1977).
9. H. J. Becher, D. Fenske, and E. Langer, *Chem. Ber.* **106,** 177 (1973).
10. H. E. Sprenger and W. Ziegenbein, *Angew. Chem., Int. Ed. Engl.* **6,** 553 (1967); **7,** 530 (1968).
11. B. Lunelli, C. Corvaja, and G. Farina, *Trans. Faraday Soc.* **67,** 1951 (1975).
12. C. Corvaja, G. Farina, and B. Lunelli, *J. Chem. Soc., Faraday Trans. 2* **71,** 1293 (1975).
13. H. Morck, R. Schmiedel, and G. Seitz, *Chem.-Ztg.* **103,** 188 (1979).
14. A. J. Fatiadi, *J. Am. Chem. Soc.* **100,** 2586 (1979), and references therein.
15. G. Maas and P. Hegenberg, *Angew. Chem., Int. Ed. Engl.* **5,** 888 (1966).
16. G. Seitz, H. Morck, K. Mann, and R. Schmiedel, *Chem.-Ztg.* **98,** 459 (1974); **99,** 332 (1975).
17. G. Seitz and R. Sutrisno, *Synthesis* p. 831 (1978).
18. G. Seitz, R. Schmiedel, and K. Mann, *Chem.-Ztg.* **99,** 463 (1975).
19. G. Seitz, R. Schmiedel, and K. Mann, *Arch. Pharm. (Weinheim, Ger.)* **310,** 549 (1977).
20. F. Götzfried, W. Beck, A. Lerf, and A. Sebald, *Angew. Chem., Int. Ed. Engl.* **18,** 463 (1979).
21. G. Le Coustumer, I. Amzil, and Y. Mollier, *J. Chem. Soc., Chem. Commun.* p. 353 (1979).
22. R. Allmann, *in* "Structural Chemistry of Hydrazo, Azo and Azoxy Compounds" (S. Patai, ed.), Chapter 2, p. 23. Wiley (Interscience), New York, 1975.
23. N. Trinajstic, *Tetrahedron Lett.* p. 1529 (1968).
24. G. Seitz and H. Morck, *Chimia* **26,** 368 (1972).
25. D. Eggerding, J. L. Straub, and R. West, *Tetrahedron Lett.* p. 3589 (1975).
26. I. Shahak and J. Sasson, *J. Am. Chem. Soc.* **95,** 3440 (1973).
27. G. Seitz, R. Schmiedel, and K. Mann, *Synthesis* p. 578 (1974).
28. R. C. De Selms, C. J. Fox, and R. C. Riordan, *Tetrahedron Lett.* p. 781 (1970).
29. G. Seitz and B. Gerecht, unpublished results.
30. D. Coucouvanis, D. G. Holak, and F. J. Hollander, *Inorg. Chem.* **14,** 2657 (1975).
31. J. W. Scheeren, P. H. J. Ooms, and R. J. F. Nivard, *Synthesis* p. 149 (1973), and references therein.
32. H. Z. Lecher, R. A. Greenwood, K. C. Whitehouse, and T. H. Chao, *J. Am. Chem. Soc.* **78,** 5018 (1956).
33. B. S. Pedersen, S. Scheibye, N. H. Nilsson, and S.-O. Lawesson, *Bull. Soc. Chim. Belg.* **87,** 223 (1978).
34. W. Kusters and P. de Mayo, *J. Am. Chem. Soc.* **96,** 3502 (1974), and references therein.
35. H. E. Simmons, D. C. Blomstrom, and R. D. Vest, *J. Am. Chem. Soc.* **84,** 4782 (1962).
36. G. Seitz and K. Mann, unpublished results, Dissertation K. Mann, Marburg, Germany, 1977.
37. Private communication from Prof. Dr. R. Mattes, University of Münster, Germany.
38. W. Walter and J. Voss, *in* "The Chemistry of Thioamides" (J. Zabicky, ed.), p. 415. Wiley (Interscience), New York, 1970.

39. R. J. Cole, J. W. Kirksey, H. G. Cutler, B. L. Doupnik, and J. C. Peckham, *Science* **179,** 1324 (1973).
40. For a review of moniliformin and related cyclobutenediones see D. Bellus and H. P. Fisher, *in* "Advances in Pesticide Science" (H. Geissbühler, ed.), Part 2, p. 373. Pergamon, Oxford, 1979.
41. G. Seitz, R. Schmiedel, and K. Mann, *Arch. Pharm. (Weinheim, Ger.)* **310,** 991 (1979).
42. G. Seitz, R. Matusch, and K. Mann, *Chem.-Ztg.* **101,** 557 (1977).
43. T. Fukunaga, *J. Am. Chem. Soc.* **98,** 610 (1976).
44. A. J. Fatiadi, *Synthesis* p. 165 (1978).
45. G. Seitz and R. Sutrisno, unpublished results.
46. H. Ehrhardt, S. Hünig, and H. Pütter, *Chem. Ber.* **110,** 2506 (1977).
47. C. U. Pittmann, A. Kress, and L. D. Kispert, *J. Org. Chem.* **39,** 378 (1974), and references therein.
48. H. Ehrhardt and S. Hünig, *Tetrahedron Lett.* p. 3515 (1976).
49. G. A. Olah and J. S. Staral, *J. Am. Chem. Soc.* **98,** 6290 (1976), and references therein.
50. For comparison, see J. E. Thorpe, *J. Chem. Soc. B* pp. 435, 1534 (1968).
51. G. Seitz and R. Schmiedel, unpublished results, Dissertation R. Schmiedel, Marburg, Germany, 1977.
52. G. Seitz, H. Morck, R. Schmiedel, and R. Sutrisno, *Synthesis* p. 361 (1979).
53. G. Seitz and R. Sutrisno, unpublished results.
54. J. F. W. McOmie and D. H. Perry, *J. Chem. Soc., Chem. Commun.* p. 248 (1973).
55. H. Ulrich, "Cycloaddition Reactions of Heterocumulenes," pp. 160ff. Academic Press, New York, 1967.
56. L. A. Paquette, T. J. Barton, and N. Horton, *Tetrahedron Lett.* p. 5039 (1967).
57. D. Eggerding and R. West, *J. Am. Chem. Soc.* **98,** 3641 (1976), and references therein.
58. M. A. Pericás and F. Serratosa, *Tetrahedron Lett.* p. 4437 (1977).
59. G. Seitz and G. Arndt, unpublished results.
60. R. Weiss, C. Schlierf, and K. Schloter, *J. Am. Chem. Soc.* **98,** 4668 (1976).
61. R. Gompper and U. Jersak, *Tetrahedron Lett.* p. 3409 (1973). Z. Yoshida, S. Miki, and S. Yoneda, *ibid.* p. 4731.
62. For a recent review, see Z. Yoshida, *Top. Curr. Chem.* **40,** 47 (1973).
63. For a review of arylsubstituted C_4-thioxocarbons, see H. Schmidt and W. Ried, *Synthesis* p. 1 (1978).
64. G. Seitz and G. Arndt, *Synthesis* p. 693 (1976).
65. M. T. Wu, D. Taub, and A. A. Patchett, *Tetrahedron Lett.* p. 2405 (1976).
66. J. U. Lerch, *Justus Liebigs Ann. Chem.* **124,** 39 (1862).
67. R. Nietzki and R. Benckiser, *Ber. Dtsch. Chem. Ges.* **19,** 299 (1886).
68. R. F. X. Williams, *Phosphorus Sulfur* **2,** 141 (1976).
69. G. Seitz, K. Mann, and R. Matusch, *Arch. Pharm. (Weinheim, Ger.)* **308,** 792 (1975).

3

Physical Chemistry of Aqueous Oxocarbons

Lowell M. Schwartz, Robert I. Gelb, and Daniel A. Laufer

I. Introduction 43
II. Survey of Oxocarbon Acid pK Determinations 44
III. Thermodynamic Data 46
IV. Structures of Aqueous Oxocarbons 49
A. Deltic Acid 49
B. Squaric Acid 50
C. Croconic Acid 52
D. Rhodizonic Acid 53
V. A New Type of Aqueous Complex 56
References 57

I. Introduction

This chapter surveys the current state of understanding of the physical chemistry of aqueous oxocarbon species and equilibria. The study focuses on the oxocarbon acids deltic, squaric, croconic, and rhodizonic and their anions and only mentions the existence of derivatives of these made by replacing hydroxyls with other functional groups or by replacing oxygens with sulfur or selenium atoms. Nor is any consideration given here to crystalline substances or to solutions or oxocarbons in mixed or nonaqueous solution.

For convenience several abbreviations are used: Sq^{2-} for squarate, Cr^{2-} for croconate (this cannot be confused with the element chromium which is never involved in these reactions), and R^{2-} for rhodizonate. It is expected that discrimination between the latter symbol and the gas constant should be obvious from context. Protonated forms of these dianions are written by attaching appropriate H's.

OXOCARBONS

ISBN 0-12-744580-3

Virtually all reactions in this chapter involve transfer of protons and hence are acid ionizations. Normally, the equilibrium constants for such reactions are subscripted, i.e., K_a or pK_a, to designate "acid," but in this chapter the subscripts a are omitted for conciseness. The term "ionization" is preferred here because some ionizations are accompanied by other forms of fragmentation of the principal species, and the term "dissociation" would be ambiguous in such cases. Finally, because our readers are likely to be more familiar with the traditional thermodynamic units, we do not use the SI system and abbreviate cal $mole^{-1}$ K^{-1} as e.u. for entropy units.

The parent series of oxocarbon acids having nominal formulas $H_2C_nO_n$ are dibasic as aqueous solutes yielding monoanions $HC_nO_n^-$ and dianions $C_nO_n^{2-}$ upon proton ionization. The strength of an acid is commonly understood to refer to the magnitude of the ionization equilibrium constant. The primary ionization constant K_1 or pK_1 corresponding to

$$H_2C_nO_n + H_2O = H_3O^+ + HC_nO_n^- \quad (1)$$

is a measure of the parent acid strength, and the secondary ionization constant K_2 or pK_2 corresponding to

$$HC_nO_n^- + H_2O = H_3O^+ + C_nO_n^{2-} \quad (2)$$

is a measure of the monoanion acid strength. Recent studies have shown however, that these reactions written as uncomplicated proton transfers do not adequately reflect structural changes that occur in the oxocarbon species in several cases. These changes are discussed in Section IV.

II. Survey of Oxocarbon Acid p*K* Determinations

A survey of oxocarbon acid ionization constant determinations is summarized in Table I. This survey is restricted to reports that give some indication of the methodology used and to those published within the previous two decades. References to earlier works can be found in papers by Alexanderson and Vannerberg [*1*] and Patton and West [*2*]. Table I includes only the parent series $H_2C_nO_n$ with n = 3–6. Acidities have been found for several derivatives formed by replacing one of the parent hydroxyls with some other functional group, and Patton and West [*3*] list those reported to 1973. Each entry in the second column of Table I is the pK value at ambient temperature and the investigator's estimate of the statistical uncertainty if quoted. Equilibrium constants are variously reported as "thermodynamic" values by extrapolation to zero ionic strength or "conditional" values appropriate to a fixed nonzero ionic strength medium. The third column of Table I gives the ionic strength corresponding to the pK, and the fourth column states the method used to extrapolate to zero ionic strength or the solvent medium if a conditional constant is reported. A number of substances

Table I. Oxocarbon Acid Strengths at 25°C

Ionization	pK	Ionic strength μ (M)	Medium/ extrapolation[a]	Experimental technique[b]	Ref.
Primary, pK_1					
$H_2C_3O_3$, deltic acid	2.57 ± 0.04	0	D–H	Pot.	*12*
$H_2C_4O_4$, squaric acid	1.7 ± 0.3	0.1	KCl	Pot.	*13*
	1.2 ± 0.2[c]	0	Davies	Pot.	*14*
	0.55 ± 0.15	0	Davies	Pot.	*15*
	0.51 ± 0.02	0	Davies	Cond.	*17*
	0.54 ± 0.06	0	Davies	Cond.	*16*
	0.96 ± 0.03	3	$NaClO_4$	Pot.	*1*
$H_2C_5O_5$,[d] croconic acid	0.68	?		Spect.	*18*
	0.32	2	HCl, NaCl	Spect.	*19*
	0.50 ± 0.05	0	H_2SO_4/a.f.	Spect.	*20*
	0.7 ± 0.03	0	HCl/H–E	Pot.	*21*
	0.80 ± 0.08	0	HCl/H–E	Spect.	*21*
	0.75 ± 0.02	0	HCl, KCl/D–H	Color.[e]	*22*
$H_2C_6O_6$,[d] rhodizonic acid	4.20	?		Spect.	*23*
	4.25 ± 0.05	0	KCl/D–H	Spect.	*2*
	4.02 ± 0.05	?		Spect.	*24*
	3.9 ± 0.5[f]	3	$NaClO_4$	Spect.	*1*
	3.45 ± 0.11	3	$NaClO_4$	Pot.	*1*
	4.378 ± 0.009	0	D–H	Pot.	*25*
Secondary, pK_2					
$HC_3O_3^-$	6.03 ± 0.06	0	D–H	Pot.	*12*
$HC_4O_4^-$	3.21	0.1	KCl	Pot.	*13*
	3.48 ± 0.02	0	Davies	Pot.	*14*
	2.89	0.5	NaCl, $NaClO_4$	Pot.	*26*
	3.48 ± 0.02	0	Davies	Pot.	*15*
	3.19 ± 0.01	3	$NaClO_4$	Pot.	*1*
$HC_5O_5^-$	1.97	?		Spect.	*18*
	1.51	2	HCl, NaCl	Spect.	*19*
	1.65 ± 0.10	0.5	H_2SO_4, Na_2SO_4	Spect.	*20*
	2.28 ± 0.14	0	HCl/H–E	Pot.	*21*
	2.24 ± 0.01	0	HCl/H–E	Spect.	*21*
$HC_6O_6^-$[d]	4.65	?		Spect.	*23*
	5.06	0.1	$NaClO_4$	Pot.	*27*
	4.72 ± 0.07	0	KCl	Spect.	*2*
	4.62 ± 0.05	?		Spect.	*24*
	3.1 ± 0.09[f]	3	$NaClO_4$	Spect.	*1*
	3.58 ± 0.11	3	$NaClO_4$	Pot.	*1*
	4.65 ± 0.01	0	D–H	Pot.	*25*

[a] Extrapolation methods: D–H, Debye–Hückel activity coefficient correlation [*28*]; Davies, activity coefficient [*29*]; H–E, Harned–Ehlers activity coefficients of HCl [*30*]; a.f., activity function.

[b] Experimental techniques: Pot., potentiometric; Cond., conductimetric; Spect., spectrophotometric; Color., colorimetric.

[c] Miscalculation of pK_1 (private communication).

[d] Nominal formula.

[e] Cresol red indicator.

[f] Note pK_1 > pK_2.

that are not members of the cyclic $H_2C_nO_n$ series might be classified as oxocarbon acids if the definition is extended to include those acidic compounds having a carbon–oxygen skeleton and whose anions are believed to be stabilized by π-electron delocalization. Ionization constants for some of these are as follows: carbonic acid H_2CO_3 : $pK_1 = 3.88$ [*4*], $pK_2 = 10.329$ [*5*]; oxalic acid $H_2C_2O_4$: $pK_1 = 1.271$ [*6*], $pK_2 = 4.266$ [*7*]; acetylenedicarboxylic acid $H_2C_4O_4$ (an isomer of squaric acid) : $pK_1 = 0.656$, $pK_2 = 2.336$ [*8*]; and tetrahydroxy-*p*-benzoquinone $H_4C_6O_6$: $pK_1 = 4.8$, $pK_2 = 6.8$ [*7,9*].

Although it is widely held that oxocarbon acids are quite strong, Table I shows that only squaric and croconic acids have pK_1 values less than unity, but these are greater than zero. In comparison, the "strong" mineral acids HNO_3 and $HClO_4$ have pK values of -1.45 and -1.59, respectively [*10*]. Several organic cyanocarbon acids, such as cyanoform, $pK = -5.1$, and hexacyanoisobutylene, $pK_1 < -8.5$, $pK_2 \simeq -2.5$ [*11*], also are much stronger than the strong oxocarbon acids. Therefore, the descriptor "moderately strong" applied to squaric, croconic, and acetylenedicarboxylic acidities puts these in a more appropriate perspective. These three acids are much stronger than enols and most carboxylic acids, which, of course, are "weak," having acid ionization pK values ranging upward from about 2.

It is clear from Table I that neither pK_1 nor pK_2 of the cyclic $H_2C_nO_n$ series varies in a regular way as n changes from 3 to 6. Such a trend might be expected if the energetics of the acids and anions varied in a simple way with the size of the ring skeleton and if the proton dissociations took place *in vacuo* in the absence of solvent interactions. Clues to the role of solvent in the irregularity of oxocarbon acid pK values can be found by comparing the thermodynamic parameter values of ionization, by ^{13}C-NMR structural studies, and to some extent by the ultraviolet–visible spectra of the aqueous oxocarbon species.

III. Thermodynamic Data

Structural information about the species involved in a reaction can be inferred from the thermodynamic parameters ΔH^0 and ΔS^0 associated with the reaction. Values for these parameters are obtained either by temperature variation of the equilibrium constants or by direct calorimetry. The former method depends on the fact that each K or pK for Eq. (1) or (2) is measured for an equilibrium mixture at fixed temperature and pressure so that the following well-known thermodynamic relationships apply:

$$\Delta G^0 = -RT \ln K = 2.303RT\, pK \tag{3}$$

$$\Delta G^0 = \Delta H^0 - T\,\Delta S^0 \tag{4}$$

$$-R\left[\frac{\partial \ln K}{\partial(1/T)}\right]_{\mathrm{p}} = 2.303R\left[\frac{\partial \mathrm{p}K}{\partial(1/T)}\right]_{\mathrm{p}} = \Delta H^{0} \tag{5}$$

$$R\left[\frac{\partial(T \ln K)}{\partial T}\right]_{\mathrm{p}} = -2.303R\left[\frac{\partial(T\,\mathrm{p}K)}{\partial T}\right]_{\mathrm{p}} = \Delta S^{0} \tag{6}$$

The temperature variation of the equilibrium constant at some given temperature thus is used to calculate ΔH^0 from the appropriate slope via Eq. (5) and ΔS^0 from Eq. (6). Alternatively, either parameter can be combined with K at that same temperature to find the other parameter by eliminating ΔG^0 from Eqs. (3) and (4). When the direct calorimetric method is used to measure ΔH^0, ΔS^0 is calculated in this way.*

Those determinations of ΔH^0 and ΔS^0 for oxocarbon acid dissociations made to date are listed in Table II. Before making deductions about the structures of species involved in a reaction from ΔH^0 and ΔS^0 data, it is important to recognize the following principle, which has been quoted many times [*32,33*]. These enthalpy and entropy changes reflect energetic and configurational differences, respectively, of the reacting system as a whole, i.e., the solvent medium as well as the reacting solutes. Consequently, a given ΔH^0 datum reflects both the difference in energies between product and reactant species *in vacuo* and the difference of energetic interaction between the product–solvent system and the reactant–solvent system. Similarly, a given ΔS^0 value can be attributed to several effects: (a) the intrinsic configurational differences between product and reactant species *in vacuo,* (b) the configurational changes induced in these species by the solvent environment, and (c) the configurational difference between the solvent surrounding the product species and that of the reactant species. Although many statistical mechanical and semiempirical theories [*34*] have been applied to the understanding of these phenomena, this discussion takes a completely empirical approach. Thermodynamic data for oxocarbon systems are compared to corresponding data for other systems, and structural explanations are offered which are consistent with results of nonthermodynamic experiments. In adopting this empirical approach, no attempt is made to explain why, for example, a given oxocarbon acid ionization has the specific ΔS^0 value observed, but rather why this ΔS^0 is similar to or different from the corresponding value of some reference acid. The reference chosen for this purpose is not a specific molecule but rather the average of all comparable carboxylic acids for which thermodynamic data of aqueous ionization are conveniently available. Eberson and Wadsö [*35*] compiled such a list of about 60 ionizations in 1963, and Larson and Hepler [*31*] gathered about 90 in 1967 and listed these in their Tables 1-2 and 1-3. Neither list

*Larson and Helper [*31*] have noted that ΔH^0 values for the same reaction measured by calorimetry and by temperature variation of K are frequently in conflict even though the two methods are equally valid in principle. Both experimental methods are subject to sources of systematic error, and so only when utmost care is taken in both experiments can agreement be expected.

Table II. Thermodynamic Data[a] for Oxocarbon Acid Ionizations

Acid	Primary ionization		Secondary ionization		Method[b]	Ref.
	ΔH_1^0	ΔS_1^0	ΔH_2^0	ΔS_2^0		
Deltic acid	~0	−12	~0	−27	T.V./Pot.	*12*
Squaric acid	−1.5 ± 0.1	−7.5 ± 0.7			T.V./Cond.	*16*
			−3.0 ± 0.5	−26 ± 2	T.V./Pot.	*15*
			−1.75 ± 0.05	−19.8 ($pK_2 = 3.05$)[c]	Cal./$\mu = 0.1\ M$	*36*
Croconic acid	3.46	8.9 ($pK_1 = 0.60$)[c]	−2.39	−16.2 ($pK_2 = 1.80$)[c]	Cal./$\mu = 0.1\ M$	*36*
	3.9 ± 0.2	9.5 ± 0.7	−3.0 ± 0.1	−20.1 ± 0.4	T.V./Spect.	*21*
	3.9 ± 0.3[d]	9.8 ± 0.7			T.V./Color.	*21*
Rhodizonic acid	4.6 ± 0.3	−4.6 ± 0.8	9.6 ± 0.3	11 ± 1	T.V./Pot.	*25*

[a] The ΔH^0 values are in kilocalories $mole^{-1}$; ΔS^0 in calories $mole^{-1}\ K^{-1}$ (entropy units); all data at 25°C.

[b] Abbreviations: T.V., temperature variation of K using technique designated; Cal., calorimetric at ionic strength designated. For other abbreviations, see footnote *b*, Table I.

[c] Calorimetric ΔS^0 values calculated from pK values in parentheses.

[d] Calculated from data given in the reference.

is up to date, but there is little likelihood that additions would be so abnormal as to change the averages very much. The Larson and Hepler tables include data for 14 dibasic acids ranging in size from oxalic to suberic acid and some including hydroxy, aromatic, or unsaturated functionality. Using these 14 as a basis, the ranges of variability are $-1.3 < \Delta H_1^0 < 0.8$ and $-1.6 < \Delta H_2^0 < 0.3$ kcal/mole and $-22 < \Delta S_1^0 < -8$ and $-36 < \Delta S_2^0 < -19$ e.u. These ranges also may be expressed as excursions from the midpoint: $\Delta H_1^0 = -0.2 \pm 1.1$, $\Delta H_2^0 = -0.6 \pm 1.0$ kcal/mole and $\Delta S_1^0 = -15 \pm 7$, $\Delta S_2^0 = -27 \pm 9$ e.u. A hypothetical typical acid, which shall be called "typic acid" and written H_2Ty, will be assigned thermodynamic data corresponding to the midpoints of the ranges, and so, by Eqs. (3) and (4), typic acid has $pK_1 = 3.1$ and $pK_2 = 5.5$ at 25°C.

A cursory examination of the same tables indicates that ΔH^0 values exist in a narrow range about zero, but ΔS^0 values are more correlated with the large variation of observed pK values. These observations led Eberson and Wadsö [*35*] to develop the empirical relationship

$$\Delta S^0 = -(0.3 + 4.94\ pK) \tag{7}$$

to fit their tabulated data, which included both primary and secondary ionizations and spanned the range $1.9 < pK < 8.3$. The quality of the fit of the data to this correlation is ± 2 e.u. It is significant that Eberson and Wadsö could not find a similar relationship between ΔH^0 and pK. Thus, in addition to "typic" acid being used for a reference level, ± 2 e.u. deviation about Eq. (7) will serve as a rough range of normal variability for ΔS^0, and similarly -0.4 ± 1.2 kcal/mole will serve for ΔH^0.

IV. Structures of Aqueous Oxocarbons

It appears that the aqueous equilibria involving the oxocarbon acid systems becomes progressively more complicated as the ring size increases from three to six carbon atoms. Each system is discussed separately.

A. DELTIC ACID

Relatively little work has been done on the physical chemistry of deltic acid. Both the parent and monoanion are weak acids ($pK_1 = 2.6$, $pK_2 = 6.0$) [*12*], somewhat similar in strength to the reference acid H_2Ty ($pK_1 = 3.1$, $pK_2 = 5.5$). Little temperature variation of the pK values was observed, and, because the scatter of data plotted versus temperature was relatively large, only rough thermodynamic values could be estimated. Nevertheless, each datum listed in Table II for deltic acid is well within the corresponding range for typical carboxylic acids mentioned in Section III. Thus, the simplest picture for the deltic acid system is that no unusual changes occur as deltate dianion protonates to monoanion and then to parent acid. If, on the other hand, deltate dianion is hypothesized to be more stabilized by π-electron delocalization than the protonated forms, this effect must be small or must be offset by some other factor tending to increase the stabilities of the protonated forms.

B. SQUARIC ACID

With $pK_1 = 0.5$ and $pK_2 = 3.5$, squaric acid and its monoanion are each stronger than the reference acid, having $pK_1 = 3.1$ and $pK_2 = 5.5$. Comparisons are more conveniently made using the magnitude of the standard Gibbs free-energy change of ionization, which, like pK, is an inverse measure of the acid strength. This parameter more directly reflects the relative contributions of energetic and configurational factors to the acid strength. At any given temperature ΔG^0 is composed of ΔH^0 and $-T\,\Delta S^0$, and Table III shows the magnitudes of these quantities for both the oxocarbon acids and the reference acid. The superior strength of H_2Sq relative to "H_2Ty" is clearly seen by the difference of 3.6 kcal/mole in ΔG_1^0. Contributing to this difference are 1.3 from ΔH_1^0 and 2.3 from $-T\,\Delta S_1^0$. Thus, H_2Sq is as strong as it is for both energetic and configurational reasons, with the latter dominating. The thermodynamic data for the primary dissociation are, nevertheless, not unusual. The ΔH_1^0 value is somewhat low but is within the normal range, and the observed ΔS_1^0 is some 4.5 e.u. more negative than predicted by Eq. (7) for $pK_1 = 0.5$, but, since no other acids of moderate strength were included in the correlation, the explanation is uncertain. An explanation for the dominant role played by $T\,\Delta S^0$ in determining acid strengths can be given by quoting Eberson and Wadsö [*35*], who in turn were reflecting earlier statements by others: "The strength of a given acid is determined to a large extent by the change in orientation and compression of solvent molecules around

Table III. Thermodynamic Contributions[a] to Acid Strengths

Ionization	ΔG_1^0	ΔH_1^0	$-T\,\Delta S_1^0$	Ref.
Primary				
"H_2Ty" reference	4.3	−0.2	4.5	—
H_2Sq	0.7	−1.5	2.2	*16*
H_2Cr	1.1	3.9	−2.8	*21*
H_2R	6.0	4.6	1.4	*25*
Secondary				
"HTy^-"	7.4	−0.6	8.0	—
HSq^-	4.7	−3.0	7.7	*15*
	4.2	−1.7	5.9	*36*
HCr^-	3.0	−3.0	6.0	*21*
HR^-	6.3	9.6	−3.3	*25*

[a] All entries in kilocalories per mole.

the solute which is appreciable even before ionization." Considering that entropic effects are so important in making H_2Sq moderately strong, this explanation seems appropriate.

There is some discrepancy in the enthalpy of secondary ionization of squaric acid. If the data of Schwartz and Howard [*15*] are correct, the monoanion is stronger than the reference "HTy^-," primarily for energetic reasons. Of the difference of 2.7 kcal/mole in ΔG_2^0, 2.4 is due to ΔH_2^0 and only 0.3 is due to ΔS_2^0. At variance with these are the calorimetric results of Orebaugh [*36*], which give more weight to entropic factors in causing the enhanced strength of HSq^- and which, if correct, would negate the following interpretations. It is hoped that future experimentation will resolve this discrepancy. The observed $\Delta H_2^0 = -3.0$ kcal/mole of Schwartz and Howard [*15*] is substantially more negative than normal, but the observed ΔS_2^0 of -26 e.u. is quite similar to that of the reference "typic" acid. It is therefore clear that energetic factors must be responsible for the strength of HSq^-. A reasonable and attractive explanation is that the stability of Sq^{2-} is enhanced relative to HSq^- due to π-electron delocalization, although another possibility is that the solvent interaction with Sq^{2-} is atypically stronger than with HSq^-. It may be significant that ΔH_2^0 for the ionization of HCr^- is also -3.0 kcal/mole, and the coincidence of these two results suggests that the same factors are responsible for both ionizations.

The ultraviolet–visible spectra of aqueous oxocarbon species should be sensitive to differences in the π-electron structures and thus should shed light on this problem. As shown in Table IV the lowest-energy part of the spectrum of each of the oxocarbon dianions [*1,2,18,19,21,37*] is characterized by two overlapping peaks the maxima of which are separated by 20–30 nm. It is observed in aqueous

Table IV. Low-Energy Electronic Spectra of H_2Sq, H_2Cr, H_2R, and Their Anions

	Peak 1		Peak 2	
Species	$\epsilon_{max} \times 10^{-4}$	λ_{max} (nm)	$\epsilon_{max} \times 10^{-4}$	λ_{max} (nm)
H_2Sq [a]	1.34	230	1.90	251
HSq^-	2.08	241	1.81	263
Sq^{2-}	2.27	251	2.31	272
H_2Cr [b]	1.46	286	0.31	341
HCr^-	1.93	309	1.88	355
Cr^{2-}	2.48	334	3.61	365
H_2R [c]				
HR^- [d]				
R^{2-}		~430	3.3	483

[a] Schwartz and Howard [*37*].
[b] Schwartz *et al.* [*21*].
[c] Single peak at 320 nm, $\epsilon_{max} = 1.1 \times 10^4$ [*2*].
[d] Probable single peak near 370 nm.

squaric acid that protonation of Sq^{2-} to HSq^- and then to H_2Sq shifts the two-peak system to slightly higher energies and diminished intensities [*37*]. If the two-peak system results from π-electron delocalization around the ring skeleton, then it is clear that some changes occur upon successive protonation of Sq^{2-}, but it is not clear how to correlate such spectral perturbations with enthalpic changes.

Carbon-13 NMR studies [*22,38*] have shown only a single resonance for squaric acid containing various proportions of the three species. This indicates that the four carbons are equivalent on the NMR time scale, showing (a) that proton exchange with the solvent is rapid and (b) that there are no hydration equilibria similar to those observed in croconic and rhodizonic acid systems.

C. CROCONIC ACID

Both ionizations of croconic acid are substantially stronger ($pK_1 = 0.8$, $pK_2 = 2.2$) than those of the reference acid ($pK_1 = 3.1$, $pK_2 = 5.5$). The thermodynamic data shown in Table II indicate that both ionizations are grossly abnormal. Whereas typical ΔH^0 values are in the range -0.4 ± 1.2 kcal/mole, the ΔH_1^0 of croconic acid is 3.9 and ΔH_2^0 is -3.0 kcal/mole. Also, the observed ΔS_1^0 of about 9.5 e.u. and ΔS_2^0 of -20 e.u. are each far from predictions based on Eq. (7). As mentioned above, ΔH_2^0 of HCr^- ionization is the same as ΔH_2^0 of HSq^-, but HCr^- is stronger as an acid because ΔS_2^0 is less negative than $\Delta S_2^0 = -26$ e.u. for HSq^- ionization. If π-electron delocalization upon ionization explains the abnormally negative ΔH_2^0 values, then perhaps the greater extent of charge distribution about the larger croconate dianion ring is less effective than the squarate ring in orienting solvent molecules. This would explain the differences in entropies observed.

The thermodynamic data for the primary ionization of H_2Cr are so abnormal that the reaction is not believed to be given by Eq. (1), which implies structure **1**

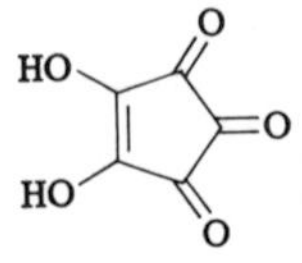

(1)

for the parent acid. There is firm evidence that the diprotonated species is covalently attached to one water molecule and has structure **2**. Carpentier *et al.*

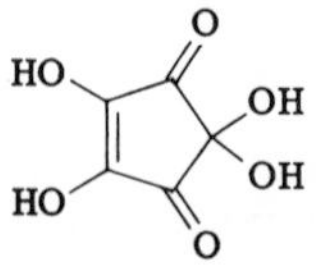

(2)

[*20*] first deduced this fact from the similarity of the UV spectrum of related compounds, and Gelb *et al.* [*22*] provided confirmation by observing three

^{13}C-NMR resonances attributable to un-ionized croconic acid and arising from the three nonequivalent carbons of **2**. It was concluded from both these investigations that the monoanion HCr^- is not hydrated, so that the primary ionization also involves a dissociation of water. The net reaction is given by Eq. (8).

$$H_2C_5O_5(H_2O) = HC_5O_5^- + H_3O^+ \quad (8)$$

Actually, Gelb *et al.* [*22*] combined the ^{13}C-NMR resonance displacements with the measured ionization constant to calculate that the equilibrium partition ratio of species **2** to species **1** is about 9 : 1 and that pK_1 for unhydrated **1** ionizing to HCr^- is nearly zero. Therefore, when 1 mole of croconic acid ionizes yielding the thermodynamic data values given in Table II, the reaction is effectively 0.9 mole by Eq. (8) and 0.1 mole by Eq. (1). Dehydration of aqueous aldehydes and ketones typically involves ΔH^0 values of ~5 kcal/mole [*39*], and similarly the dehydration of the α-keto acid pyruvic acid has $\Delta H^0 = 5.5$ kcal/mole and $\Delta S^0 = 16$ e.u. [*40*]. Therefore, by analogy, a reaction according to Eq. (8) is expected to have ΔH_1^0 some 5 kcal/mole more positive than according to Eq. (1) and ΔS_1^0 some 16 e.u. more positive. These clearly account for the observed anomalous thermodynamic results and also suggest that the primary ionization of unhydrated H_2Cr would be quite similar to that of H_2Sq.

D. RHODIZONIC ACID

The aqueous dissociation of rhodizonic acid is, like that of croconic acid, complicated by hydration, and several investigators have studied the phenomenon recently [*2,24,25*]. All agree that R^{2-} dianion is unhydrated at equilibrium but that both the monoanion and neutral acid are dihydrated. Patton and West [*2*] assumed that the dihydrated system had an ortho geometry, i.e.,

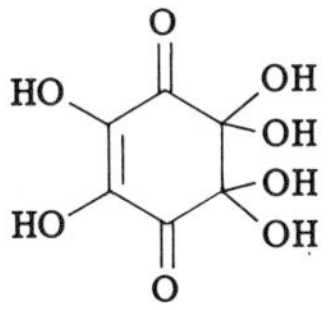

(3)

but Gelb *et al.* [*25*] interpreted ^{13}C-NMR evidence given in Tables V and VI to suggest that the ortho dihydrate system isomerizes slowly to the more stable para dihydrate **4.** Because of rapid proton exchange between the enols and carbonyls,

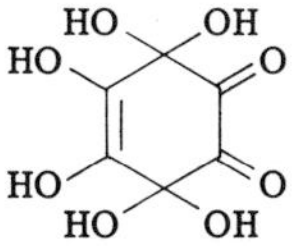

(4)

Table V. ^{13}C-NMR Spectra of Rhodizonic Acid and Its Anions in 5% D_2O/H_2O at 25°C[a]

Mole ratio LiOH/R[b]	^{13}C-NMR (ppm relative to Me_4Si)				Acquisition time (hr)
0	191.2		142.8	95.0	0.1
0.15	191.1		143.3	94.9	0.1
0.57	191.3		~144[c]	95.1	0.4
0.75	191.3		~145[c]	95.1	0.6
1.05	191.3			95.1	2.6
1.29	191.3	177.88		95.0	1.6
1.51	191.2	177.87		95.0	2.1
2.00		177.88			2.5

[a] From Gelb *et al.* [25]. Reprinted with permission. Copyright (1978) American Chemical Society.
[b] Here R represents total rhodizonate species in solution.
[c] Broad peak.

this structure has only two NMR-equivalent carbons and so is consistent with the two resonances that are observed after solutions of H_2R and HR^- are allowed to equilibrate for several days.

The existence of two isomeric forms of aqueous H_2R and HR^- introduces the additional complication as to which of these forms is present in any of the solutions examined in the various spectrophotometric or pH potentiometric studies reported to date. This ambiguity can be resolved by the following observations, however. First, the data in Table V indicate that little or no para dihydrate species are formed for ~2 hr when solid $H_2R \cdot 2H_2O$ is used in solution preparation. Second, Gelb *et al.* [25] found essentially identical pK_1 values when

Table VI. ^{13}C-NMR Spectra of H_2R and Its Anions in 5% D_2O/H_2O at 30°C with Prior Equilibration[a]

Mole ratio LiOH/R[b]	^{13}C-NMR (ppm relative to Me_4Si)					Equilibration time[c] (hr)
0	191.1		142.8	95.0		0
0	191.3	172.7	142.9	95.1	92.7	50
0.07	191.2	172.8	142.9	95.0	92.6	50
0.07		172.4			92.6	75
0.10		172.7			92.7	50
0.27		173.0			92.8	50
0.50		173.9			93.1	50

[a] From Gelb *et al.* [25]. Reprinted with permission. Copyright (1978) American Chemical Society.
[b] Here R represents total rhodizonate species in solution, which was ~0.5*F* throughout.
[c] The equilibration times correspond to standing times at ~21°C after solution preparation until the start of data acquisition, which required about 20 hr.

solutions of H_2R and K_2R were titrated: replicates 4.377 ± 0.005, 4.372 ± 0.007, 4.391 ± 0.008 for K_2R titrations with HCl and 4.372 ± 0.004 for H_2R titrated with NaOH. These observations seem to indicate that the ortho dihydrated species are kinetically favored when R^{2-} solutions are acidified and that the stable para dihydrated species form only after several days of standing. Thus, it appears that studies reported in the literature to date deal with the ortho dihydrated H_2R and HR^- species since the general experimental practice has involved acidification of R^{2-} solutions and prompt measurement.

Visible spectral evidence summarized in Table IV also suggests that substantial changes in π-electron structure accompany protonation and hydration of R^{2-}. The dianion spectrum shown by Patton and West [*2*] has two overlapping peaks near 430 and 480 nm, which are characteristic of π-electron delocalization around the ring, but these two peaks diminish in intensity without shifting to higher energy as the pH decreases. This is in contrast to the corresponding spectral behavior observed in the squarate and croconate systems. The double hydration of R^{2-} evidently perturbs the ring molecular orbitals into a more localized configuration, which yields a spectrum of substantially higher energy for the monoanion. Thus, decreasing the pH of a solution of R^{2-} simply diminishes the characteristic R^{2-} peak intensities in the visible, but decreasing the pH of a solution of, say, Sq^{2-} not only diminishes the Sq^{2-} spectrum but increases the HSq^- spectrum amplitude, which is only slightly shifted to higher energy.

From Table I it can be seen that the apparent pK values of rhodizonic acid are quite close. The thermodynamic pK_1 found by Patton and West [*2*] is 4.25, whereas that measured by Gelb *et al.* [*25*] is slightly different at 4.38. Both groups of investigators measured pK_2 values agreeing within statistical uncertainty near 4.7. Only a single thermodynamic study has been made [*25*], and that showed very unusual results, as seen in Table II. The secondary ionization enthalpy change is approximately 10 kcal/mole more positive than normal and no doubt reflects the energetically unfavorable dissociation of two water molecules along with the proton. The corresponding entropy change is also abnormally positive, and this also reflects the double dehydration which is entropically favorable, perhaps by as much as 30 e.u. However, the thermodynamic data for the primary ionization are more difficult to interpret. The observed ΔH_1^0 of 4.6 kcal/mole shows that the reaction is energetically unfavorable, but in this case dehydration cannot be the source. The ΔS_1^0 of -4.6 e.u. is much less negative than would be expected from the correlation of Eq. (7). These discrepancies may result from changes in bonding caused by the ionization reaction. Thus, bonding in neutral rhodizonic acid might involve a structure (**5**) having a bond across the ring, similar to that suggested many years ago by Carpeni [*41*]. The existence of such a bond would impart energetic stability and rigidity to the neutral molecule such that its disruption upon ionization would tend to make both enthalpy and

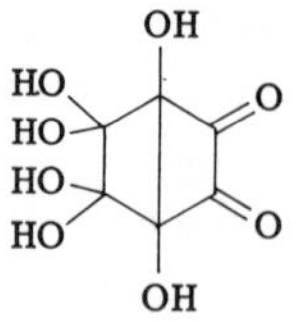

(5)

entropy changes more positive. Structure **5** is also consistent with the three ^{13}C-NMR resonances observed for the neutral acid with ortho dihydration.

The physical chemistry of rhodizonic acid in solution is far from being well understood, and it is hoped that other studies will be undertaken at least to verify the thermodynamics if not the kinetics involved. The experiments are difficult not only because the solutions are sensitive to air oxidation but because the reactions of hydration and isomerization are slow, so that attainment of equilibrium is difficult to ascertain.

V. A New Type of Aqueous Complex

In Section IV,B thermodynamic data led to the suggestion that the enhanced acid strength of H_2Sq could be attributed to the orientation of solvent molecules around the neutral but polar acid before ionization. Other important configurational evidence about aqueous H_2Sq was the discovery of complexes formed between this molecule and unprotonated forms of the colorimetric acid–base indicators cresol red and 4-phenylazodiphenylamine (PDPA) [*42*]. The visible spectra of these complexes are virtually the same as those of the protonated indicators, which indicates that, as far as its reaction with the unprotonated indicator species is concerned, the squaric acid molecule is essentially indistinguishable from an aqueous proton. This further implies that one of the protons on H_2Sq is virtually transferred to a neighboring solvent molecule, and so aqueous H_2Sq probably has a structure resembling an associated ion pair $H_3O^+HSq^-$. Such an ion pair dipole would interact strongly with the solvent, and hence the configuration of solvent water around the H_2Sq molecule has an unusually high degree of ordering compared to carboxylic acids.

Gelb and Schwartz [*42*] measured the formation constants of these H_2Sq–indicator complexes and found values of $\sim$30 and $\sim$10 for the complexes with cresol red and PDPA, respectively. These values compare with protonation constants of $\sim$19 and $\sim$7 for the two indicators, respectively, and show that the affinities of both indicators are stronger for H_2Sq than for H_3O^+. These investigators searched for similar complexes in aqueous solutions of a few other moderately strong acids. Oxalic acid solutions showed a slight discrepancy between pK_1 measured conductometrically and colorimetrically, and this discrepancy might be due to a very weak complex or might result from another minor source of experimental error [*42*]. On the other hand, it was shown unambiguously [*21*] that croconic acid does not form such a complex. It is now understood, as discussed in Section IV,C, that croconic acid is moderately strong not so much because the neutral acid is highly polar but because of the entropic factors associated with the removal of hydrated water. It is conceivable that unhydrated neutral H_2Cr forms such complexes with indicators but this species cannot reach a sufficiently high concentration to detect it. This assertion is a consequence of the hydration equilibrium of H_2Cr, which at 25°C leaves only about 7–14% of the acid in the unhydrated form. This concentration of H_2Cr corresponds to only 1–3% of the H^+ concentration in the solutions reported by Gelb *et al.* [*22*], so that a complex between unhydrated H_2Cr and cresol red might easily have gone undetected.

Recently, a complex has been found between the moderately strong acetylenedicarboxylic acid and cresol red [*43*]. This discovery indicates that the phenomenon with squaric acid is not an isolated example. The use of colorimetry in solutions containing un-ionized, moderately strong acids is now open to question, and further experimentation to characterize these complexes is likely to be undertaken in the future.

References

1. D. Alexanderson and N.-G. Vannerberg, *Acta Chem. Scand.* **26,** 1909 (1972).
2. E. Patton and R. West, *J. Phys. Chem.* **74,** 2512 (1970).
3. E. Patton and R. West, *J. Am. Chem. Soc.* **95,** 8703 (1973).
4. D. Berg and A. Patterson, *J. Am. Chem. Soc.* **75,** 5197 (1953).
5. H. S. Harned and S. R. Scholes, *J. Am. Chem. Soc.* **63,** 1706 (1941).
6. L. S. Darken, *J. Am. Chem. Soc.* **63,** 1007 (1941).
7. P. W. Preisler, L. Berger, and E. S. Hill, *J. Am. Chem. Soc.* **70,** 871 (1948).
8. L. M. Schwartz, R. I. Gelb, and D. A. Laufer, *J. Chem. Eng. Data* **25,** 95 (1980).
9. P. W. Preisler, L. Berger, and E. S. Hill, *J. Am. Chem. Soc.* **69,** 326 (1947).
10. G. C. Hood and C. A. Reilly, *J. Chem. Phys.* **32,** 127 (1960).
11. R. H. Boyd, *J. Phys. Chem.* **67,** 737 (1963).
12. R. I. Gelb and L. M. Schwartz, *J. Chem. Soc., Perkin Trans. 2* p. 930 (1976).
13. D. T. Ireland and H. F. Walton, *J. Phys. Chem.* **71,** 751 (1967).
14. D. J. MacDonald, *J. Org. Chem.* **33,** 4559 (1968).
15. L. M. Schwartz and L. O. Howard, *J. Phys. Chem.* **74,** 4374 (1970).
16. L. M. Schwartz and L. O. Howard, *J. Phys. Chem.* **75,** 1798 (1971).
17. R. I. Gelb, *Anal. Chem.* **43,** 1110 (1971).
18. P. Souchay and M. Fleury, *C. R. Hebd. Seances Acad. Sci.* **252,** 737 (1961).
19. B. Carlquist and D. Dyrssen, *Acta Chem. Scand.* **16,** 94 (1962).
20. J. M. Carpentier, M. B. Fleury, and J. F. Verchere, *Bull. Soc. Chim. Fr.* **4,** 1293 (1972).
21. L. M. Schwartz, R. I. Gelb, and J. O. Yardley, *J. Phys. Chem.* **79,** 2246 (1975).
22. R. I. Gelb, L. M. Schwartz, D. A. Laufer, and J. O. Yardley, *J. Phys. Chem.* **81,** 1268 (1977).
23. H. Takahashi, A. Kotaki, and K. Yagi, *Seikagaku* **37,** 413 (1965).

24. M. F. Fleury and G. Molle, *C. R. Hebd. Seances Acad. Sci., Ser. C* **273,** 605 (1971).
25. R. I. Gelb, L. M. Schwartz, and D. A. Laufer, *J. Phys. Chem.* **82,** 1985 (1978).
26. P. H. Tedesco and H. F. Walton, *Inorg. Chem.* **8,** 932 (1969).
27. A. Banerjee, S. Mandal, T. Singh, and A. Dey, *Indian J. Chem.* **7,** 733 (1969).
28. R. A. Robinson and R. H. Stokes, "Electrolyte Solutions," 2nd rev. ed. Butterworth, London, 1965.
29. G. W. Davies, "Ion Association." Butterworth, London, 1962.
30. H. S. Harned and R. W. Ehlers, *J. Am. Chem. Soc.* **55,** 2179 (1933).
31. J. W. Larson and L. G. Hepler, *in* "Solute-Solvent Interations" (J. F. Coetzee and C. D. Ritchie, eds.), p. 1. Dekker, New York, 1969.
32. E. J. King, "Acid-Base Equilibria." Pergamon, Oxford, 1965.
33. L. G. Hepler and E. M. Woolley, *in* "Water: A Comprehensive Treatise" (F. Franks, ed.), Vol. 3, Chapter 3. Plenum, New York, 1973.
34. H. L. Friedman and C. V. Krishnan, *in* "Water: A Comprehensive Treatise" (F. Franks, ed.), Vol. 3, Chapter 1. Plenum, New York, 1973.
35. L. Eberson and I. Wadsö, *Acta Chem. Scand.* **17,** 1552 (1963).
36. E. Orebaugh, Ph.D. Dissertation, Florida State University, Tallahassee, 1972.
37. L. M. Schwartz and L. O. Howard, *J. Phys. Chem.* **77,** 314 (1973).
38. W. Stadeli, R. Hollenstein, and W. von Philipsborn, *Helv. Chim. Acta* **60,** 948 (1977).
39. R. P. Bell, *Adv. Phys. Org. Chem.* **4,** 1 (1966).
40. G. Ojelund and I. Wadsö, *Acta Chem. Scand.* **21,** 1408 (1967).
41. G. Carpeni, *J. Chim. Phys.* **35,** 193 (1938).
42. R. I. Gelb and L. M. Schwartz, *Anal. Chem.* **44,** 554 (1972).
43. R. I. Gelb, L. M. Schwartz, and D. A. Laufer, unpublished work.

4

New Bond-Delocalized (Dicyanomethylidene)croconate Derivatives: "Croconate Violet" and "Croconate Blue"

Alexander J. Fatiadi

I. Introduction 59
II. Reaction of Croconates with Malononitrile 62
A. 1,3-Bis(dicyanomethylidene)croconate Salts: Croconate Violet . 62
B. 2-(Dicyanomethylidene)croconate Salts 65
C. 1,2,3-Tris(dicyanomethylidene)croconate Salts: Croconic Acid Blue 68
D. Attempted Condensation of Six-Carbon Oxocarbons with Malononitrile 72
III. Electrical Conductivity of Some Bond-Delocalized Salts 73
IV. Summary 76
References 76

I. Introduction

The first oxocarbons, croconic acid (**1**) and the dipotassium salt of croconate anion (**2**), were obtained by Gmelin as early as 1825 [*1*]. Recognition that the oxocarbons $C_nO_n{}^{m-}$ were members of a class of hitherto unknown aromatic substances in the 1960's [*2*] led to greatly renewed interest in this area [*3*]. In recent years several species have been synthesized in which the carbonyl oxygen atoms have been partially or completely replaced by other atoms; these are called pseudo-oxocarbons. They include nitrogen [*4*], sulfur [*5,6*], and selenium [*7*] analogs of squarate ion and a sulfur analog of croconate [*8*] (Fig. 1). This chapter

OXOCARBONS

ISBN 0-12-744580-3

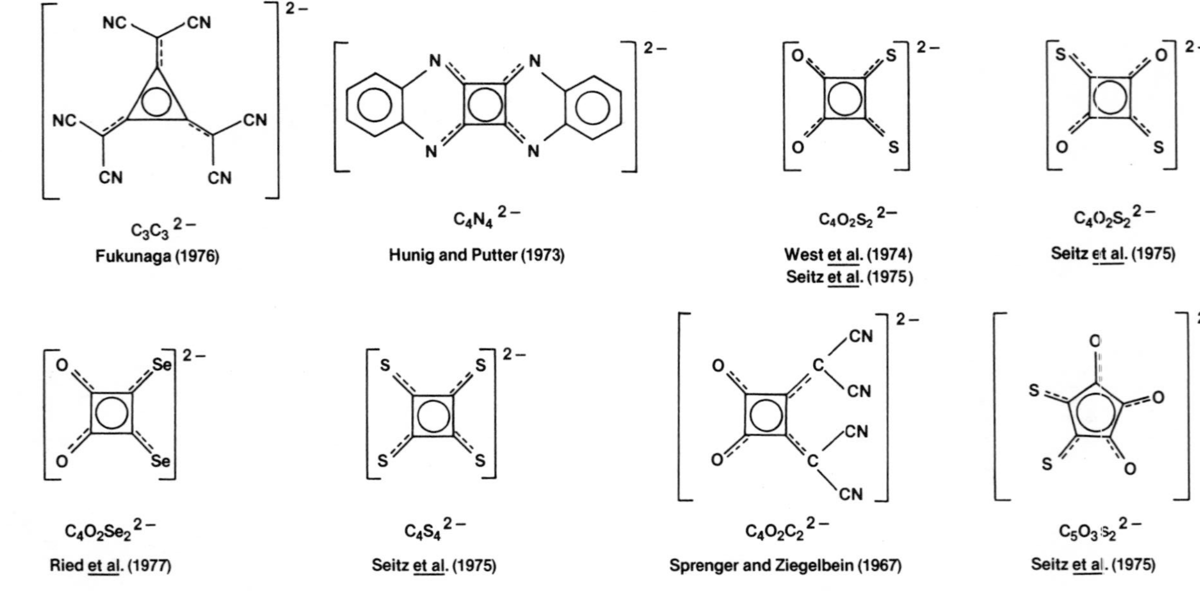

Fig. 1. Some aromatic pseudo-oxocarbon dianions.

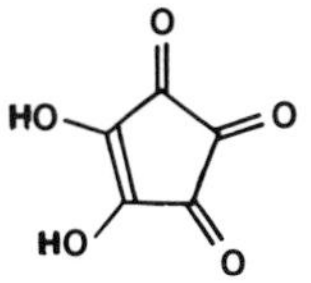

(1) 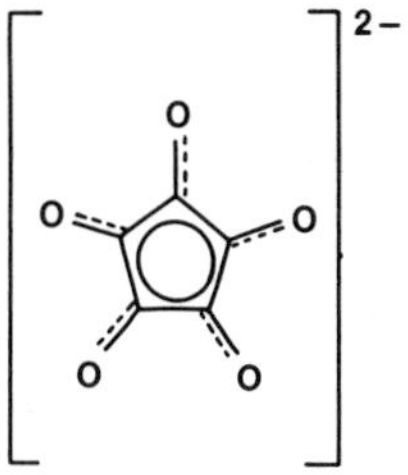 (2)

is concerned with new oxocarbon derivatives in which one or more of the oxygen atoms are replaced with dicyanomethylidene groups $=C(CN)_2$.

Our interest in these substances grew from the observation that 1,2,3-indanetrione reacts readily with malononitrile, with replacement of the most reactive 2-carbonyl oxygen [*9,10*]. Brief refluxing (5–10 min) of indanetrione with malononitrile in aqueous solution produced 2-(dicyanomethylidene)-1,3-indanedione (**3**) in almost quantitative yield [*11*].

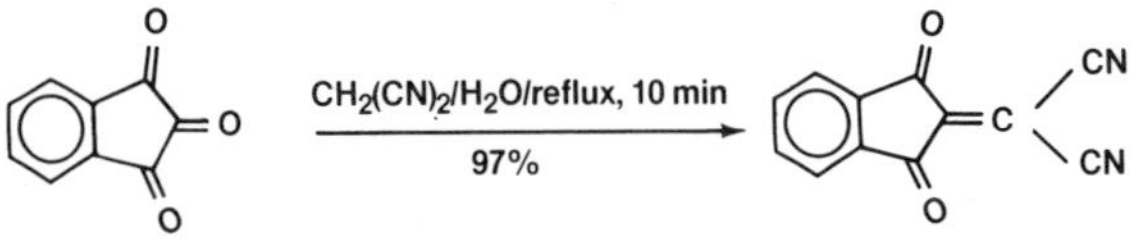

$CH_2(CN)_2/H_2O$/reflux, 10 min
97%

(**3**)

The reviews on oxocarbons [*3*] or vincinal polyketones [*12*] discuss only briefly the reaction with malononitrile, and a literature survey showed only three reports on this topic. When dibutyl squarate was treated with malononitrile in the presence of sodium butoxide in butanol, Sprenger and Ziegenbein [*13*] obtained a deep yellow disodium salt of 1,2-bis(dicyanomethylidene)-3-cyclobutene-3,4-dione, a new analog of the squarate dianion (Fig. 1). On treatment of the bis(dimethylamide) of squaric acid with malononitrile in the presence of triethylamine in dichloromethane, Seitz *et al.* [*14*] isolated a red triethylammonium salt of 2,4-bis(dicyanomethylidene)-3-(dimethylamino)-1-oxo-3-cyclobutene; the bond-delocalized salt is a new example of an aromatic pseudo-oxocarbon. A salt of

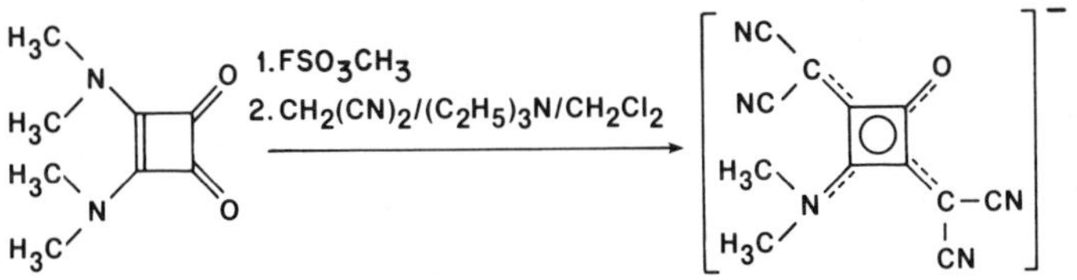

1. FSO_3CH_3
2. $CH_2(CN)_2/(C_2H_5)_3N/CH_2Cl_2$

1,2,3,-tris(dicyanomethylidene)deltate has also been synthesized by Fukunaga [*15*] using an indirect method. When tetrachlorocyclopropene was treated with 3 equivalents of malononitrile in 1,2-dimethoxyethane in the presence of 6 equivalents of sodium hydride, hexacyanotrimethylenecyclopropanediide (Fig. 1) was obtained in almost quantitative yield. The product was isolated as the bis(tetrabutylammonium) salt.

Still another recent report [*15*a] described a synthesis of a nitrogen analog of deltic acid, a new addition to the pseudo-oxocarbon family.

The most interesting and best studied dicyanomethylidene derivatives to date are derivatives of croconic acid and the croconate dianion. We have found that one, two, or three oxygens in $C_5O_5^{2-}$ can be replaced by $=C(CN)_2$ groups to give the dianions **4**, **5**, and **6** respectively [*11,16,17*]. Salts of these anions are

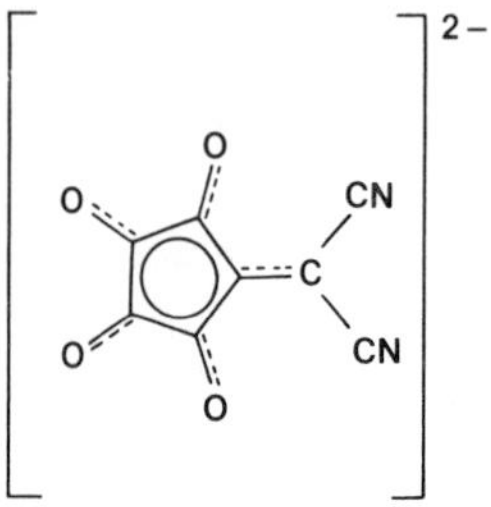

(4) $C_5O_4C_1^{2-}$

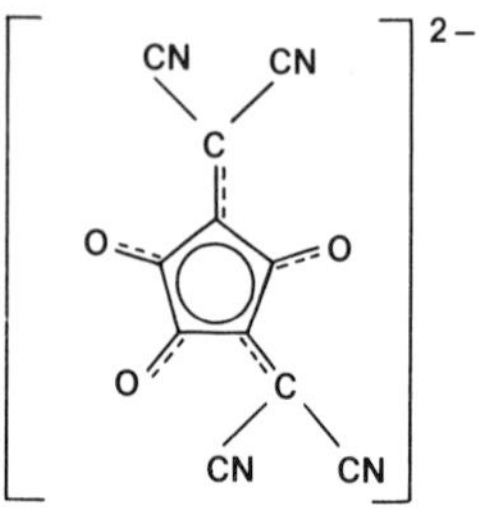

(5) $C_5O_3C_2^{2-}$

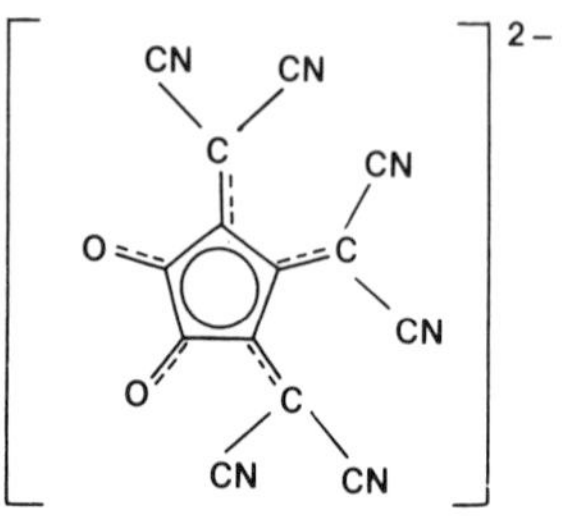

(6) $C_5O_2C_3^{2-}$

brilliantly colored and show remarkable electrical conductivity. These species and the few mentioned above are members of a new class of pseudo-oxocarbons containing dicyanomethylidene groups with the general formula $C_nO_n[C(CN)_2]^{r-}_{n-m}$.

II. Reaction of Croconates with Malononitrile

A. 1,3-BIS(DICYANOMETHYLIDENE)CROCONATE SALTS: CROCONATE VIOLET

Treatment of dipotassium croconate (**7**, R = K; $R^1 = R^2$ = O) [*18*] with a 1–2 molar excess of malononitrile in aqueous solution at 85°–90°C yields the dipotassium salt of 1,3-bis(dicyanomethylidene)-2-oxo-4-cyclopentene-4,5-diol (**5**) (Scheme 1) [*16*]. Recrystallization from hot water gives the salt of **5** as deep blue metallic needles of the dihydrate. This dipotassium salt of **5** is a dye for which we have suggested the name "croconate violet" because of its intense violet color in solution [$\lambda_{max}(H_2O)$ 533 nm (ϵ = 100,000)]. The infrared spectrum of the dipotassium salt of **5** is relatively simple, consistent with a symmetric structure for the dianion of the salt. Four strong bands are observed at 1680, 1620, 1580,

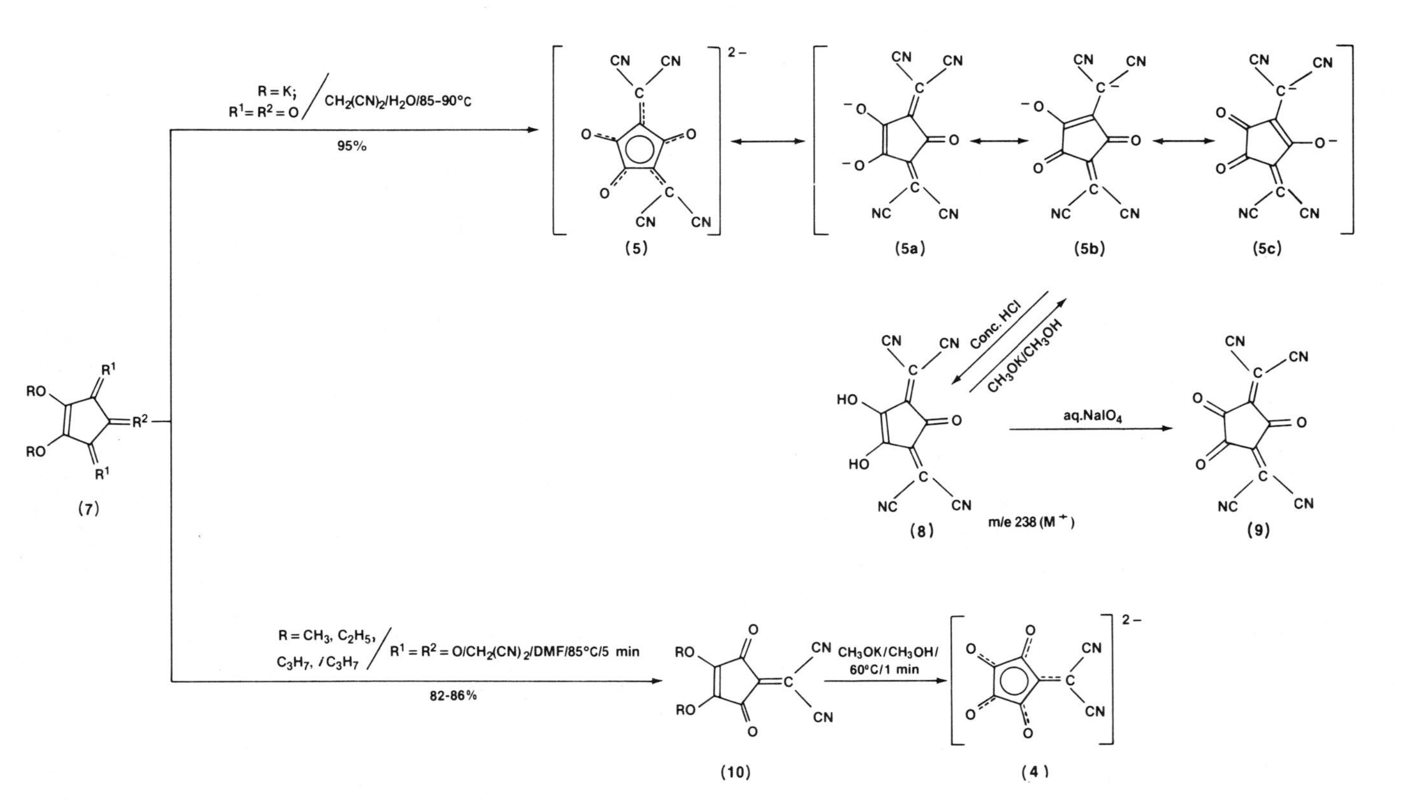
R = K;
R¹ = R² = O
CH₂(CN)₂/H₂O/85-90°C
95%
(5)
(5a)
(5b)
(5c)
(7)
Conc. HCl
CH₃OK/CH₃OH
aq.NaIO₄
(8)
m/e 238 (M⁺)
(9)
R = CH₃, C₂H₅,
C₃H₇, /C₃H₇
R¹ = R² = O/CH₂(CN)₂/DMF/85°C/5 min
82-86%
CH₃OK/CH₃OH/
60°C/1 min
(10)
(4)

Scheme 1

and 1520 cm^{-1}, probably due to combinations of C=O, C=C, and $C=C(CN)_2$ vibrations.

The ^{13}C-NMR spectrum of **5** shows three ^{13}C resonances for the ring carbon atoms at δ = 181.4 (C=O), 172.0 (C=O), and 147.8 ppm [**C**=C(CN)$_2$], as well as resonances assigned to exocyclic carbons at 119.5 (CN) and 51.4 [C=**C**(CN)$_2$].* This spectrum is also consistent with a symmetric structure for the dianion.

A nearly complete ionization of the dipotassium salt of **5** in aqueous solution is evidenced from the specific conductance of **5** (235.7147 μS/cm) (1 m*M*, 22.8°C), as compared to potassium chloride (147 μS/cm) (1 m*M*, 22.8°C). Subtraction of the ionic conductance of the dipotassium ions ($2K^+ \times 73.5$ μS/cm, 25°C) leaves the ionic conductance (~88 μS/cm) of the bulky 1,3-bis(dicyanomethylidene)croconate ion, and it is only slightly higher from the ionic conductance (76.3 μS/cm, 25°C) of the chloride ion.

The structure of **5** has been definitely established by an X-ray crystallographic study [*19*]. As shown in Fig. 2, compound **5** has the 1,3 structure with D_{2d} symmetry. The dianion is planar, with mean C—O and ring C—C bond lengths of 1.244 and 1.450 Å, respectively. These can be compared with the C—O and C—C distances for croconate ion (in diammonium croconate) of 1.262 and 1.457 Å [*20*]. The internal CCC bond angles are all within 1 standard deviation of 108°, the internal angle for a regular pentagon. Thus, the dicyanomethylidene groups do not greatly change the electronic distribution or disturb the D_{5h} symmetry of the croconate ring.

Dianion **5** can be represented in resonance terms as the hybrid of canonical forms **5a**↔**5b**↔**5c** (Scheme 1). The bond lengths in Fig. 2 suggest that, consistent with the very high electronegativity of the dicyanomethylidene group, forms such as **5b** and **5c** make a slightly greater contribution than **5a**.

When the dipotassium salt of **5** is treated with strong aqueous acid, the purple color of the solution gradually changes to orange as protonation to the acid form **8** takes place [*16*]. Brief warming of dipotassium **5** with concentrated HCl produces orange crystals of the free acid **8** (Scheme 1). In water, **8** undergoes immediate ionization to **5**. Compound **8** is perhaps the strongest acid yet known in the oxocarbon or pseudo-oxocarbon series, having an immeasurably large pK_1 and pK_2 of 0.07 ± 0.02 [*21*]. Careful neutralization converts **8** back to **5**, and such neutralization is the best way of making the other alkali metal salts of **5**. Oxidation of **8** with aqueous sodium periodate followed by extractions with ethyl

*The unusual high-field ^{13}C chemical shift at 51.4 ppm, assigned to the exocyclic olefinic carbon atoms in **5**, can be explained as being due to shielding by the triple bond of the cyano group. Similar resonances are observed for other compounds in this series (**4** and **6**) and for model compounds. The peak assigned to cyano-group carbons at 119.5 is a probable multiplet, consistent with structure **5**, which contains two nonequivalent kinds of CN groups.

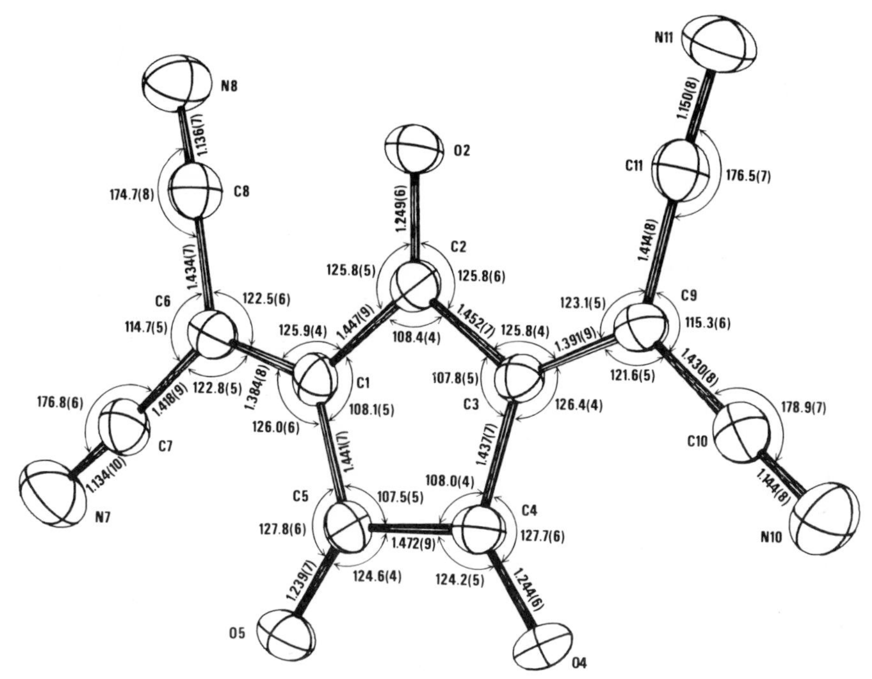

Fig. 2. Bond distances (angstroms) and angles for the dianion of **5**.

acetate gave an orange product (λ_{mas} 430–440 nm), which may have the unprecedented structure **9** (Scheme 1).

Cyclic voltammetry of the salt of **5** in water revealed an irreversible oxidation potential at +0.52 V versus SCE. (However, a reversible oxidation process for the salt of **5** was observed on a glassy carbon electrode.) The dianion **5** (and its conjugate acid **8**) apparently both undergo two-electron reduction showing two irreversible peak potentials at −0.88 and −1.33 V (Fig. 3). The two-electron reduction probably involves a dicyanomethylidene group and a keto group, e.g., $C{=}C(CN)_2 + C{=}O \rightarrow HO{-}C{=}C{-}CH(CN)_2$. The redox reaction is pH dependent; the reduction step proceeds faster in alkaline media, as stepwise hydrolysis of **5** takes place via the intermediate dianion **4** to croconate ion (**2**) (see Fig. 4). The oxidation and reduction potentials of **5** are rather similar to those of croconate ion, which shows an oxidation at $E_{\frac{1}{2}} = +0.59$ V and reduction steps at −0.48 and −1.33 V.

B. 2-(DICYANOMETHYLIDENE)CROCONATE SALTS

When esters of croconic acid are allowed to react with malononitrile, only *one* of the oxygens is replaced by a dicyanomethylidene group. When dimethyl, diethyl, or di-*n*-propyl croconate is warmed with malononitrile in DMF, golden

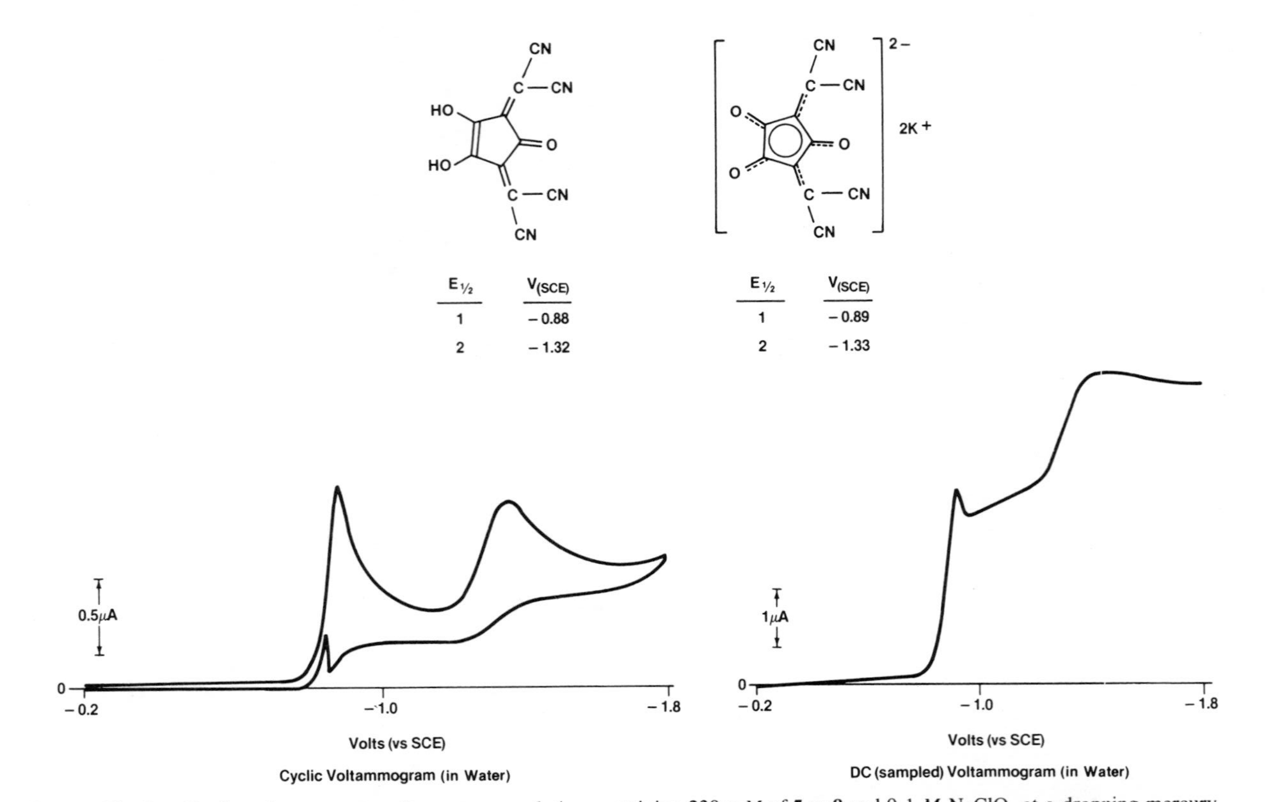

Fig. 3. Cyclic voltammograms of an aqueous solution containing 230 m*M* of **5** or **8** and 0.1 *M* $NaClO_4$ at a dropping mercury electrode. Scan rate 50 mV/sec.

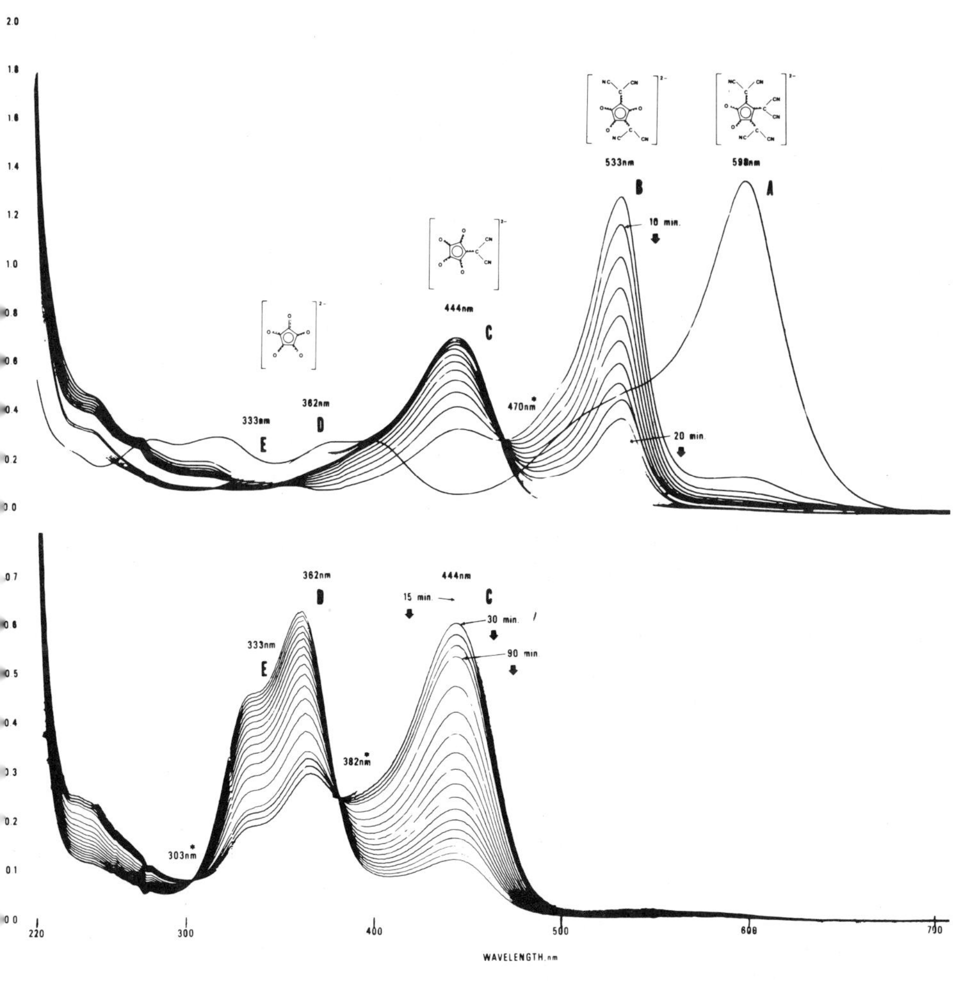

Fig. 4. Spectrophotometric monitoring of the rate of alkaline (pH 13.2) hydrolysis of the bis(tetramethylammonium) salt of **6** (croconate blue). Concentration 22.9 m*M*; asterisks indicate isosbestic point.

plates of the corresponding 2-(dicyanomethylidene)dialkyl croconates (**10**) are produced. The ^{13}C-NMR spectra of these compounds are consistent with the symmetric D_{2d} structures shown in Scheme 1. These compounds are efficient electron acceptors, forming deeply colored charge-transfer complexes with polycyclic aromatic hydrocarbons such as pyrene, benzo[*a*]pyrene, or anthracene.

The crystal structure of the red 1 : 1 complex (pyrene–**10**, R = C_2H_5) has been determined [*22*]; a novelty of the structure is that its crystal unit cell contains two pyrene molecules differing in symmetry. However, with tetrathiafulvalene as the donor, 2-(dicyanomethylidene)alkyl croconates (**10**, R = CH_3, C_2H_5, n-C_3H_7, i-C_3H_7) form charge-transfer salts that are semiconductors [*11*] (see Section III). Chemically, 2-(dicyanomethylidene) alkyl croconates react readily with aniline in warm alcohol to give deep red dyes with structures similar to that described for the product of reaction of 2-(dicyanomethylidene)-1,3-indanedione with aniline [*23*].

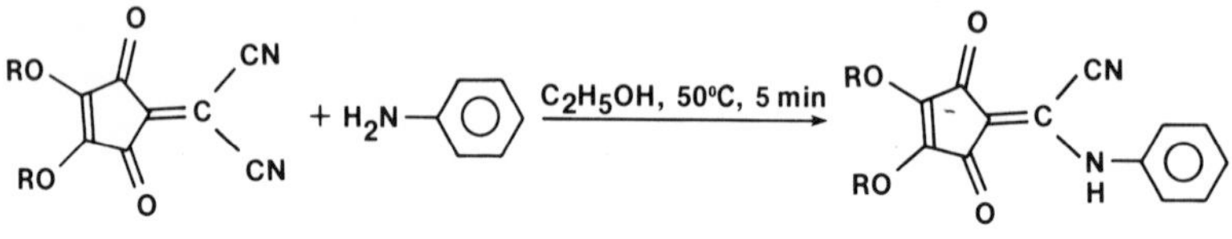

C_2H_5OH, 50°C, 5 min

R = CH_3; R = C_2H_5

When the esters (**10**) are carefully hydrolyzed with potassium methoxide in methanol, the dipotassium salt of **4** can be isolated. Recrystallization from water yields cherry red crystals of the dihydrate. The infrared spectrum of **4** is similar to that of **5**, which suggests that **4** may also be represented as the hybrid of resonance structures resembling those for **5** (Scheme 1). Anion **4** gives intensely red-violet solutions in water, with λ_{max} 533 (ϵ = 70,000). When the dipotassium salt is warmed with HCl, it forms orange crystals of the conjugate acid (**10**, R = H).

C. 1,2,3-TRIS(DICYANOMETHYLIDENE)CROCONATE SALTS: CROCONIC ACID BLUE

In Sections II, A and B we have seen that croconate anions and esters give quite different products upon reaction with malononitrile. Yet another result is observed when croconic acid (**11**) (Scheme 2) is treated with malononitrile. In this case *three adjacent* carbonyl groups are replaced to give the new oxocyanocarbon acid **12** [1,2,3-tris(dicyanomethylidene)-4-cyclopentene-4,5-diol] [*17*]. Shining purple plates of **12** crystallize from solution when croconic acid is simply warmed briefly with malononitrile in aqueous solution at 85°–90°C. The product, obtained in over 90% yield, is intensely blue in water (λ 600 nm; ϵ = 55,000), and we have termed it croconic acid blue. However, **12** is very strongly solvatochromic, giving red solutions with λ 475–480 nm in anhydrous acetone or alcohol. As expected, **12** is a strong oxocarbon acid; its pK_2 value is ~1, placing it intermediate in acid strength between **8** and croconic acid [*21*].

The infrared spectrum (KBr) of **12** shows strong bands at 1755, 1700(sh), 1650, 1590, and 1520 cm^{-1}. The band at 1755 cm^{-1} suggests that tautomers with C—H structure (**12a** or **12b**) may be present in the solid.

As shown in Scheme 2, neutralization of **12** with potassium methoxide in

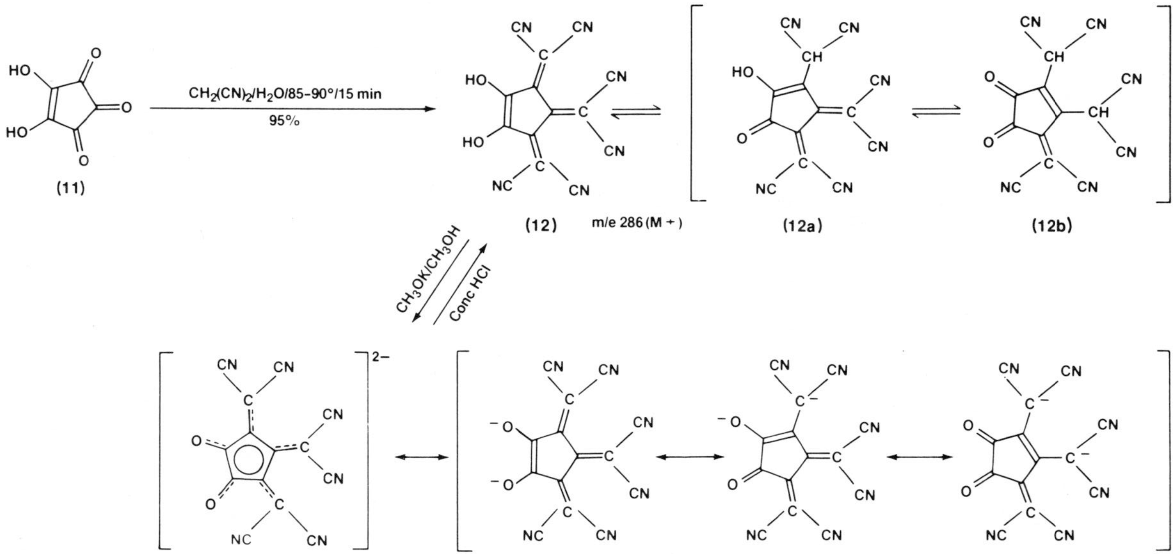

$CH_2(CN)_2/H_2O/85–90°/15\ min$

95%

(11)

(12) m/e 286 (M +)

(12a)

(12b)

CH_3OK/CH_3OH

Conc HCl

(6)

(6a)

(6b)

(6c)

Scheme 2

methanol gives the dipotassium salt of **6**, croconate blue [*17*]. The salt is obtained as shining blue-green crystals of the trihydrate upon recrystallization from water. Dianion **6** has its principal absorptions in water at 599 (ϵ = 54,600) and 538 nm [ϵ = 32,000(sh)]. The 1,2,3 structure is confirmed by the ^{13}C-NMR spectrum, which shows resonances at δ = 178.0 ppm assigned to *C*=O, 147.3 and 139.1 ppm [**C**=C(CN)$_2$], 120.2, 118.7, and 118.6 ppm (CN), and 53.1 ppm [*C*=**C**(CN)$_2$]. The three CN peaks are expected for structure **6**, which has three intrinsically different cyano groups. Additional fine structure may also be present in this cluster, possibly reflecting different conformations in solution. The single peak at 53.1 probably results from accidental overlap of resonances for the two kinds of C=*C*(CN)$_2$ groups.

The properties of **6** resemble generally those of **4** and **5**, suggesting that the croconate blue anion also has a bond-delocalized structure. The probable contributors to the resonance hybrid structure for **6** are shown as **6a**↔**6b**↔**6c** in Scheme 2. However a stereomodel of **6** indicates that, in order to relieve steric crowding, a staggered conformation of the dicyanomethylidene groups is required. The twist angle for the central dicyanomethylidene group sufficient to relieve steric strain is estimated to be about 30° [*17*]. The ^{13}C-NMR spectrum for the central C≡N groups may show evidence for this lowered symmetry.

The acid **12** in water at room temperature (sensitive to UV irradiation) hydrolyzes slowly (95% in 85 days) to yield the more thermodynamically stable croconic acid violet **8** [λ_{max}(H_2O) 534 nm]. In warm (95°C) 10 *M* hydrochloric acid, the acid **12** hydrolyzes completely in a few minutes to **8**. However, excessive heating of **12** in water causes apparent polymerization to give as product deep green, lustrous plates.

Sever crowding and lowered symmetry in **6** compared to **5** may be responsible for the fact that **6** is more easily reduced than **5**. Cyclic voltammetry of **6** revealed the first reduction wave at −0.69 V versus SCE (Fig. 5) compared with −0.88 V for **5**.* Crowding in **6** is probably also associated with its relative instability in aqueous alkaline media. At pH 13.2 the 2-dicycanomethylidene group is split from **6** within 2–3 sec, giving the croconate violet anion **5** (Fig. 4). This rapid change is followed by much slower, stepwise hydrolysis of the remaining 1,3-bis(dicyanomethylidene) groups to give, first, the orange dianion **4** (band at 444 nm, a change from B to C in 4 hr, Fig. 4) and then, as the final product, yellow croconate dianion **2** [bands at 362 and 333(sh) nm, a change from C to D in 40 hr]. The decomposition of **6** and formation of new products are indicated by three isosbestic points, at 303 (ϵ = 3800), 382 (ϵ = 11,400), and 470 nm (ϵ = 13,000). Note the large bathochromic shift in the visible spectrum of the dianion **2** (363 → 600 nm) due to consecutive addition of the dicyanomethylidene chromophores.

*Further electrochemical study of both **5** and **6**, including isolation and identification of their redox products, is in progress.

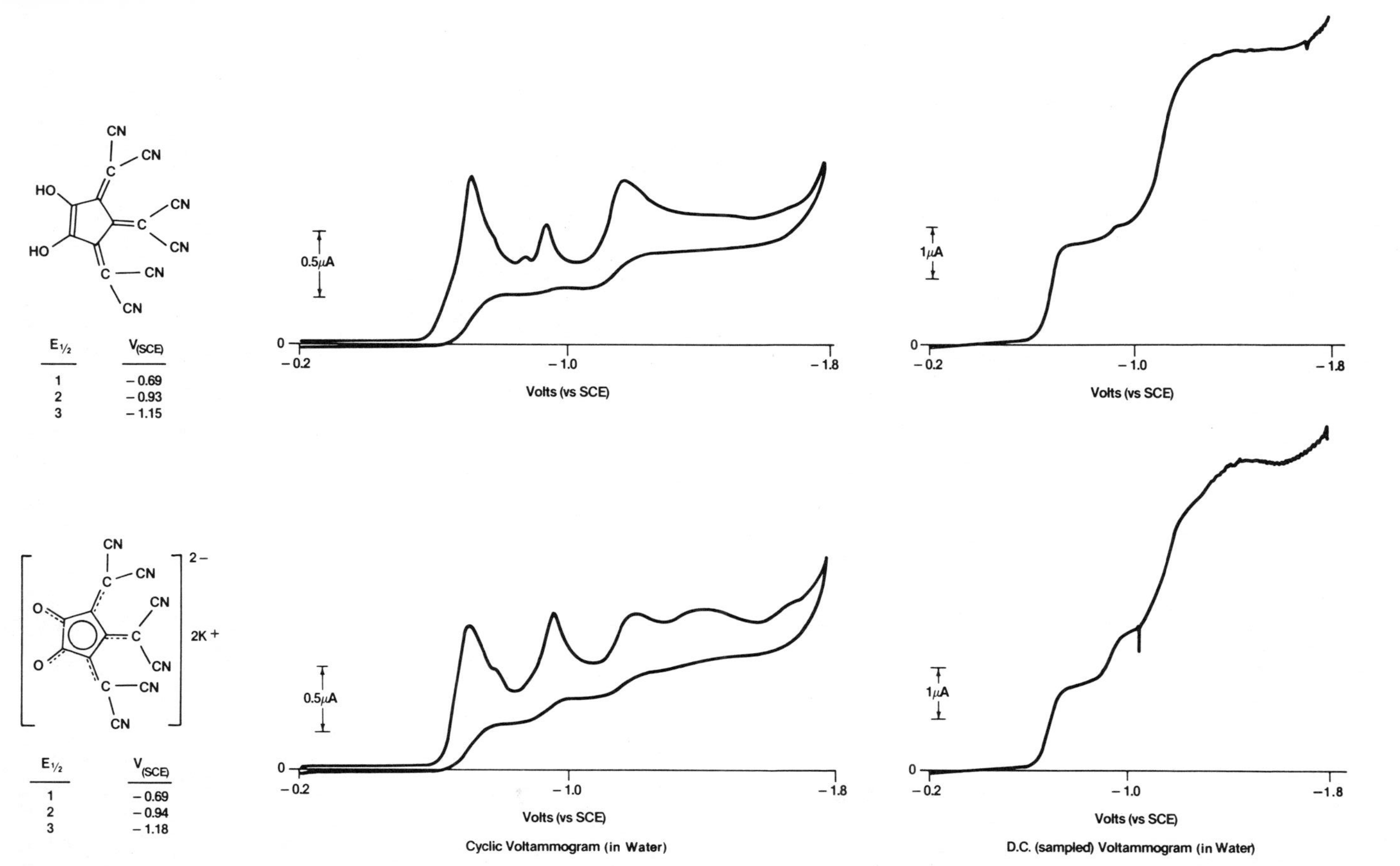

Fig. 5. Cyclic voltammograms of an aqueous solution containing 240 m*M* of **12** or **6** and 0.1 *M* $NaClO_4$ at a dropping mercury electrode. Scan rate 50 mV/sec.

The kinetics of alkaline hydrolysis of the bis(tetramethylammonium) salt of **6** are first order, with an apparent rate-constant of $K_b = 0.0136\ \text{min}^{-1}$ from B to C (Fig. 4); the rate was 10 times slower going from C to D (e.g., $K_c = 0.00135\ \text{min}^{-1}$).

D. ATTEMPTED CONDENSATION OF SIX-CARBON OXOCARBONS WITH MALONONITRILE

Unlike the reaction of croconic acid or its salts with malononitrile to give crystalline products (Sections IIA, B, and C), treatment of six-carbon oxocarbons, e.g., tetrahydroxy-*p*-benzoquinone (THQ), rhodizonic acid (1,2-dihydroxy-1-cyclohexene-3,4,5,6-tetrone) or their alkali salts with malononitrile in aqueous media produces colored products, the structures of which are under study [*24*]. For example, THQ [*25*] and malononitrile in *N,N*-dimethylformamide (DMF) (or water) at 80° yielded a benzodifuran derivative (λ_{max}^{DMF} 626–642 nm) [*25a*]. However, a gentle treatment [*24*] (50° for 10–15 min) of 1 mole of the disodium salt of THQ with 2 moles of malononitrile gave a somewhat differently colored product, with λ_{max}^{DMF} 562–565 nm. Prolonged treatment of the same salt with an excess of the reagent under reflux yielded a colorless trimer of malononitrile (2-cyanomethyl-4,6-diamino-3,5-dicyanopyridine) in high yield. Similar treatment of disodium rhodizonate or disodium squarate with an excess of malononitrile also gave a trimer of malononitrile, but formation of colored products was again observed under milder reaction temperature. The mechanism of the trimerization reaction can be explained as being due first to the hydrolysis of the salt, followed by a base-catalyzed aldol reaction of malononitrile [*11*].

NaO, NaO, O, O, O, O
H_2O / $CH_2(CN)_2$ /reflux/4 hr
90%
NH_2, NC, CN, H_2N, N, CH_2CN

Also: Na – THQ
Na – squarate

A dark green compound is also formed on treatment of benzenehexol with malononitrile, presumed to be a benzodifuran derivative. Controlled reaction or triquinoyl (cyclohexanenexone octahydrate) or leuconic acid (cyclopentanepentone pentahydrate) [*18*] with the reagent gave lustrous black microcrystalline products that showed identical infrared spectra different from that of the product from benzenehexol [*24*].

For the monomeric series of the six-carbon oxocarbons, a direct route for the preparation of the malononitrile adducts should be avoided, unless the reaction is thoroughly controlled. Instead, an indirect method similar to that described for the synthesis of the 1,2,3-tris(dicyanomethylidene)deltate salt [*15*] (Fig. 1) should be explored. The synthesis of novel, six-carbon pseudo-oxocarbons wherein the original carbonyl oxygen atoms are either partially or completely substituted with nitrogen, sulfur, selenium, phosphorus, etc., offers an attractive area of study.

III. Electrical Conductivity of Some Bond-Delocalized Salts

The charge-transfer salts of the π donor tetrathiafulvalene (TTF) [*26*] or its analog tetraselenafulvalene [*27*] with tetracyanoquinodimethane (TCNQ) as the acceptor are the best known examples of a new class of organic solids having high electrical conductivity (the quasi-one-dimensional "organic metals") [*28*]. The search for more effective organic metals has led to the production of a vast number of TTF–TCNQ derivatives that differ primarily in the nature of substituents on either the donor (TTF) or the acceptor (TCNQ) unit. However, only a limited number of such charge-transfer salts show high conductivity. Despite extensive recent work, it is still difficult to specify detailed molecular requirements that would give rise to, and increase conductivity in, organic metals [*27–29*]. High symmetry, high polarizability, small molecular size, and molecular planarity [*26,29*] are among the factors that are generally considered to favor high conductivity. However, equally important factors for such conducting salts appear to be (a) an alternating, donor–acceptor stacking arrangement [*26*], (b) crystal packing [*30*], and (c) molecular volume [*30,31*].

Using TTF as a model symmetric π donor in reactions with unsymmetric acceptors (e.g., croconic acid and its esters) in acetonitrile (or methanol-acetonitrile), we have prepared a series of new TTF salts that show all of the properties of semiconductors, having a conductivity of 2.5×10^{-3} to 10^{-4} Ω^{-1} cm^{-1} (pellet, 300 K) [*11*].

Dipotassium tetrathiosquarate [*6*] (Fig. 1) has been reported [*32*] to react with 1,2-dithiolynium salts to give shining black 2 : 1 complexes; these have, however, been found to be insulators, with a room-temperature compaction conductivity of 10^{-10} Ω^{-1} cm^{-1}. The metathesis between donor and acceptor salts probably involves almost complete electron transfer, leading to considerable electrostatic repulsion in the solid state, which may partially explain the low conductivity of these salts [*32*]. Indeed, the most recent report [*33*] described the preparationof the squaric acid complex bis(squarato)platinum(II) salt, a new example of a one-dimentional electrical conductor. We have found that the 1 : 1 adduct between dimethyl sulfoxide and tetrahydroxyquinone (which is a proton-transfer complex) is also a poor conductor with a conductivity of 10^{-9} Ω^{-1} cm^{-1} (pellet, room temperature) [*24*].

However, croconate violet, the dipotassium salt of **5** (Table I, entry 2) has a room temperature single-crystal conductivity of $2 \times 10^{-6}\ \Omega^{-1}\ cm^{-1}$, comparable to that of the potassium salt of the TCNQ anion-radical (Table I, entry 1). Measurements of magnetic susceptibility indicate that the salt of **5** is diamagnetic, and the salt exhibits no ESR spectrum. An unusually large increase in the

Table I. Electrical Conductivities of Some Oxocarbons and Pseudo-Oxocarbons

Entry	Salt	Conductivity ($\Omega^{-1}\ cm^{-1}$)	Entry	Salt	Conductivity ($\Omega^{-1}\ cm^{-1}$)
1	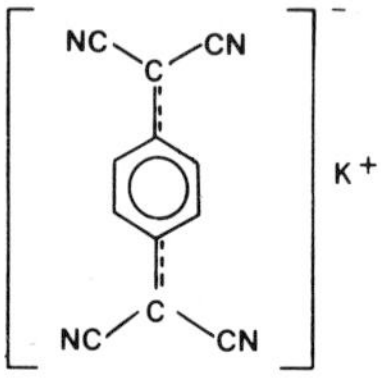	K^+ 2×10^{-4}[a]	5	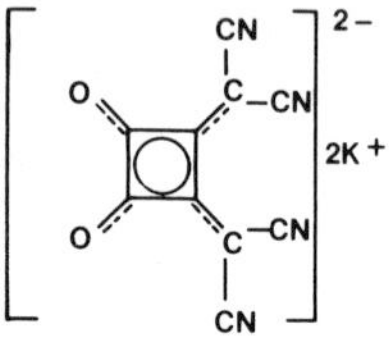	$2K^+$ 8×10^{-8}[a]
2	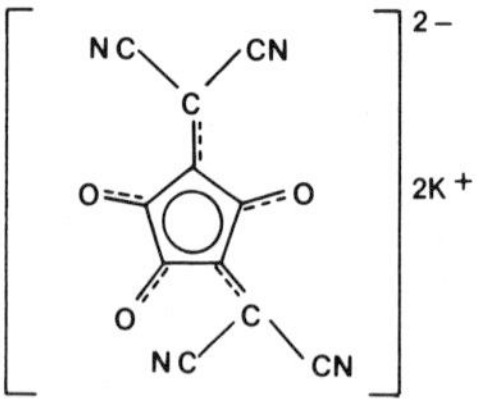	$2K^+$ 2×10^{-6}[b]	6	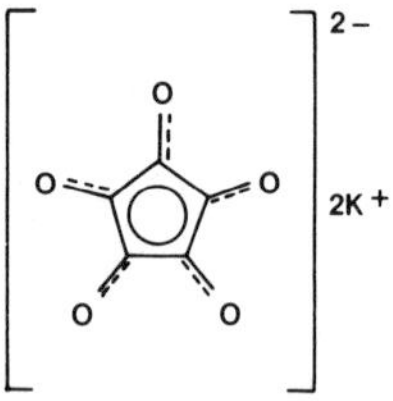	$2K^+$ 5×10^{-9}[a]
3	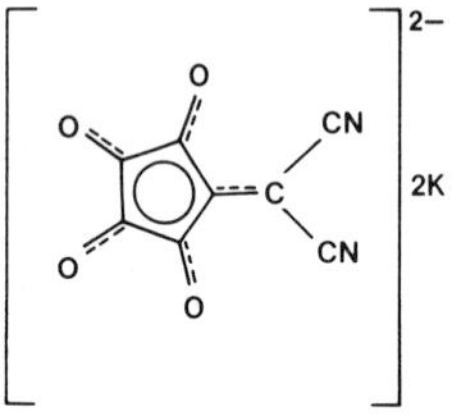	$2K^+$ 1.5×10^{-7}[a]	7	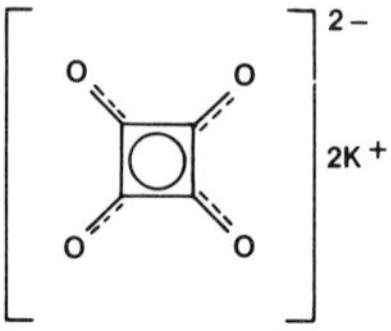	$2K^+$ 4×10^{-9}[a]
4	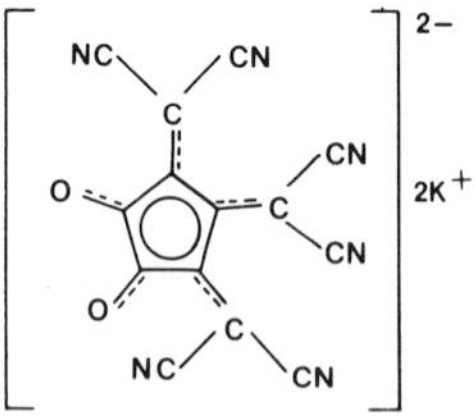	$2K^+$ 1×10^{-7}[a]	8	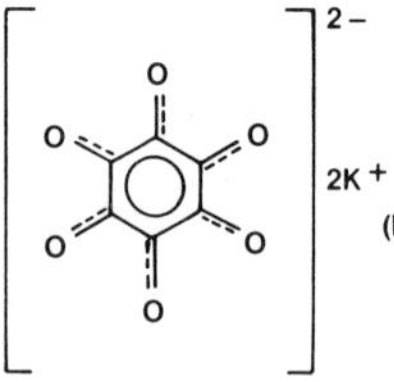	$2K^+$ 6×10^{-10}[a] (Rhodizonate)
			9	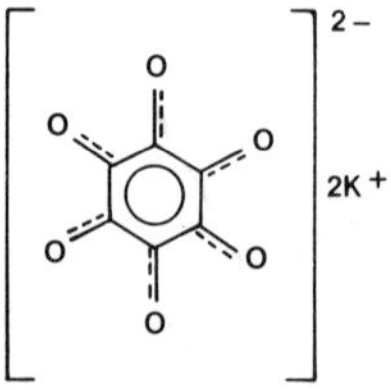	$2K^+$ 2×10^{-10}[a] (THQ)

[a] Compressed pellet, 300 K.
[b] Single crystal, 300 K.

electrical conductivity, by a factor of 10^4–10^5, was observed on heating the dipotassium salt of **5** from 25° to 400°, and this is typical behavior of a semiconductor. The dipotassium salt of **5** showed a remarkable thermal stability since the conductivity was not altered appreciably on repeating the heating–cooling cycle (25°–400°) (Fig. 6) [*22*].

The intrinsic conductivity of the dipotassium salts of **4**, **5**, and **6** (Table I, entries 2–4) is partially due to the electronic conductivity that is associated with polarizability (via the strongly electronegative dicyanomethylene groups), structural symmetry, extensive electronic delocalization, and appreciable aromatic character of the dianion. However, in the case of the dipotassium salt of **5**, the ionic contribution to the conductivity may be due to the unique crystal structure and crystal packing. Structurally there are columns of the cyclopentane ring anions parallel to the cation columns, with only 3.32 Å separation between adjacent molecular sites [*19*] (as compared to 3.30 Å found for the interplanar spacing for TTF–TCNQ). Hence, the (potassium) cation, by virtue of partial occupancy of sites in a channel, is capable of relatively high ionic conductivity.

Much lower conductivities are found for oxocarbons containing no di-

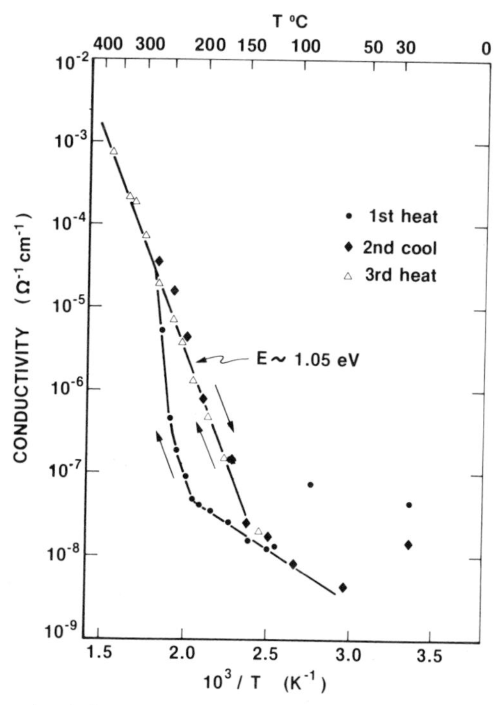

Fig. 6. Rise in the electrical conductivity on heating of the dipotassium salt **5** (croconate violet) from 25° to 400°.

cyanomethylene group. The diamagnetic dipotassium salts of croconic, squaric, and rhodizonic acids and of tetrahydroxyquinone (Table I, entries 6–9) are all poor conductors, and the single-crystal (four-probe), room-temperature ionic conductivity of dilithium croconate trihydrate [*18*] was found to be $1 \times 10^{-8}\ \Omega^{-1}$ cm^{-1}, a value within the semiconductor range [*24*]. However, metal complexes of oxocarbons may be much better conductors. A recent report [*33*] described the preparation of the bis(squarato)platinum(II) salt $K_2[Pt(C_4O_4)_2]\cdot 2H_2O$, a new example of a one-dimensional electrical conductor. The observed conductivity of the salt at 25°C was $5 \times 10^{-3}\ \Omega^{-}\ cm^{-1}$, suggesting a metallic state. Metal complexes of **4, 5,** and **6,** when they are prepared, may also show higher conductivity.

IV. Summary

Either partial or complete replacement of the original carbonyl oxygen atoms in three-, four-, five-, or six-membered oxocarbon anions $C_nO_n^{m-}$ with the C=O equivalent, π-isoelectronic groups C=C, C=N, C=P, C=S, C=Se, etc., yields a series of unusual oxocarbon analogs, the pseudo-oxocarbons. Many of these new oxocarbons may be of considerable interest to the theoretical chemist with regard to their molecular symmetry, the planarity of their anions, and their novel, nonbenzenoid aromaticity. Studies of their chemistry (e.g., esters, amides, and salts), their physical properties (e.g. electronic, vibrational, and NMR spectroscopic, dipole moment, force constant, X-ray crystal structure, and coordinate analysis, and MO + LCAO + MO calculations), their analytical applications (e.g., electrochemistry, photolysis, and solution chemistry), and their biomedical applications offer new directions in the chemistry of oxocarbons. In their 1969 review on oxocarbons, West and Niu [*3*] stated that the chemistry of the more complex oxocarbons was only beginning and that hundreds of possible stable structures awaited synthesis by the organic chemists of tomorrow. This prediction seems valid today.

Acknowledgment

The author expresses appreciation to Dr. R. S. Tipson for reading the manuscript and wishes to thank Dr. L. M. Doane for the cyclic voltammetry, Dr. B. Coxon for the ^{13}C-NMR spectrum, Dr. L. R. Hilpert for the mass spectra, Dr. C. K. Chiang for some electrical conductivity measurements, Dr. R. Cook for the specific conductance measurements, and F. J. Savluk for some spectral measurements

References

1. L. Gmelin, *Ann. Phys. (Leipzig)* [2] **4,** 31 (1825).
2. R. West, H. Y. Niu, D. L. Powell, and M. A. Evans, *J. Am. Chem. Soc.* **82,** 6204 (1960).

3. R. West and J. Nieu, *in* "Nonbenzenoid Aromatics" (J. P. Snyder, ed.), Vol. 1, Chapter 6. Academic Press, New York, 1969; *in* "The Chemistry of the Carbonyl Group" (J. Zabicky, ed.), Vol. 2, Chapter 4. Wiley (Interscience), New York, 1970; R. West, *Isr. J. Chem.* (in press).
4. S. Hünig and H. Putter, *Angew. Chem., Int. Ed. Engl.,* **12,** 149 (1973); *Chem. Ber.* p. 2532 (1977). See also S. Kanoktanadon and J. A. H. MacBride, *J. Chem. Res. S.* 206 (1980).
5. D. Coucouvanis, F. J. Hollander, R. West, and D. Eggerding, *J. Am. Chem. Soc.* **96,** 3006 (1974).
6. G. Seitz, K. Mann, R. Schmiedel, and R. Matusch, *Chem.-Ztg.* **99,** 90 (1975); R. Allmann, T. Debaerdemacker, K. Mann, R. Matusch, R. Schmiedel, and G. Seitz, *Chem. Ber.* **109,** 2208 (1976).
7. A. H. Schmidt, W. Ried, and P. Pustoslemsek, *Chem.-Ztg.* **101,** 154 (1977); *Chem. Abstr.* **86,** 189258 (1977).
8. G. Seitz, K. Mann, and R. Matusch, *Arch. Pharm. (Weinheim, Ger.)* **308,** 792 (1975); R. F. X. Williams, *Phosphorus and Sulfur* **2,** 141 (1976).
9. S. Chatterjee, *J. Chem. Soc. B* p. 725 (1969); *Science* **157,** 314 (1967).
10. A Schönberg and E. Singer, *Tetrahedron* **34,** 1285 (1978); *Chem. Ber.* **103,** 3871 (1970); *Tetrahedron Lett.* p. 4571 (1969).
11. A. J. Fatiadi, *Synthesis* pp. 165, 241 (1978).
12. M. B. Rubin, *Chem. Rev.* **75,** 177 (1975).
13. H. E. Sprenger and W. Ziegenbein, *Angew. Chem., Int. Ed. Engl.* **6,** 553 (1967); **7,** 530 (1968).
14. H. Morck, R. Schmiedel, and G. Seitz, *Chem.-Ztg.* **103,** 188 (1979).
15. T. Fukunaga, *J. Am. Chem. Soc.* **98,** 610 (1976); U.S. Patent 3,963,769 (1976); *Chem. Abstr.* **86,** 55052 (1977).
15a. R. Weiss and M. H. Hertel, *J. Chem. Soc.* Chem. Commun. p. 223 (1980).
16. A. J. Fatiadi, *J. Am. Chem. Soc.* **100,** 2586 (1978).
17. A. J. Fatiadi, *J. Org. Chem.* **45,** 1338 (1980); *J. Res. Natl. Bur. Stand.* **85,** 73 (1980).
18. A. J. Fatiadi, H. S. Isbell, and W. F. Sager, *J. Res. Natl. Bur. Stand., Sect. A* **67,** 153 (1963).
19. V. L. Hines, A. D. Mighell, C. R. Hubbard, and A. J. Fatiadi, *J. Res. Natl. Bur. Stand.* **85,** 87 (1980).
20. N. C. Baenziger and J. J. Hegenbarth, *J. Am. Chem. Soc.* **86,** 3250 (1964).
21. L. M. Schwartz, personal communication.
22. A. J. Fatiadi *et al.,* to be submitted.
23. H. Junek, H. Aigner, and H. Fischer-Colbrie, *Monatsh. Chem.* **103,** 639 (1972).
24. A. J. Fatiadi, unpublished observations.
25. A. J. Fatiadi and W. F. Sager, *Org. Synth.* **42,** 66, 90 (1962).
25a. R. Peltzmann, B. Unterweger, and H. Junek, *Montash. Chem.* **110,** 739 (1979).
26. A. F. Garito and A. J. Heeger, *Acc. Chem. Res.* **7,** 232 (1974); H. Meier, "Organic Semiconductors," pp. 190–211. Verlag Chemie, Weinheim, 1974; A. N. Bloch, D. O. Cowan, and T. O. Poehler, *In* "Energy and Charge Transfer in Organic Semiconductors" (K. Masuda and M. Silver, eds.), pp. 159–174. Plenum, New York, 1974.
27. E. M. Engler, *Chem. Technol.* **6,** 274 (1976).
28. J. H. Perlstein, *Angew. Chem., Int. Ed. Engl.* **16,** 519 (1977).
29. R. C. Wheland and J. L. Gillson, *J. Am. Chem. soc.* **98,** 3917 (1976).
30. A. I. Kitaigorodsky, "Molecular Crystals and Molecules," Chapter 1. Academic Press, New York, 1973.
32. D. J. Sandman, A. J. Epstein, T. J. Holmes, and A. P. Fisher, III, *J. Chem. Soc., Chem. Commun.* p. 177 (1977).
32. G. LeCoustumer, J. Amzil, and Y. Mollier, *J. Chem. Soc., Chem. Commun.* p. 353 (1979).
33. H. Toftlund, *J. Chem. Soc., Chem. Commun.* p. 837 (1979).

5

Excited States of Oxocarbon Dianions

Josef Michl and Robert West

I. Introduction 79
II. The Perimeter Model for $(4N+2)$-Electron $[n]$Annulenes 80
A. Molecular Orbitals 80
B. Electronic States and Spectra 85
III. The Perimeter Model for Oxocarbon Dianions 88
A. Molecular Orbitals 88
B. Electronic States and Spectra 91
IV. Comparison with Experiment 95
References 99

I. Introduction

The cyclic oxocarbon dianions $C_nO_n^{2-}$ ($n \geqslant 3$) represent an intriguing class of compounds. Their structure consists of a polygon of carbon atoms separated by about 1.46 Å and a concentric larger polygon of oxygen atoms, with C—O bonds radial and about 1.26 Å long [1]. The simplicity of the structures and the existence of the homologous series $n = 3, 4, 5, \ldots$ tempt one to conclude that there might be simple regularities and trends in the properties of these compounds, including the properties of their excited states. The purpose of this chapter is to investigate the degree to which such expectations hold for the excited singlet states.

We begin by considering π,π^* excited singlets. Our discussion is based on the classic perimeter model [2,3], which provides simple physical insight; but we also utilize the results of numerical calculations [4,5] as necessary. Comparison with experimental data ($n = 4, 5, 6$) is not limited to the absorption [6] and reso-

OXOCARBONS

ISBN 0-12-744580-3

nance Raman [7] spectra, which have been known for some time, but also utilizes the results of recent measurements of magnetic circular dichroism (MCD), which made possible the recognition of additional excited states [5]. Since MCD theory may not be well known to the reader, its elements are reviewed, with particular attention to their application to the perimeter model.

Finally, we briefly consider n,π^* excited singlet states for which no definitive experimental information exists at present. Our discussion is based on numerical calculations [5] using the Ridley-Zerner [8] version of the INDO/S model.

II. The Perimeter Model for (4*N*+2)-Electron [*n*]Annulenes

A. MOLECULAR ORBITALS

The regular cyclic structure of the oxocarbon dianions invites analysis in terms of the perimeter model as developed by Platt [2], Moffitt [3], and others and extended recently by one of the authors for use in MCD spectroscopy [9]. Symmetry dictates the form of the n molecular orbitals (MO's) which result from π interactions between neighbors in a regular cyclic array of n $2p_z$ atomic orbitals (AO's, ϕ). We shall number the AO's consecutively, $\phi_0, \phi_1, \ldots, \phi_{n-1}$, in counterclockwise order as viewed from the positive end of the z axis. We shall label the MO's by the irreducible representations of the C_n point group, ϵ_k and ϵ_{-k}, where k is a positive integer or zero. The coefficient of AO ϕ_j in the MO ϵ_k is equal to $n^{-\frac{1}{2}} \exp(2\pi ijk/n) = n^{-\frac{1}{2}}[\cos(2\pi jk/n) + i \sin(2\pi jk/n)]$. All coefficients are thus equal to the normalization factor $n^{-\frac{1}{2}}$ in absolute value, and they differ only in their complex phase. These phases are complex unities which can be represented as the n vertices of a regular polygon of unit radius with its center at the origin in the complex plane and oriented so that its vertex number zero coincides with +1. We refer to this polygon as the "phase polygon" (Fig. 1). As one proceeds from one AO of the real polygon-shaped molecule to the next, the coefficient of a given MO, say ϵ_k, changes its complex phase. The phase goes around the phase polygon, starting at +1, in a counterclockwise fashion for ϵ_k and in a clockwise fashion for ϵ_{-k}. The rate at which the complex phase changes from one AO to the next, as one goes around the perimeter of the molecule, is proportional to k. In the MO ϵ_0, the rate of progress around the polygon in the complex plane is zero; i.e., not only the coefficient of the AO ϕ_0, but the coefficients of all the n AO's are real, have complex phase +1, and are equal to $n^{-\frac{1}{2}}$. This orbital is therefore totally symmetric and is usually labeled a. In the MO's ϵ_1 and ϵ_{-1} the complex phase of the AO coefficient jumps from one vertex of the phase polygon to the next, counterclockwise for ϵ_1 and clockwise for ϵ_{-1}, as one moves counterlockwise from one AO in the real molecule to the next. In the MO's ϵ_2 and ϵ_{-2}, the complex phase jumps twice as fast, skipping every other

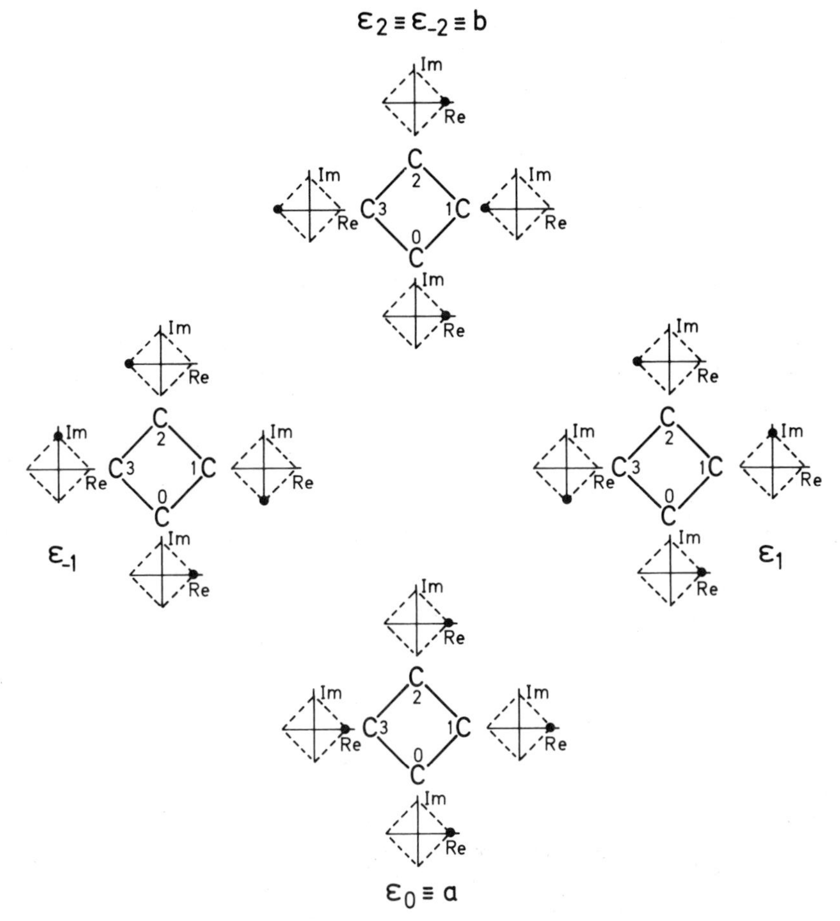

Fig. 1. The MO's of an [n]annulene. The coefficient of each AO in each MO has the absolute magnitude $n^{-\frac{1}{2}}$ and the complex phase shown by a dot in the phase polygon (Re, real; Im, imaginary). The numbering of the AO's is shown.

vertex of the phase polygon, and therefore runs twice around its perimeter as one goes around the real molecule once. In ϵ_3 and ϵ_{-3}, the complex phase changes three times as fast and skips two out of every three vertices of the phase polygon, etc.

Since n is finite, the number of values of k that lead to distinct MO's is limited. If n is even, the phase polygon has a vertex at -1. When $k = n/2$, the complex phase will reach this vertex immediately upon going from AO ϕ_0 to AO ϕ_1. As one proceeds to ϕ_2, the complex phase will jump back to $+1$, etc., and it is clear that the MO will have the coefficients $n^{-\frac{1}{2}}$, $-n^{-\frac{1}{2}}$, $n^{-\frac{1}{2}}$, $-n^{-\frac{1}{2}}$, etc., on

$\phi_0, \phi_1, \phi_2, \phi_3$, etc. For $-k = n/2$, the jumps will again be between $+1$ and -1, and the same orbital will result. Thus, if n is even, $\epsilon_{n/2} = \epsilon_{-n/2}$. This unique orbital is real, has nodes between all pairs of neighboring AO's, and is usually labeled b. If k is allowed to become larger than $n/2$, no new MO's will result, and it is readily checked that $\epsilon_{k \pm j \cdot n} = \epsilon_k$, where j is an integer. If n is odd, the phase polygon has no vertex at -1, and all orbitals ϵ_k, ϵ_{-k} are distinct up to and including $k = (n - 1)/2$. For larger values of k, however, duplication once again sets in. It is easily verified that $\epsilon_{(n+1)/2} = \epsilon_{-(n-1)/2}$, and in general again $\epsilon_{k \pm j \cdot n} = \epsilon_k$. In either case, then, there are n MO's. Their energies are split apart by amounts proportional to the resonance integral between the neighbors in a manner shown in Fig. 2. The totally symmetric orbital $\epsilon_0 \equiv$ a is the most bonding and is followed in the order of increasing energy by the degenerate pair $\epsilon_1, \epsilon_{-1}$, then $\epsilon_2, \epsilon_{-2}$, etc., until either the degenerate pair $\epsilon_{(n-1)/2}, \epsilon_{-(n-1)/2}$ (n odd) or the single orbital $\epsilon_{n/2} \equiv \epsilon_{-n/2} \equiv$ b (n even) is reached. The degeneracy of the orbitals ϵ_k and ϵ_{-k} follows from the fact that they are complex conjugates of each other, $\epsilon_{-k} = \epsilon_k^*$.

It is of interest to ask about the physical meaning of the complex phase of the AO coefficients in a given MO. Since the whole orbital can always be multiplied by any complex unity without any change in its physical significance, it is clear that only the relative change of the complex phase from one part of an MO to another, i.e., in the case at hand, from one AO coefficient to another, matters. These relative changes are a measure of net electron current in the MO. In the LCAO picture, electrons move around the molecule by jumping from one AO to another. The rate of electron flow from ϕ_0 directly to another AO, say ϕ_j, is proportional to the imaginary part of the complex phase of ϕ_j.* The proportionality constant contains the resonance integral between the two AO's and is therefore large between neighbors, an order of magnitude smaller and of the opposite sign between next-nearest neighbors, and negligible for more distant AO's. The opposite sign found for next-nearest neighbors results from the fact that, strictly speaking, our AO's are Löwdin-orthogonalized orbitals, each of which is not completely localized on a single center but contains small "wings" on its neighbors and much smaller contributions on more distant centers [9].

We now see that in the purely real MO ϵ_0, there is no net electron flow around the perimeter. This is in general true of all purely real orbitals. We also see that the currents in ϵ_k and ϵ_{-k} will be of equal magnitude but will flow in opposite sense. In order to consider this flow, we at first limit attention to its dominant part, due to electron jumps between neighboring AO's. As defined here, an

*More generally, that part of the change in the complex phase incurred upon going from a given AO to another AO within a given MO, which is perpendicular in the complex plane to the complex phase of the former AO, provides a measure of the net current of electrons occupying the MO from the former AO to the latter AO. If we choose the former AO to be our ϕ_0, which has a real coefficient, the statement in the text results.

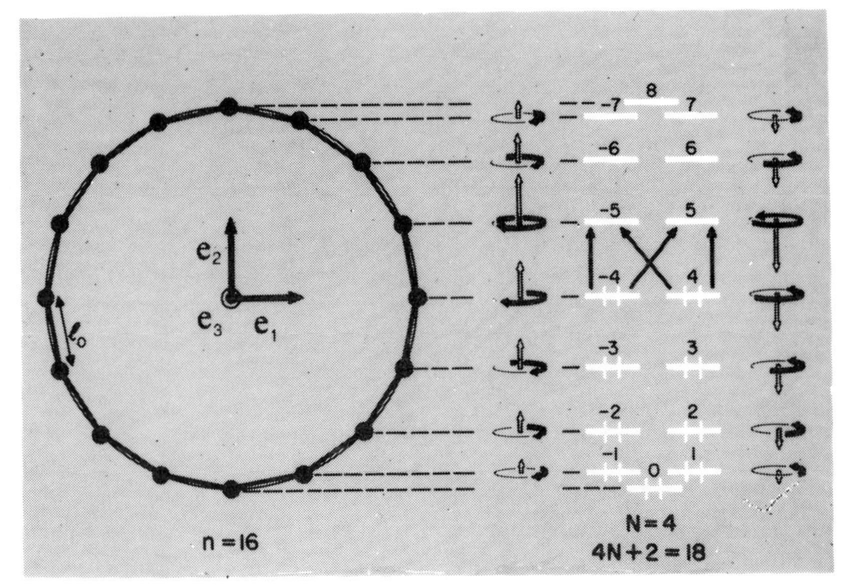

Fig. 2. Geometry and MO energies for a $(4N+2)$-electron $[n]$annulene. For each MO ϵ_k, the value of k is given and the sense and amount of electron circulation as well as the resulting magnetic moment are shown schematically. The MO's ϵ_0 and $\epsilon_{n/2} \equiv \epsilon_{-n/2} \equiv$ b have no net electron circulation and no magnetic moment. Electron occupancy in the ground configuration and the four HOMO → LUMO promotions are indicated. From Michl [9]. Reprinted with permission of the copyright holder, the American Chemical Society.

electron in ϵ_k will circulate in the counterclockwise sense, and an electron in ϵ_{-k} will circulate clockwise. The circulation of a charged particle will produce a current in a loop and thus a magnetic moment. Since the electron is negative, the moment will be negative, i.e., pointed toward the negative end of the z axis, if the electron is in the MO ϵ_k, and positive if the electron is in ϵ_{-k}. For a perimeter of a given size, the magnitude of the magnetic moment due to jumps between neighbors will be a function of the rate of electron flow, and thus of the resonance integrals between neighboring AO's, and of the imaginary part of the complex phase of ϕ_1. Since the bond lengths between neighbors are quite constant for perimeters constituted of a given type of atom, say carbon, the resonance integrals will be almost constant as well, and we need not consider their effects further at the moment. For now, we concentrate on the effects of the imaginary part of the complex phase of ϕ_1. It is apparent from Fig. 1 that this imaginary part increases at first when k increases since this makes the complex phase of ϕ_1 move counterclockwise along the perimeter of the phase polygon. However, when k has increased so much that the angle swept in the complex plane upon going from ϕ_0 to ϕ_1 is $\pi/2$ or larger, any further increase in k will reduce the imaginary part of the complex phase of ϕ_1 again and thus decrease the magnetic moment of the orbital ϵ_k. The extreme is reached for $k = n/2$ (even n), where the complex phase of ϕ_1 has changed by a full π with respect to ϕ_0. At this point, its imaginary part vanishes, and there is no net electron flow and consequently no orbital magnetic moment. Similar considerations hold for the negative values $-k$, except that the imaginary part of the complex phase of ϕ_1 is now of the opposite sign and the current will flow in the opposite direction.

When electron jumps between next-nearest neighbors are also taken into account, the result changes only slightly. The magnetic moment of the orbitals ϵ_k and ϵ_{-k} increases in absolute magnitude from the value zero for ϵ_0 until it reaches a maximum for a value of k somewhat larger than $n/4$. It then decreases rapidly and is small for the least-bonding orbital pair $\epsilon_{(n-1)/2}$, $\epsilon_{-(n-1)/2}$ if n is odd, or zero for the least-bonding orbital b $= \epsilon_{n/2}$ if n is even.

Electron transitions between the individual MO's can be induced by electromagnetic radiation. Nonvanishing electric dipole transition moments occur only for those transitions which change the orbital subscript k by 1. If we take our light as propagating in the positive direction of the z axis, the transition $\epsilon_k \rightarrow \epsilon_{k+1}$ calls for the absorption of a left-handed circular (LHC) polarized photon (angular momentum $\hbar$), and the transition $\epsilon_{-k} \rightarrow \epsilon_{-k-1}$ calls for the absorption of a right-handed circular (RHC) polarized photon (angular momentum $-\hbar$). In a sense, then, the orbital subscript k serves as a quantum number, and the selection rule is similar to that known in atomic spectroscopy, with the nodeless ϵ_0 orbital playing the role of the atomic s orbital, the ϵ_1, ϵ_{-1} pair being analogous to the atomic p_+, p_- orbital pair, etc. The most striking difference is the fact that in the cyclic perimeter absorption of an LHC photon and thus an increase in the "quantum number" k to $k + 1$ can cause either an increase or a decrease in the magnitude of the orbital magnetic moment depending on the value of k, as discussed above, whereas in an atom it always corresponds to an increase. As shown below, this fact has a profound effect on the MCD spectra of cyclic π-electron systems compared with those of atoms. The different behavior of the orbitals in atoms is due to their higher degree of symmetry, which ensures a complete transfer of the angular momentum of the photon to the orbital angular moment of the promoted electron. In the picture described above, it is due to the fact that the complex phase $\exp(ik\varphi)$ changes continuously as one goes around an atom in the xy plane, so that taking the difference of the complex phase between two points is now replaced by taking the derivative $(d/d\varphi)[\exp(ik\varphi)]$, which is always perpendicular to the phase itself and proportional to k.

In a discussion of atomic orbitals, one usually converts the complex AO's p_+ and p_- into their real linear combinations p_x and p_y, which are more conveniently visualized. A similar transformation is commonly performed on the degenerate pairs of perimeter orbitals ϵ_k, ϵ_{-k}. The real form belongs to the doubly degenerate real representation e_k and is undoubtedly familiar to the reader. Since the real MO's are mixtures of equal amounts of ϵ_k and ϵ_{-k}, an electron is just as likely to move counterclockwise as clockwise when it is described by such a real MO. It then has no net magnetic moment, and these real MO's provide less insight into the origin of MCD spectra.

B. ELECTRONIC STATES AND SPECTRA

In most cases, it is reasonable to approximate electronic states of annulenes by configurations, and we shall consider only first-order configuration interaction. If $4N + 2$ electrons are present in the perimeter (Fig. 2), its ground configuration has the orbital occupancy $\epsilon_0^2\epsilon_1^2\epsilon_{-1}^2 \ldots \epsilon_N^2\epsilon_{-N}^2$. There are two special cases. First, if there are only two π electrons ($N = 0$), the ground configuration is ϵ_0^2 and there are two degenerate low-energy excited singlet configurations, $\epsilon_0\epsilon_1$ and $\epsilon_0\epsilon_{-1}$, which we shall denote by $\epsilon_0 \rightarrow \epsilon_1$ and $\epsilon_0 \rightarrow \epsilon_{-1}$, respectively. Second, if $N = n/2 - 1$ (n even), the ground configuration is $\epsilon^2 \ldots \epsilon_{n/2-1}^2\epsilon_{-(n/2-1)}^2$, and all orbitals but the topmost one will be filled. There are now two degenerate low-energy excited singlet configurations similarly denoted $\epsilon_{n/2-1} \rightarrow \epsilon_{n/2}$ and $\epsilon_{-(n/2-1)} \rightarrow \epsilon_{-n/2}$ (note that $\epsilon_{n/2} \equiv \epsilon_{-n/2} \equiv$ b). In both of these special cases one low-energy excited state of symmetry E_1 results, and a transition into it is electric-dipole-allowed. In the general case, $0 < N < n/2 - 1$, and both the highest occupied MO's (HOMO's) ϵ_N and ϵ_{-N} and the lowest unoccupied MO's (LUMO's) ϵ_{N+1} and $\epsilon_{-(N+1)}$ of the ground configuration are degenerate. As a result, there are four low-energy excited configurations. In two of them, the sense of circulation of the excited electron is preserved, counterclockwise in $\epsilon_k \rightarrow \epsilon_{k+1}$ and clockwise in $\epsilon_{-k} \rightarrow \epsilon_{-(k+1)}$. These give rise to an excited state of symmetry E_1, a transition into which is electric-dipole-allowed. In the other two configurations, $\epsilon_k \rightarrow \epsilon_{-(k+1)}$ and $\epsilon_{-k} \rightarrow \epsilon_{k+1}$, the sense of circulation of the promoted electron is reversed, and these produce a single lower-energy excited state of symmetry E_{2N+1} if the perimeter is charged ($n \neq 4N + 2$) or two lower-energy excited states of symmetry B if it is uncharged ($n = 4N + 2$). In either case, transitions into these sense-reversing lower-energy excited states are electric-dipole-forbidden in absorption.

The magnetic moment of the ground configuration vanishes in every case, since the moments of the two electrons in the MO ϵ_k always exactly cancel those of the two electrons in the MO ϵ_{-k}. The magnetic moments of the excited configurations generally do not vanish and are given by the difference of the magnetic moment of the promoted electron after and before excitation. Since the electron has no orbital magnetic moment when it is in the MO $\epsilon_0 \equiv$ a, in the case $N = 0$ the magnetic moment of the excited configuration $\epsilon_0 \rightarrow \epsilon_1$ is equal to that of the MO ϵ_1 and is negative, whereas that of the configuration $\epsilon_0 \rightarrow \epsilon_{-1}$ is of equal size but positive. Similarly, in the case $N = n/2 - 1$, the electron has no orbital magnetic moment when in the MO $\epsilon_{n/2} \equiv \epsilon_{-n/2} \equiv$ b, whereas it had a positive moment in $\epsilon_{-(n/2-1)}$ or a negative one in $\epsilon_{n/2-1}$ before promotion, so that the magnetic moment of the excited configuration $\epsilon_{n/2-1} \rightarrow \epsilon_{n/2}$ is equal to minus the moment of the orbital $\epsilon_{n/2-1}$ and thus is positive, that of the configuration $\epsilon_{-(n/2-1)} \rightarrow \epsilon_{-n/2}$ being of equal size but negative. We shall see below how these signs are reflected in the MCD spectra.

The magnetic moments due to the sense-reversing excitations $\epsilon_k \rightarrow \epsilon_{-(k+1)}$ and $\epsilon_{-k} \rightarrow \epsilon_{k+1}$ which occur in the general perimeter ($0 < N < n/2 - 1$) are also easy to predict. That of the former is clearly large and positive, since a negative contribution is destroyed when an electron is taken out of ϵ_k and a positive one is generated when it is placed in $\epsilon_{-(k+1)}$. That of the latter is equally large but negative. The magnetic moments resulting from the sense-preserving excitations $\epsilon_k \rightarrow \epsilon_{k+1}$ and $\epsilon_{-k} \rightarrow \epsilon_{-(k+1)}$ are less easily predictable. They are clearly relatively small and can be of either sign, depending on the value of k. As we have seen above (Fig. 2), for smaller values of k, up to a little past $n/4$, the magnitude of the magnetic moment increases with k, so that the moment of the configuration $\epsilon_k \rightarrow \epsilon_{k+1}$ will be small and negative, and that of $\epsilon_{-k} \rightarrow \epsilon_{-(k+1)}$ will be equally small and positive. For large values of k, the magnitude of the magnetic moment decreases with k, so that the moment of $\epsilon_k \rightarrow \epsilon_{k+1}$ will be small and positive, and that of $\epsilon_{-k} \rightarrow \epsilon_{-(k+1)}$ will be small and negative. Numerical values for typical cases have been tabulated [9].

In the nomenclature introduced by Platt [2], the above-described excited states of E_1 symmetry are referred to as the B states and the transitions from the ground state into them as B transitions. The lower-energy excited states of E_{2N+1} (n odd) or B (n even) symmetry, which exist only if $0 < N < n/2 - 1$ and are due to sense-reversing electron promotions, are referred to as the L states and the transitions into them as the L transitions. These labels form the basis of the standard nomenclature for the spectra of aromatic molecules.

We can now summarize our expectations for the singlet π,π^* contributions to the absorption spectra of $(4N+2)$-electron [n]annulenes. First, the spectra should contain a band due to an allowed B transition into the state of E_1 symmetry and, second, if $0 < N < n/2 - 1$, they should also contain one or two symmetry-forbidden L transitions at lower energies. If $n \neq 4N + 2$, there will be one L transition into a degenerate state of symmetry E_{2N+1}; if $n = 4N + 2$, there will be two L transitions into nondegenerate states of symmetry $B \equiv E_{n/2} \equiv E_{-n/2}$. When the full D_{nh} symmetry of the [n]annulene is considered, the lower of the two L states belongs to the symmetry species B_{2u} (the L_b state) and the upper to the symmetry species B_{1u} (the L_a state). This picture agrees very well with the observed spectra, in which the symmetry-forbidden L transitions appear weakly, due to vibronic intensity borrowing, which we have ignored here.

It remains to consider how the signs of the magnetic moments of the excited states will be reflected in the MCD spectra. In these spectra, the difference of the extinction coefficients for RHC polarized light and for LHC polarized light, $\epsilon_L - \epsilon_R$, measured in the presence of unit magnetic field directed along the light propagation direction, is plotted against photon energy. More commonly, molar ellipticity $[\theta]_M$, which is proportional to $\epsilon_L - \epsilon_R$, is plotted. In the absence of magnetic field, the two components of a doubly degenerate state, one LHC polarized, i.e., reached from the ground state by absorption of LHC light, the other RHC polarized, i.e., reached by absorption of RHC light, are of equal energy and have identical spectral shapes. Therefore, $\epsilon_L = \epsilon_R$ at all wavelengths, and the [*n*]annulene exhibits no circular dichroism. However, in the presence of a magnetic field the Zeeman effect will split the two components (Fig. 3). The component with a negative magnetic moment will be shifted to a higher energy, and the one with a positive magnetic moment to a lower energy. These shifts are small so that for ordinary magnetic field strengths any splitting that would be expected in the absorption spectrum is completely buried within the line width, which is typically quite considerable for molecules of the type considered here. The existence of the shifts can, however, still be detected, since the components differ in circular polarization. Their mutual displacement on the energy scale will therefore cause the appearance of a nonvanishing $\epsilon_L - \epsilon_R$ difference and thus an MCD effect. If the LHC polarized component is the one that has a negative magnetic moment, it is reached later as the photon energy is swept to higher values, and the $\epsilon_L - \epsilon_R$ curve therefore first dips to negative values, where the absorption into the RHC polarized component dominates, then passes through zero at the center of the absorption band, and acquires positive values as the photon energy increases. Of course, it goes back to zero for wavelengths beyond the absorption band for which $\epsilon_L = \epsilon_R = 0$. The resulting S-shaped curve in the MCD spectrum is called a positive *A* term. This is the MCD sign sequence that would be expected in atoms. The above analysis for $(4N+2)$-electron [*n*]an-

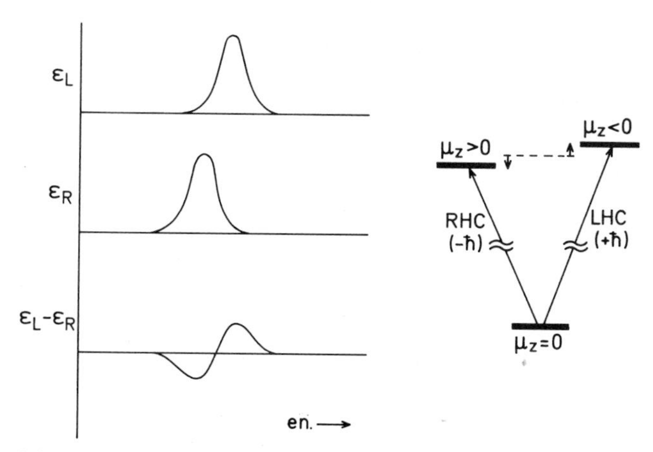

Fig. 3. Origin of the *A* term in the Zeeman effect. In the case shown, absorption of LHC light produces an excited state with a negative component of the magnetic moment along the *z* axis μ_z and thus a positive *A* term. A negative *A* term would result if LHC light produced $\mu_z > 0$.

nulenes shows that positive A terms are expected for the B transition in two-electron annulenes ($N = 0$) and that weak positive A terms are also expected for the B transition in those annulenes with $0 < N < n/2 - 1$ in which the magnitude of the magnetic moment of LUMO exceeds that of the magnetic moment of HOMO, as is usually the case. However, in practice the small A terms in the latter case are likely to be perturbed by vibronic borrowing, which admixes some of the much larger magnetic moments of the L states into the vibronic wavefunctions of the B states.

If the LHC polarized component of the degenerate excited state is the one that has a positive magnetic moment, it is reached first as the photon energy is swept to higher values, and the MCD curve therefore first rises to positive values of $\epsilon_L - \epsilon_R$ and then passes through zero to negative values at higher energies. The resulting S-shaped curve is a mirror image of that discussed above and is referred to as a negative A term. We expect it to occur for the B transition in those $(4N+2)$-electron $[n]$annulenes for which $N = n/2 - 1$. Weakly negative A terms are also expected for the B transitions in those annulenes with $0 < N < n/2 - 1$ in which the magnitude of the magnetic moment of LUMO is smaller than that of the magnetic moment of HOMO. Once again, these small A terms are likely to be perturbed by vibronic effects. The MCD effects of the L transitions in $(4N+2)$-electron $[n]$annulenes with $0 < N < n/2 - 1$ are also vibronic in origin, primarily due to borrowing of circularly polarized intensity from the B transitions, but they need not be discussed here.

A more quantitative treatment of the origin of A terms requires averaging over molecular orientations, since the molecules in an isotropic sample are not aligned with their z axes along the magnetic field, as has been assumed in our qualitative discussion for simplicity. The qualitative results remain unchanged, and it can be shown that the magnitude of the A term, defined as $A = 33.53^{-1} \int d\tilde{\nu}(\tilde{\nu} - \tilde{\nu}_0)[\theta]_M/\tilde{\nu}$, where $\tilde{\nu}$ is wavenumber, $\tilde{\nu}_0$ is band center, and $[\theta]_M$ is molar ellipticity per unit magnetic field (deg. L.m^{-1} mole^{-1} G^{-1}), is related to the magnetic moment μ of the excited state by $-2A/D = \mu$, where D is the dipole strength of the transition.

III. The Perimeter Model for Oxocarbon Dianions

A. MOLECULAR ORBITALS

The MO's of the cyclic $C_nO_n^{2-}$ dianions of interest to us here can be derived in a way similar to that used for the simple perimeters discussed above. If we consider the C_nO_n skeleton as originating in n CO units interacting through their carbon ends, it is useful to first construct the C—O π-bond orbitals (right-hand side of Fig. 4). These are the n degenerate strongly bonding π_{CO} orbitals, and the n degenerate strongly antibonding π^*_{CO} orbitals. Next, we consider only the

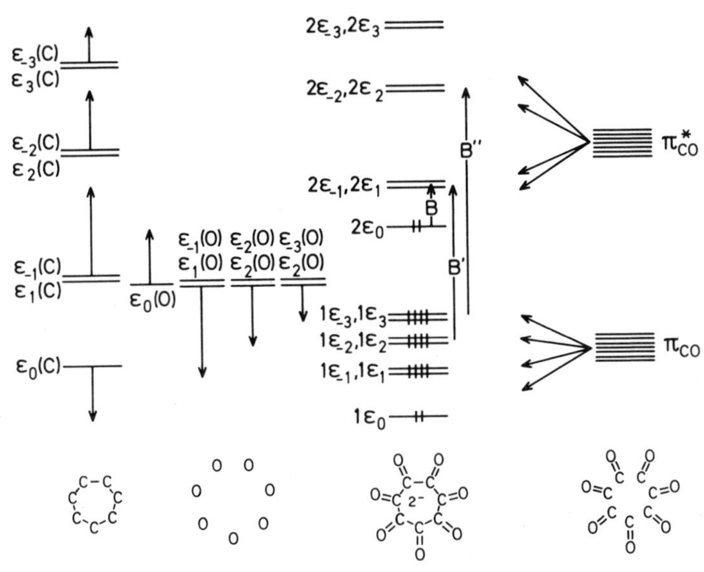

Fig. 4. On the left, qualitative derivation of the MO's of a $C_nO_n^{2-}$ dianion ($n = 7$) from those of the perimeter orbitals $\epsilon_k(C)$ and the oxygen symmetry orbitals $\epsilon_k(O)$. On the right, their relation to those of the bonding (π_{CO}) and antibonding (π^*_{CO}) bond orbitals of the constituent carbonyl groups. The first three allowed transitions, B, B', and B'', are indicated.

first-order interactions between degenerate orbitals. The n low-energy π_{CO} orbitals will interact to produce a set of relatively low energy delocalized orbitals of the C_nO_n skeleton, all of which will be C—O bonding and which we refer to as group 1. The n high-energy π^*_{CO} orbitals will similarly produce a set of relatively high energy delocalized orbitals, all of which will be C—O antibonding and which we refer to as group 2. The course of these interactions is governed by symmetry exactly as was the case for the simple perimeter in which a set of n nonbonding AO's was allowed to interact: Within each group, the most stable orbital is of symmetry $\epsilon_0 \equiv$ a and is followed by a degenerate pair of symmetry ϵ_1, ϵ_{-1}, etc., and eventually by the least stable orbital of the group, either the $\epsilon_{n/2-1}$, $\epsilon_{-(n/2-1)}$ pair (n odd) or the $\epsilon_{n/2} \equiv \epsilon_{-n/2} \equiv$ b orbital (n even).

Finally, second-order mixing between the resulting bonding and antibonding orbitals is allowed to occur. This will clearly not change the qualitative picture of the orbital arrangement but will permit the degree of C—O bonding or antibonding to vary from one MO to another. Still, the MO's of the low-energy group 1 will be generally C—O bonding, and those of the high-energy group 2 will be generally C—O antibonding. Within each group, the lowest a orbital will be purely C—C bonding and the highest $\epsilon_{n/2-1}$, $\epsilon_{-(n/2-1)}$ pair (n odd) or $\epsilon_{n/2} \equiv \epsilon_{-n/2} \equiv$ b orbital (n even) strongly C—C antibonding, with a gradual transition from one extreme to the other.

An alternative way of deriving this orbital scheme is shown on the left-hand side of Fig. 4. It is useful since it allows us to make easy estimates of the magnetic moments associated with the various orbitals. Here, we first construct the ordinary perimeter orbitals from the carbon AO's in the usual fashion and label them $\epsilon_0(C)$, $\epsilon_1(C)$, $\epsilon_{-1}(C)$, etc., and combine the $2p_z$ orbitals on the oxygens into symmetry orbitals. Since the resonance integrals between the oxygen atoms are negligible due to their large distance, these symmetry orbitals will all be of about the same energy and will remain nonbonding, but their symmetry properties will be the same as those of the strongly bonding and antibonding carbon orbitals, and we shall label them $\epsilon_0(O)$, $\epsilon_1(O)$, $\epsilon_{-1}(O)$, etc. It is important to note that, unlike the carbon perimeter orbitals, the analogous oxygen symmetry orbitals will have negligible magnetic moments, since the resonance integrals within even the nearest pairs of oxygens will be negligibly small.

Finally, the carbon perimeter orbitals and the nonbonding symmetry orbitals on the oxygens are permitted to interact. Since there is only one delocalized orbital of each symmetry species in each group, the consideration of this interaction is very simple. Effectively, orbitals of like symmetry will act as if their energies repelled, and the degree of the effect will be dominated by their separation on the energy scale. The lower-energy combination, $\epsilon_k(C) + \epsilon_k(O)$, will be identical to one of the MO's of the above group 1. It will be bonding along the C—O bonds corresponding to the new interactions and will consist largely of that original orbital $\epsilon_k(C)$ or $\epsilon_k(O)$ which was of lower energy. The higher-energy combination, $\epsilon_k(C) - \epsilon_k(O)$, will correspond to one of the MO's of the above group 2. It will be C—O antibonding and will consist predominantly of the original orbital that was of higher energy. If the two orbitals are of roughly equal energy, they will mix in a ratio of approximately one to one.

Consideration of the relative electronegativity of carbon and oxygen suggests, and both π-electron and all-valence-electron calculations using standard parameters confirm [5] that the energy of the nonbonding oxygen orbitals is comparable to that of the ϵ_1, ϵ_{-1} bonding pair of the carbon perimeter orbitals, so that they mix in a roughly one-to-one ratio. This is clearly somewhat dependent on the value of n. In very small rings, the ϵ_1, ϵ_{-1} pair is nonbonding ($n = 4$) or even antibonding ($n = 3$), and then the oxygen orbitals are of lower energy. In general, however, the $\epsilon_1(C) - \epsilon_1(O)$ mixing is stronger than any other and produces two widely split ϵ_1, ϵ_{-1} pairs of MO's, $1e_1$ and $2e_1$. The ϵ_0 symmetry orbital of the oxygens $\epsilon_0(O)$ is pushed up in energy by interaction with $\epsilon_0(C)$, which is lower but close enough for a reasonably effective mixing. Calculations suggest that the weight of $\epsilon_0(C)$ in the lower of the resulting MO's, $1\epsilon_0$, is somewhere between two-thirds and three-quarters of the total, whereas its weight in the upper of the resulting MO's, $2\epsilon_0$, is between one-quarter and one-third. The remaining symmetry orbitals of the oxygens, $\epsilon_2(O)$, $\epsilon_{-2}(O)$, $\epsilon_3(O)$, $\epsilon_{-3}(O)$, etc., are all pushed down in energy by interaction with the higher carbon perimeter orbitals $\epsilon_2(C)$, $\epsilon_{-2}(C)$, $\epsilon_3(C)$, $\epsilon_{-3}(C)$, etc., in progressively smaller amounts as the energy separation $\epsilon_k(O) - \epsilon_k(C)$ increases with increasing k. At the same time, the resulting bonding orbitals $1\epsilon_2$, $1\epsilon_{-2}$, $1\epsilon_3$, $1\epsilon_{-3}$, etc., become more purely localized on oxygen atoms alone. The calculations suggest that the weight of $\epsilon_2(C)$ in $1\epsilon_2$ is only about one-fifth and that the weight of the $\epsilon_k(C)$'s decreases further for higher $1\epsilon_k$'s. These results are in qualitative agreement with those of ESR measurements on the related radical-anions [*10*]. The g value for $C_6O_6^{3-}$ is lower than that of $C_6O_6^-$, as expected if the weight of $\epsilon_1(C)$ in $2\epsilon_1$ is larger than the weight of $\epsilon_0(C)$ in $2\epsilon_0$. The ^{13}C coupling constant in $C_5O_5^-$ suggests that $\epsilon_0(O)$ dominates over $\epsilon_0(C)$ in $2\epsilon_0$.

The resulting orbital pattern is the same as that deduced by constructing n CO fragments first and consists of two full sets of orbitals ϵ_0, ϵ_1, $\epsilon_{-1}, \ldots, \epsilon_{n/2-1}$, $\epsilon_{-(n/2-1)}$ (n odd) or $\ldots, \epsilon_{n/2}$ (n even), one bonding (group 1), the other antibonding (group 2). The most stable among the orbitals of group 2, $2\epsilon_0$, must actually become at least weakly bonding if the molecule is in a polar solvent, or ion-paired, since the dianions $C_nO_n^{2-}$ do not autoionize and are stable in solution, but this was not considered in the calculations. The advantage of the second method of constructing the MO's of $C_nO_n^{2-}$ is that it provides an estimate of the relative weight of the carbon perimeter orbitals $\epsilon_k(C)$ and oxygen symmetry orbitals $\epsilon_k(O)$ in the MO's $1\epsilon_k$ and $2\epsilon_k$. Since the contribution of the $\epsilon_k(O)$ orbitals to the magnetic moments of the resulting MO's is virtually nil, we can estimate the relative magnitude of the magnetic moments of the final MO's. In the case of $n = 5$–7, the magnetic moments of the orbitals $\epsilon_1(C)$ and $\epsilon_2(C)$ do not differ much, and then the relative weight of $\epsilon_k(C)$ and $\epsilon_k(O)$ alone permits an estimate to be made. The magnetic moments of MO's $1\epsilon_1$ and $2\epsilon_1$ will be comparable and quite large, those of higher orbitals of group 1, such as $1\epsilon_2$, will be small, those of higher orbitals of group 2, such as $2\epsilon_2$, will be large, until one gets near the upper end of the MO spectrum.

B. ELECTRONIC STATES AND SPECTRA

Once again, we approximate electronic states by single configurations. In a more quantitative analysis, it would be necessary to allow for configuration mixing. The mixing of the ground configuration with doubly excited configurations must be considered if the singlet–triplet splitting is to be understood [*4*], but this does not concern us here.

The $C_nO_n^{2-}$ dianion contains $2n+2$ π electrons. In the ground configuration, all n MO's of group 1 and the lowest MO of group 2, $2\epsilon_0$, will be doubly occupied. There will be a considerable number of low-energy excited singlet configurations, corresponding to excitations from the various orbitals of group 1 into the LUMO, $2\epsilon_1$, $2\epsilon_{-1}$, into the next higher vacant orbital $2\epsilon_2$, $2\epsilon_{-2}$, etc. However, relatively few of these will be of E_1 symmetry, which is required if transitions into them are to be symmetry-allowed. To produce an E_1 configuration, the excitation has to be of the type* $\epsilon_k \rightarrow \epsilon_{k\pm1}$. The lowest-energy excited configuration, $2\epsilon_0 \rightarrow 2\epsilon_1, 2\epsilon_1$, is of the type $\epsilon_k \rightarrow \epsilon_{k\pm1}, \epsilon_{-k} \rightarrow \epsilon_{-(k\pm1)}$ and is completely analogous to the configuration $\epsilon_0 \rightarrow \epsilon_1, \epsilon_{-1}$ in a simple two-electron [n]annulenylium cation such as the hypothetical cyclobutadiene dication. Among configurations obtained by transitions from orbitals of group 1 into orbitals $2\epsilon_1$, $2\epsilon_{-1}$, two are of this type. One of these, $1\epsilon_0 \rightarrow 2\epsilon_1, 2\epsilon_{-1}$, is at very high energies, but the other, $1\epsilon_2 \rightarrow 2\epsilon_1$, $1\epsilon_{-2} \rightarrow 2\epsilon_{-1}$, is more likely to be observed. This configuration has no analogy in simple $(4N+2)$-electron [n]annulenes discussed previously, in which the order of MO's is such that, if $\epsilon_k, \epsilon_{-k}$ is vacant in the ground configuration, all $\epsilon_j, \epsilon_{-j}$ pairs for which $j > k$ are also vacant since they are higher in energy. The next group of configurations corresponds to promotions into $2\epsilon_2, 2\epsilon_{-2}$. The allowed transitions will originate in $1\epsilon_1, 1\epsilon_{-1}$ and $1\epsilon_3, 1\epsilon_{-3}$ (if $n = 6$, $1\epsilon_3 \equiv 1\epsilon_{-3}$; if $n < 6$, ϵ_3 does not exist) but are likely to be at high energies and difficult to observe for small values of n.

The forbidden transitions are of two types. Some originate in electron promotions from the higher orbitals of group 1 into the $2\epsilon_1, 2\epsilon_{-1}$ pair and possibly also into the $2\epsilon_2, 2\epsilon_{-2}$ pair. Others are analogous to the L transitions in the simple perimeters in that they are the sense-reversing excitations between a pair of degenerate orbitals which also produces an allowed transition: $\epsilon_k \rightarrow \epsilon_{-(k+1)}, \epsilon_{-k} \rightarrow \epsilon_{k+1}$. The upper states will be degenerate and of symmetry E_{2k+1}. If n is even, this representation may not exist, and the upper states will then be nondegenerate and of symmetry A or B.

For larger values of n, a fairly complex absorption spectrum is thus expected. This will be simplified, at least apparently, by the fact that most transitions are forbidden and probably difficult to observe. The spectrum should start with an allowed transition into the degenerate excited state $2\epsilon_0 \rightarrow 2\epsilon_1, 2\epsilon_{-1}$ of E_1 symmetry, analogous to the B transition expected for two-electron [n]annulenylium ions. We refer to this characteristic first π,π^* transition as the B transition of the $C_nO_n^{2-}$ dianion. It may be followed by forbidden transitions, depending on the value of n. For $n > 3$, the next allowed transition, which we shall label B', should correspond to the two degenerate B-type excitations $1\epsilon_2 \rightarrow 2\epsilon_1$ and $1\epsilon_{-2} \rightarrow 2\epsilon_{-1}$ (if $n = 4$, $\epsilon_2 \equiv \epsilon_{-2}$ and the corresponding forbidden transition of the L type, $1\epsilon_2 \rightarrow 2\epsilon_{-1}$ and $1\epsilon_{-2} \rightarrow 2\epsilon_1$, will be missing). If $n \geqslant 6$, the allowed absorption due to the degenerate B-type excitations $1\epsilon_3 \rightarrow 2\epsilon_2$ and $1\epsilon_{-3} \rightarrow 2\epsilon_{-2}$ will come at even higher energies (transition B''), etc.

The effect of an increase in n should be twofold. First, as already mentioned, a larger number of orbitals will be available, and a larger number of both allowed

*Strictly speaking, the electron promotion $1\epsilon_{(n-1)/2} \rightarrow 2\epsilon_{(n-1)/2}$ (n odd) also produces an allowed transition into a configuration of symmetry E_1', but this is likely to lie at high energy and to be of little interest in the present context.

and forbidden transitions will be possible. In particular, an increasing number of forbidden transitions should occur in the experimentally interesting region between B and B'. Calculations [4,5] suggest that in $C_4O_4^{2-}$, this region will contain none, in $C_5O_5^{2-}$, one (E_2': $1\epsilon_2 \rightarrow 2\epsilon_{-1}$, $1\epsilon_{-2} \rightarrow 2\epsilon_1$), and in $C_6O_6^{2-}$, three or four ($2 \times E_{2g}$ from $1b \rightarrow 2\epsilon_1, 2\epsilon_{-1}$ and $2a \rightarrow 2\epsilon_2, 2\epsilon_{-2}$; B_{1u} and B_{2u} from $1\epsilon_2 \rightarrow 2\epsilon_{-1}$, $1\epsilon_{-2} \rightarrow 2\epsilon_1$). Second, even if it is assumed that the C—O and C—C bond lengths, and thus the resonance integrals, remain constant, the excitation energies of transitions of the various types should still decrease with increasing n, just as they do in the simple $(4N+2)$-electron [n]annulenes. This happens because the splitting between the individual MO's decreases as n grows. For each of the two groups, a total of n orbitals must be packed into an energy region spanned by four times the effective resonance integral between the π_{CO} or π^*_{CO} bond orbitals on neighboring CO groups (see the right-hand side of Fig. 4). This effect should be most striking for the allowed transitions B, B', etc., the presence of which should dominate the absorption spectra.

MCD spectroscopy is likely to be of use in two ways. First, it may allow the detection of some of the forbidden transitions, particularly those of the L type, which should have very large magnetic moments, if they acquire some intensity by vibronic interactions. Second, it may confirm our deductions concerning the relative size of the magnetic moments associated with the various orbitals. For this purpose, it is useful to translate these deductions into predictions of the signs of A terms in the MCD spectra.

From our previous discussion of the origin of these signs in simple $(4N+2)$-electron [n]annulenes, it is already clear that the degenerate excitations $\epsilon_k \rightarrow \epsilon_{k+1}$, $\epsilon_{-k} \rightarrow \epsilon_{-(k+1)}$ produce a positive A term if the magnetic moment of an electron in orbital ϵ_{k+1} is larger in absolute value than that of an electron in orbital ϵ_k, and a negative A term if it is smaller. Since we are now also faced with excitations of the type $\epsilon_{k+1} \rightarrow \epsilon_k$, $\epsilon_{-(k+1)} \rightarrow \epsilon_{-k}$, we must consider them as well. We first note that since the promotion $\epsilon_k \rightarrow \epsilon_{k+1}$ required an LHC photon and $\epsilon_{-k} \rightarrow \epsilon_{-(k+1)}$ an RHC photon, the promotion $\epsilon_{k+1} \rightarrow \epsilon_k$ will require an RHC photon and $\epsilon_{-(k+1)} \rightarrow \epsilon_{-k}$ an LHC photon. Our old argument (Fig. 3) then shows that, if the magnetic moment of ϵ_k is less negative than that of ϵ_{k+1}, the excited configuration $\epsilon_{k+1} \rightarrow \epsilon_k$ will have a positive magnetic moment and will represent the lower Zeeman component in magnetic field. Since it is produced by an RHC photon, the values of $\epsilon_L - \epsilon_R$ will be negative at lower photon energies and positive at higher photon energies, which corresponds to a positive A term. Similarly, if the magnetic moment of ϵ_k is more negative than that of ϵ_{k+1}, a negative A term will result.

In summary, the sign of the A term of the pair of electron transitions between ϵ_k and ϵ_{k+1} and between ϵ_{-k} and $\epsilon_{-(k+1)}$ does not depend on whether the excitation is from k to $k+1$ or from $k+1$ to k. In either case, the A term will be positive if the absolute value of the magnetic moment of an electron in the orbital

ϵ_{k+1} (or $\epsilon_{-(k+1)}$) is larger than that of an electron in the orbital ϵ_k (or ϵ_{-k}), and negative if the opposite is true.

Since the orbital $2\epsilon_0$ has no magnetic moment, whereas the orbitals $2\epsilon_1$ and $2\epsilon_{-1}$ do, there is no doubt about the sign of the *A* term of the *B* transition in the $C_nO_n^{2-}$ dianions, represented by the $2\epsilon_0 \rightarrow 2\epsilon_1$, $2\epsilon_0 \rightarrow 2\epsilon_{-1}$ pair of excitations; it must be positive. This prediction was already reached in the original analysis of the MCD spectra of $(4N+2)$-electron [*n*]annulenes in terms of the perimeter model [9].

According to our analysis, the next allowed transition, B', should correspond to the pair of excitations $1\epsilon_2 \rightarrow 2\epsilon_1$, $1\epsilon_{-2} \rightarrow 2\epsilon_{-1}$. This transition does not exist in $C_3O_3^{2-}$. In $C_4O_4^{2-}$, $1\epsilon_2 \equiv 1\epsilon_{-2} \equiv 1b$. This orbital has a zero magnetic moment, whereas the pair $2\epsilon_1$, $2\epsilon_{-1}$ has a nonzero moment, so that the above argument leaves no doubt that the *A* term of the B' transition must be negative. In higher members of the $C_nO_n^{2-}$ series, both the starting ($1\epsilon_2$, $1\epsilon_{-2}$) and ending ($2\epsilon_1$, $2\epsilon_{-1}$) orbitals involved in the B' transition are degenerate and have nonvanishing magnetic moments. It then becomes more difficult to make a firm prediction of the sign of the *A* term. Our analysis suggested very strongly that the orbital pair $2\epsilon_1$, $2\epsilon_{-1}$ has a large magnetic moment since it contains a heavy contribution from the carbon perimeter orbitals $\epsilon_1(C)$, $\epsilon_{-1}(C)$, whereas the orbital pair $1\epsilon_2$, $1\epsilon_{-2}$ has a smaller magnetic moment since it is composed primarily of the oxygen symmetry orbitals $\epsilon_2(O)$, $\epsilon_{-2}(O)$, which carry essentially no magnetic moment. There is little doubt that this conclusion is correct for $n = 5$–7, but, as n increases, it may become questionable, both because of the increasing content of $\epsilon_2(C)$, $\epsilon_{-2}(C)$ in the orbital pair $1\epsilon_2$, $1\epsilon_{-2}$ (Fig. 4) and because the magnetic moment of $\epsilon_2(C)$, $\epsilon_{-2}(C)$ gradually becomes substantially larger than that of $\epsilon_1(C)$, $\epsilon_{-1}(C)$ as n increases (in the limit of large n, it is twice as large), whereas they were comparable for $n = 5$–7.

The next allowed transition should be B'', corresponding to the pair of excitations $1\epsilon_3 \rightarrow 2\epsilon_2$, $1\epsilon_{-3} \rightarrow 2\epsilon_{-2}$. It appears only if $n \geqslant 6$. In $C_6O_6^{2-}$, $1\epsilon_3 \equiv 1\epsilon_{-3} \equiv 1b$. This orbital has zero magnetic moment whereas the members of the pair $2\epsilon_2$, $2\epsilon_{-2}$ have a nonzero moment, so that there is no doubt that the *A* term of the B'' transition must be negative. In $C_7O_7^{2-}$ and higher members of the series, both the starting and the ending orbitals of the B'' transition are degenerate and have nonvanishing magnetic moments. The arguments given above indicate very strongly that the orbital pair $2\epsilon_2$, $2\epsilon_{-2}$ is primarily carbon-based and has a large magnetic moment, whereas the $1\epsilon_3$, $1\epsilon_{-3}$ pair is primarily oxygen-based and therefore has a small magnetic moment. We conclude that the *A* term of the B'' transition will again be negative. As was the case for the B' transition, this prediction will be on progressively shakier grounds as n increases but should be quite safe for $C_7O_7^{2-}$.

There is little point in continuing the analysis for even higher transitions, since these will not descend into an observable region until n becomes quite large, and there are presently no experimental data for such dianions. However, the principles of the analysis remain the same, and it will be necessary only to take increasingly into account the effects of configuration interaction as excited configurations of like symmetry come closer together in energy and interact more strongly.

The neglect of vibronic interactions throughout our discussion makes it impossible to predict the signs of the MCD effects of forbidden transitions that acquire intensity only by vibronic coupling. It also makes it impossible to discuss the Jahn–Teller distortions possibly present in the degenerate excited states. There is presently good evidence from resonance Raman spectroscopy [7] that the B excited state of the ions $C_4O_4^{2-}$, $C_5O_5^{2-}$, and $C_6O_6^{2-}$ is Jahn–Teller-distorted. The MCD spectra [5] are certainly compatible with this in that the two peaks observed for the transition in absorption spectra correspond jointly to a single S-shaped MCD curve, rather than separately to one S-shaped MCD curve each. The Jahn–Teller distortion will not affect our arguments concerning the sequence of absolute signs in the MCD spectrum.

IV. Comparison with Experiment

The absorption and MCD spectra of the squarate ($C_4O_4^{2-}$), croconate ($C_5O_5^{2-}$), and rhodizonate ($C_6O_6^{2-}$) dianions, taken from West *et al.* [5], are shown in Figs. 5–7. A summary of the experimental data is provided in Table I.

The spectral patterns observed are exactly those anticipated from our previous discussion. There are two intense absorption bands, decreasing in energy with increasing n in $C_nO_n^{2-}$, the lower with a positive A term, the upper with a negative A term (it is out of range of our instrument for $C_4O_4^{2-}$). Accordingly, we assign the former as due to the B transition and the latter as due to the B' transition. In the spectra of croconate and rhodizonate dianions, very weak transitions are observed in the region between the B and B' bands. We assign these as the expected forbidden π,π^* transitions. In $C_5O_5^{2-}$, their expected symmetry is E_2', and a weak A term should be observed. This appears to be present but is partially obscured by overlap with the B' transition. In $C_6O_6^{2-}$, we expect two forbidden transitions into degenerate states of E_{2g} symmetry as well as forbidden transitions into B_{2u} and B_{1u} states. Two weak A terms are observed and must correspond to the degenerate transitions. One or both expected non-degenerate transitions may also be present if they are of very weak intensity.

The assignment of the excited states in terms of dominant electron configurations is also shown in the table and the figures. In $C_6O_6^{2-}$, considerable mixing of the configurations $2a_{2u} \rightarrow 2e_{2u}$ and $1b_{2g} \rightarrow 2e_{1g}$, which are responsible for the two E_{2g} states, is calculated to occur.

The magnitudes of the magnetic moments of the B states in the three dianions

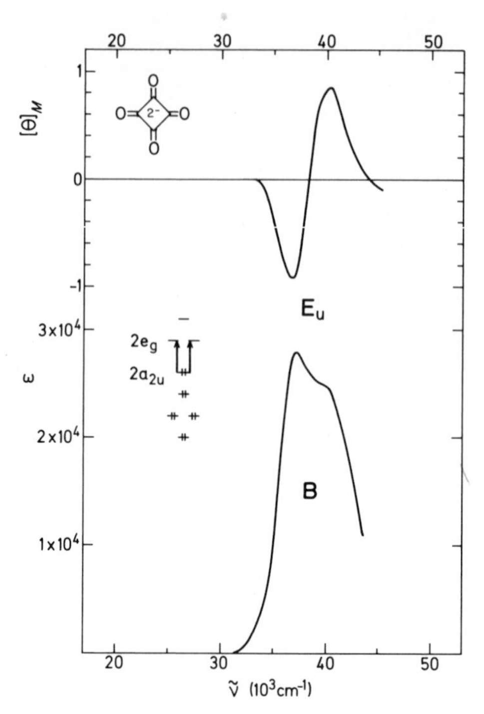

Fig. 5. Sodium squarate. Top, MCD spectrum; bottom, absorption spectrum (in water). The observed transition is of type *B*. Excited-state symmetry and a schematic representation of the dominant configuration are also shown. The D_{4h} symmetry labels of initial and final MO's are given; the correspondence to the C_4 group labels used in the text is $a_{2u} \rightarrow \epsilon_0$; $e_g \rightarrow \epsilon_1, \epsilon_{-1}$; $b_{1u} \rightarrow \epsilon_2 \equiv \epsilon_{-2}$.

are quite similar. If we take the magnetic moments of the pure perimeter orbitals $\epsilon_1(C)$ to be given by the estimates of Michl [*9*], $-0.6\beta_e$ in $C_4H_4^{2+}$, $-0.7\beta_e$ in $C_5H_5^{3+}$, and $-0.8\beta_e$ in $C_6H_6^{4+}$ (β_e is Bohr magneton), we would anticipate a gradual, slow increase of the observed magnetic moments as *n* increases, provided that the weight of the perimeter orbitals in the MO's $2\epsilon_0$, $2\epsilon_1$, and $2\epsilon_{-1}$, involved in the *B* transition, remained constant along the series. Calculations [*5*] indicate that the weight of the carbon perimeter orbital pair $\epsilon_1(C)$, $\epsilon_{-1}(C)$ in the MO's $2\epsilon_1$, $2\epsilon_{-1}$ is about one-half. To account for the observed values in $C_5O_5^{2-}$ and $C_6O_6^{2-}$, the weight of the carbon perimeter orbital $\epsilon_0(C)$ in the MO $2\epsilon_0$ should be less than half, in accordance with expectations based on Fig. 4 and with the ESR spectrum of $C_5O_5^-$ [*10*]. The larger value observed for $C_4O_4^{2-}$ runs against the trend in the moments of the perimeter orbitals $\epsilon_1(C)$ and indicates an increased weight of at least one of the perimeter orbitals $\epsilon_0(C)$ and $\epsilon_1(C)$ in the MO's involved in the *B* transition. We believe that the weight of the $\epsilon_0(C)$ perimeter orbital in the MO $2\epsilon_0$ is not likely to change much with *n*, although the

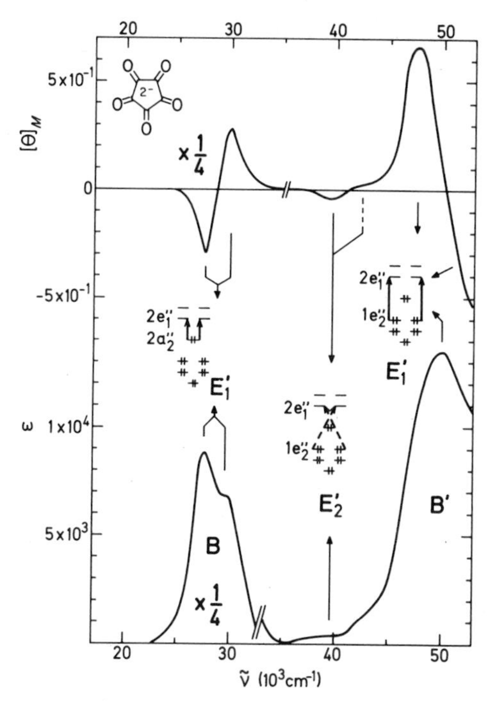

Fig. 6. Potassium croconate. Top, MCD spectrum; bottom, absorption spectrum (in water). The assignment of the allowed transitions B and B' is shown. Center, excited-stated symmetry and schematic representation of the dominant configuration. Allowed electron transitions are shown as full lines, forbidden ones as dashed lines. The D_{5h} symmetry labels of initial and final MO's are shown. They correspond to the C_5 group labels used in the text as follows: $a_2'' \rightarrow \epsilon_0$; $e_1'' \rightarrow \epsilon_1, \epsilon_{-1}$; $e_2'' \rightarrow \epsilon_2, \epsilon_{-2}$.

presence of some change has been inferred from the ESR spectra of radical monoanions [*10*]; in the simple Hückel picture, it would not change at all (Fig. 4). However, there is a good reason to expect an increase in the weight of the $\epsilon_1(C)$, $\epsilon_{-1}(C)$ orbital pair in the MO pair $2\epsilon_1$, $2\epsilon_{-1}$ in $C_4O_4^{2-}$, since $\epsilon_1(C)$ is nonbonding in the parent perimeter if $n = 4$, whereas it is bonding and of lower energy if $n > 4$. According to Fig. 4, an increase in the energy of $\epsilon_1(C)$, $\epsilon_{-1}(C)$ will increase its weight in the final MO's $2\epsilon_1$, $2\epsilon_{-1}$. This is also observed in the results of numerical calculations.

Finally, the question of the possible presence of n,π^* transitions in the spectra must be addressed. It is clear from the above that their presence need not be postulated in order to account for our absorption and MCD data, but it is not possible to exclude their presence in the energy region observed. A very weak absorption ($\epsilon = 174$) was reported [*11*] at 31,000 cm^{-1} in $C_4O_4^{2-}$. If not due to an impurity, this might be an n,π^* band, as suggested by the authors. We

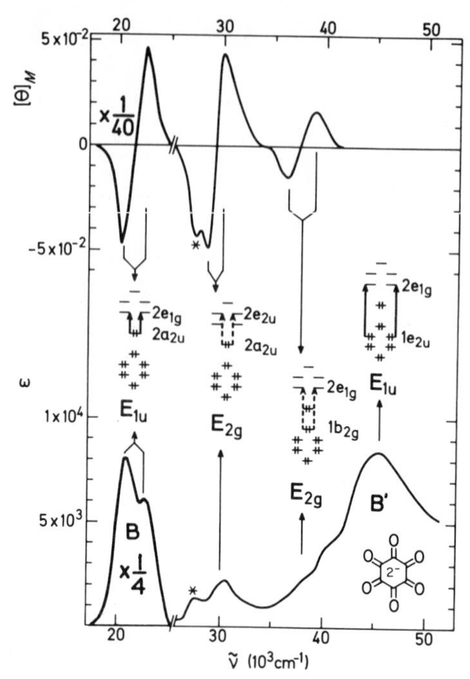

Fig. 7. Sodium rhodizonate. Top, MCD spectrum; bottom, absorption spectrum (in water). The asterisks mark peaks due to a $C_5O_5^{2-}$ impurity. The assignment of the allowed transitions B and B' is shown. Center, excited-state symmetry and schematic representation of the dominant configuration. Allowed electron transitions are shown as full lines, forbidden ones as dashed lines. The D_{6h} symmetry labels of initial and final MO's are shown. They correspond to the C_6 group labels used in the text as follows: $a_{2u} \rightarrow \epsilon_0$; $e_{1g} \rightarrow \epsilon_1, \epsilon_{-1}$; $e_{2u} \rightarrow \epsilon_2, \epsilon_{-2}$; $b_{2g} \rightarrow \epsilon_3 \equiv \epsilon_{-3}$.

Table I. Excited-State Properties

Species	E (cm^{-1})	Osc. str.	μ (β_e)	Symmetry	Type	Dominant configuration
$C_4O_4^{2-}$	39,000	0.9	−0.3	E_u	B	$2\epsilon_0 \rightarrow 2\epsilon_1, 2\epsilon_{-1}$
$C_5O_5^{2-}$	28,500	0.7	−0.25	E_1'	B	$2\epsilon_0 \rightarrow 2\epsilon_1, 2\epsilon_{-1}$
	41,000			E_2'		$1\epsilon_2 \rightarrow 2\epsilon_{-1}, 1\epsilon_{-2} \rightarrow 2\epsilon_1$
	50,000	0.4	>0	E_1'	B'	$1\epsilon_2 \rightarrow 2\epsilon_1, 1\epsilon_{-2} \rightarrow 2\epsilon_{-1}$
$C_6O_6^{2-}$	21,500	0.5	−0.35	E_{1u}	B	$2\epsilon_0 \rightarrow 2\epsilon_1, 2\epsilon_{-1}$
	30,000			E_{2g}		$2\epsilon_0 \rightarrow 2\epsilon_2, 2\epsilon_{-2}$
	38,000			E_{2g}		$1\epsilon_3 \equiv 1\epsilon_{-3} \rightarrow 2\epsilon_1, 2\epsilon_{-1}$
	45,500	ca. 0.4		E_{1u}	B'	$1\epsilon_2 \rightarrow 2\epsilon_1, 1\epsilon_{-2} \rightarrow 2\epsilon_{-1}$

observe some tailing on the long-wavelength side of the first band in absorption but not in MCD in all three dianions, and this could perhaps be due to n,π^* absorption. Overall, it seems that the experimental evidence is rather inconclusive. A calculation of the energies of n,π^* states is of limited value, given the generally increased difficulty in handling electrons of σ symmetry and given the undoubtedly large effects of hydrogen bonding in aqueous solutions of the dianions. The reported extended Hückel calculations [*11*] suggest that the lowest-energy transitions are of n,π^* type, but this method is well known for placing orbitals of σ symmetry too high in energy. We have attempted to derive additional qualitative insight into the problem [*5*] and have used the method developed by Ridley and Zerner [*8*], which has given good results for n,π^* states of neutral aza heterocycles. For $C_nO_n^{2-}$, $n = 3$–6, the lowest calculated excited state again was of the n,π^* type and was located a few thousand wavenumbers below the calculated energy of the *B* state. A total of about a dozen n,π^* states were calculated to lie below the *B*′ state. However, in each dianion, only one of the n,π^* transitions calculated to lie in the accessible region (below 200 nm) was symmetry-allowed, and only very weakly at that. Its energy was almost constant at 43,000 cm^{-1}. Inspection of Figs. 5–7 reveals no convincing evidence for its presence, neither in absorption, nor in MCD, where it could have a *B* term. Of course, n,π^* transitions are generally quite weak in MCD spectra of cyclic conjugated molecules relative to the π,π^* transitions, just as they are in absorption, and, besides, they should be blue-shifted considerably in aqueous solution. In conclusion, we believe that some n,π^* absorption is present underneath the observed π,π^* absorption of the dianions and possibly even at lower energies, but proving its presence unequivocally will require additional work.

Acknowledgment

The authors gratefully acknowledge support from U.S. Public Health Service (GM 21153) and the National Science Foundation (CHE 76-80374). We are grateful to the editor of the *Journal of the American Chemical Society* for kind permission to reproduce Fig. 2.

References

1. N. C. Baenziger, J. J. Hegenbarth, and D. G. Williams, *J. Am. Chem. Soc.* **85,** 1539 (1963); W. M. Macintyre and M. S. Werkema, *J. Chem. Phys.* **40,** 3563 (1964).
2. J. R. Platt, *J. Chem. Phys.* **17,** 484 (1949).
3. W. Moffitt, *J. Chem. Phys.* **22,** 320, 1820 (1954).
4. K. Sakamoto and Y. J. I'Haya, *J. Am. Chem. Soc.* **92,** 2636 (1970), and references therein.
5. R. West, J. W. Downing, S. Inagaki, and J. Michl, in preparation.
6. R. West and J. Niu, *in* "Non-Benzenoid Aromatics" (J. Snyder, ed.), Vol. 1, p. 163. Academic Press, New York, 1969.

7. M. Takahashi, K. Kaya, and M. Ito, *Chem. Phys.* **35,** 293 (1978); M. Iijima, Y. Udagawa, K. Kaya, and M. Ito, *ibid.* **9,** 229 (1975); S. Muramatsu, K. Nasu, M. Takahashi, and K. Kaya, *Chem. Phys. Lett.* **50,** 284 (1977); cf. S. Muramatsu and K. Nasu, *J. Phys. Soc. Jpn.* **46,** 189 (1979).
8. J. Ridley and M. Zerner, *Theor. Chim. Acta* **32,** 111 (1973).
9. J. Michl, *J. Am. Chem. Soc.* **100,** 6801 (1978).
10. E. Patton and R. West, *J. Phys. Chem.* **77,** 2652 (1973).
11. K. Sakamoto and Y. I'Haya, *Bull. Chem. Soc. Jpn.* **44,** 1201 (1971).

6

The Mycotoxin "Moniliformin" and Related Substances

H.-D. Scharf and H. Frauenrath

I. Moniliformin 101
A. Discovery 101
B. Structure and Physical Properties 105
C. Biological and Toxicological Properties 108
II. Syntheses and Properties of Semisquaric Acid and Its Derivatives . . 109
A. Most Efficient Synthetic Routes 109
B. Chemical and Physical Properties 111
III. The Series of Semioxocarbons $CH(CO)_m^{(-)}M^{(+)}$ 114
IV. Naturally Occurring Phenylog Substances 116
References 117

I. Moniliformin

A. DISCOVERY

In 1970 a strain of the mold *Fusarium moniliforme* was isolated from corn seed damaged by southern leaf blight [*1*]. The mold was found to produce a water-soluble mycotoxin, which was given the trivial name "moniliformin" (Fig. 1). This compound has growth-regulating and phytotoxic effects on plants and also has a toxic effect on mammals. Moniliformin from *Gibberella fujikuroi* (a stage of *Fusarium moniliforme*) was found by X-ray analysis to be the potassium salt of semisquaric acid (**1a**) [*2*]. The original toxin is the corresponding sodium salt (**1b**) [*1*].

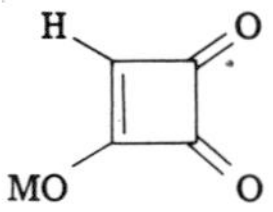

1. M = H
1a. M = K
1b. M = Na
1c. M = CH_3

OXOCARBONS

ISBN 0-12-744580-3

Fig. 1A. Photograph of a moniliformin crystal.

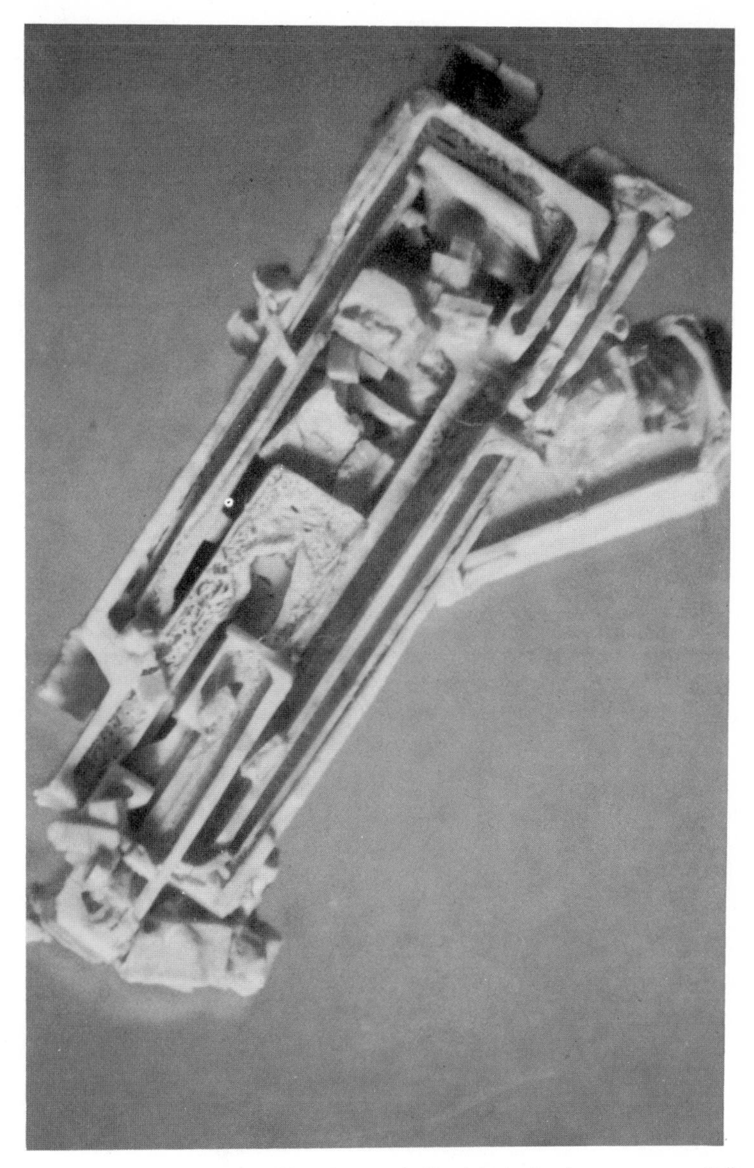

Fig. 1B. Photograph of a moniliformin crystal after dehydrating in vacuum over P_2O_5.

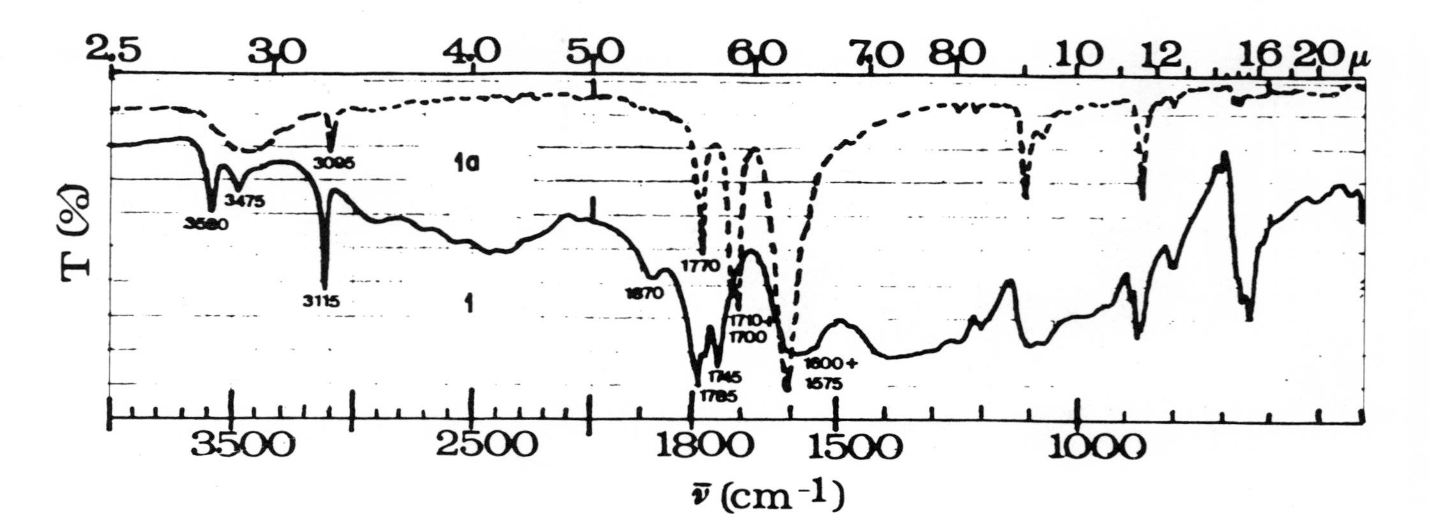

Fig. 2. Infrared spectra of **1** (——) and **1a** (- - -) in KBr.

Semisquaric acid itself (**1**) had been prepared independently in the laboratory of Hoffman *et al.* [*3*] and in our laboratory [*4,5*] years before its discovery in nature. Compound **1** may be regarded as the first member of a new class of reduced oxocarbons. Moreover, it is a parent compound for a series of biologically active substances the structure–activity relationships of which are only now being studied.

B. STRUCTURE AND PHYSICAL PROPERTIES

The semisquaric acid molecule contains a vinylogous acid function linked to a carbonyl group in a four-membered ring [*6*]. Therefore, in correspondence to squarate ion [*7*], the ground state of the semisquarate ion can be represented as the resonance hybrid A↔B↔C. However, the IR and ^{13}C-NMR spectra of **1**

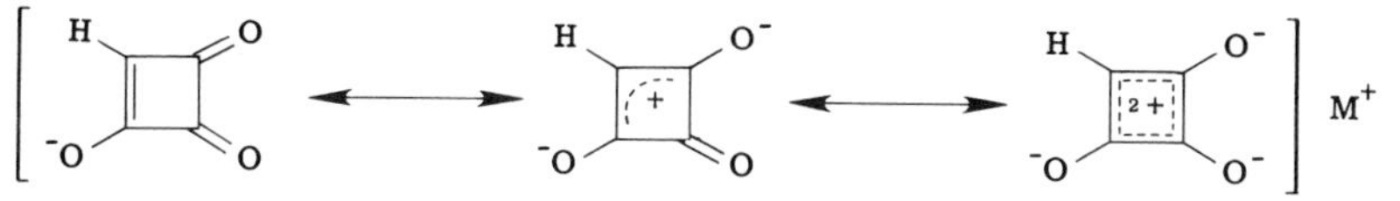

(A) (B) (C) M^+

indicate that the contribution of resonance form C is low, and therefore the main contributors to the ground state of **1** are canonical forms A and B. The IR spectrum, shown in Fig. 2, contains C=O and C=C stretching absorption bands in the region from 1600 to 1800 cm^{-1}, similar to those of other cyclobutenone systems [*8–11*]. The ^{13}C-NMR spectra of both **1a** and **1** show three signals (Table I) corresponding to the hybrid structure shown above, with C_s symmetry. However, the methyl ester **1c** shows resonances for four different ring carbon atoms.

The mass spectrum of semisquaric acid shows strong peaks at *m*/*e* 70 and 42 as well as at the parent molecular ion peak. The fragmentation pattern is very likely as follows:

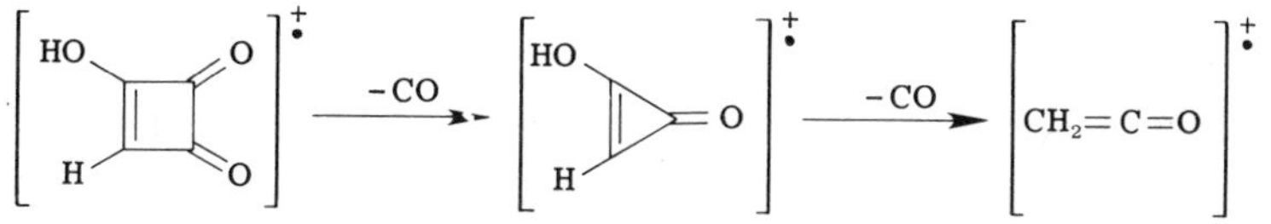

m/e = 98 (25%) —(−CO)→ m/e = 70 (85%) —(−CO)→ $[CH_2{=}C{=}O]^{+\bullet}$ m/e = 42 (100%)

Ultraviolet spectra in water and sulfuric acid at different concentrations are plotted in Fig. 3 (see also Section II,B) [*6*].

Emission spectra of an early sample of moniliformin (**1a**) [*12*] showing both fluorescence and phosphorescence were reported [*14*]. On excitation at 300 nm, fluorescence emission was detected at 360 nm with a quantum efficiency of 5.32

Table I. ^{13}C-NMR Data (ppm) for Compounds **1**, **1a**, and **1c**[a,b]

Structure	C-1	C-2	C-3	C-4	C-5	Solvent
(1a)	203.6	171.0	214.8	—	—	D_2O
(1)	199.0	166.4	202.2	—	—	DMSO-d_6
(1c)	194.6	166.0	203.4	196.4	63.6	DMSO-d_6

[a] Tetramethylsilane as internal standard; frequency 20 MHz.
[b] From Scharf [6].

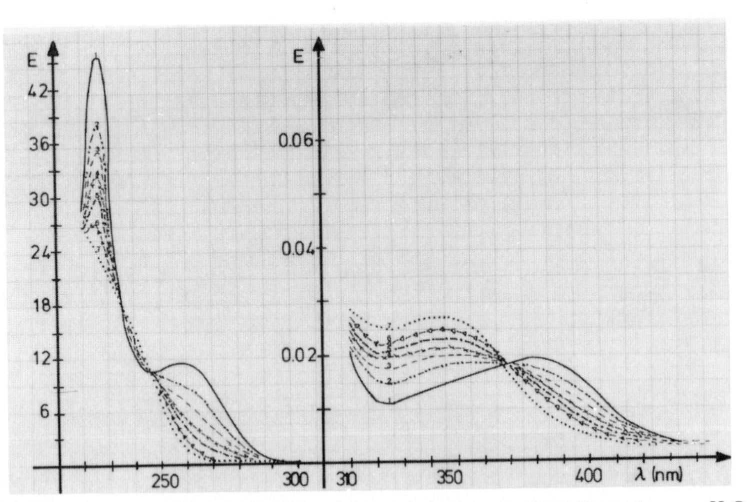

Fig. 3. Ultraviolet spectra of **1** in sulfuric acid of various concentrations. Curve 1: ——, H_2O; 2: —··—, 0.1 *M* H_2SO_4; 3: - - -, 0.25 *M* H_2SO_4; 4: -x-, 0.5 *M* H_2SO_4; 5: -··-, 1.01 *M* H_2SO_4; 6: -Δ- 4.07 *M* H_2SO_4; 7: ··· 7.11 *M* H_2SO_4.

$\times\ 10^{-2}$ and a mean lifetime of 0.165 nsec. This fluorescence emission was said to be insensitive to oxygen quenching. Luminescence experiments in frozen solutions revealed phosphorescence with a maximal emission at 420 nm upon excitation at 305 nm. Phosphorescence decay measurements showed the possible existence of two separate emissions with lifetimes of 5.7 and 2.6×10^{-2} sec. However, when these experiments were repeated [*15*] with a synthetic sample of **1a** in doubly distilled water [*16*], no UV absorption maximum was found in the range of 300 nm (see Figs. 1A and 4). Moreover, no emission was observed on excitation at this wavelength [*14*] except for a weak emission caused by an

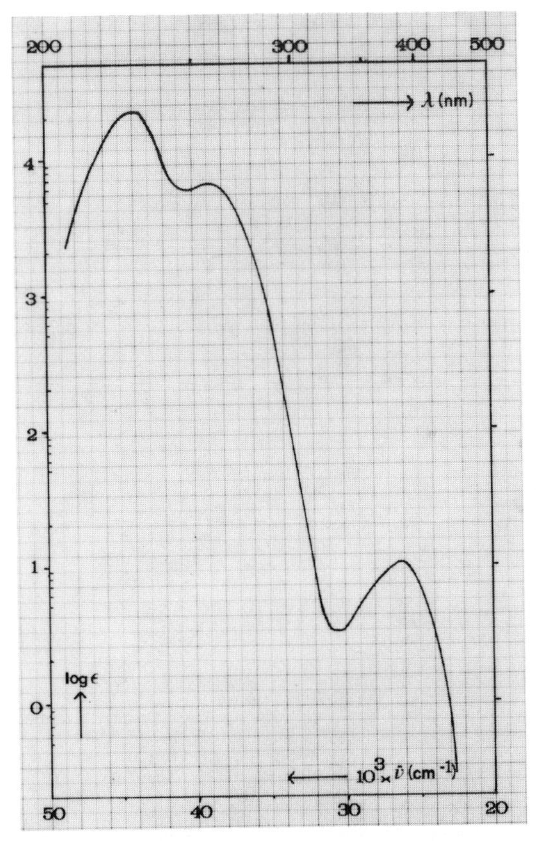

Fig. 4. Ultraviolet absorption spectrum of **1a** in doubly distilled H_2O. Recording conditions: Varian Cary 14 spectrometer (500–290 nm: $c = 3.7 \times 10^{-3}$ M, d = 10 cm, d = 1 cm; 290–200 nm: $c = 3.7 \times 10^{-4}$ M, d = 0.1 cm). Aging of the compound as well as dehydrating in vacuum over P_2O_5 results in the appearance of new absorption maxima at 350 and 360 nm.

impurity of the solvent MeOH/EtOH at 77 K. Thus, the question of emission by **1** is still unresolved.

C. BIOLOGICAL AND TOXICOLOGICAL PROPERTIES

Moniliformin has marked growth-regulating effects on many plants. In such plants as corn and tobacco [*1*], and even wheat and barley [*17*], the growth regulation is accompanied by thin stalk growth and considerable cell destruction. Among related substances, compound **2** has a slight inhibiting effect on growth

H_5C_6 O

KO O

(2)

of grasses [*17*]. Herbicidal effects of semisquaric and squaric acid derivatives on weeds have also been demonstrated and compared [*18*].

Moniliformin is quite toxic to mammals, with an oral LD_{50} of 4.0 mg/kg body weight in 1-day-old cockerels [*1*]. Histological lesions in cockerels were ascites with edema of the mesenteries and small hemorrhages in the proventriculus, gizzard, small and large intestine, and skin. Furthermore, leukoencephaly, liver damage, and/or esophagal carcinoma have been observed in mammals fed with corn contaminated with *Fusarium moniliforme* [*19*]. Biochemically, moniliformin causes selective inhibition of the mitochondrial pyruvate and α-ketoglutarate oxidation (Fig. 5).

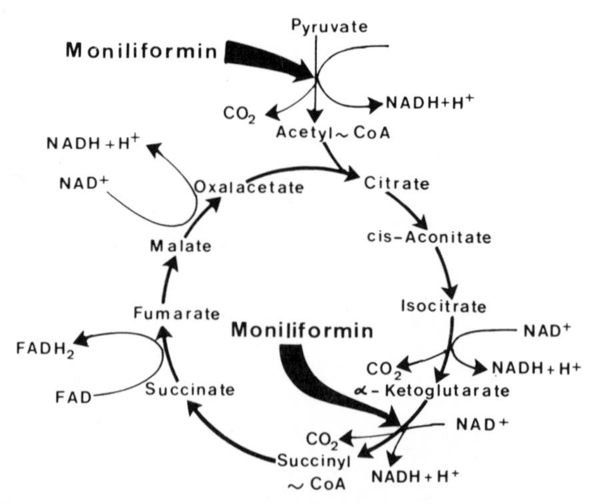

Fig. 5. Points of intervention in the Krebs cycle by moniliformin.

II. Syntheses and Properties of Semisquaric Acid and Its Derivatives

A. MOST EFFICIENT SYNTHETIC ROUTES

In 1971 Hoffman *et al.* [3] synthesized semisquaric acid by [2+2] cycloaddition of ketene to tetramethoxyethene (**3**):

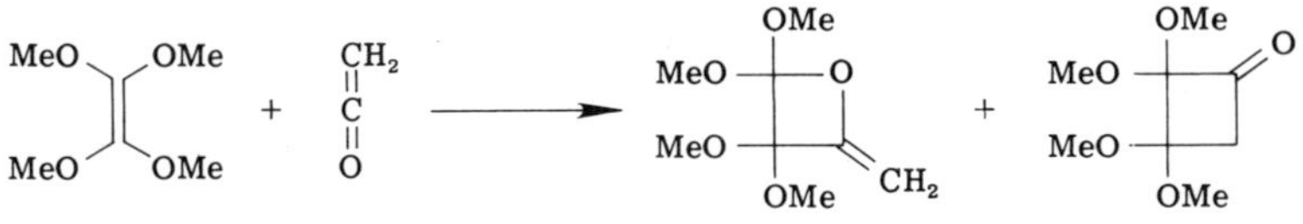

(3) (4) (5)

This reaction yields a 1 : 1 ratio of **4** to **5**. Cyclobutanone (**5**) was hydrolyzed to semisquaric acid:

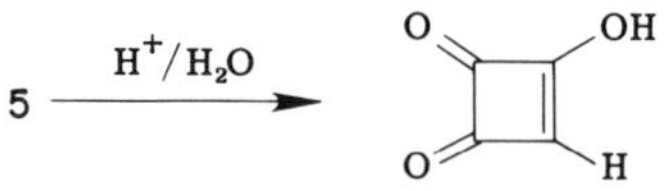

$$5 \xrightarrow{H^+/H_2O}$$

(1)

Recently, a number of more or less successful syntheses of **1** [*2,4–6,13*] as well as its derivatives [*18,20–24*] have been published. Only those syntheses that have proved to be particularly useful are specified below.

1. Cycloaddition of Tetraethoxyethene and Ketenes

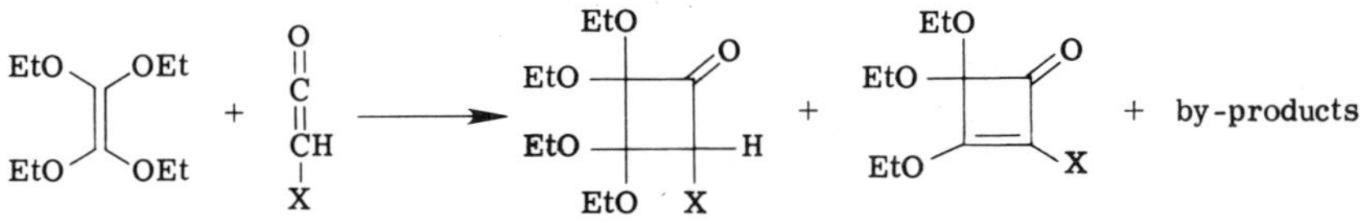

(6) (7) (8) (9) + by-products

$$8 \xrightarrow[\text{hexane}]{SiO_2/Et_3N,\ Al_2O_3} 9 + \text{by-products} \xrightarrow{HCl/H_2O}$$

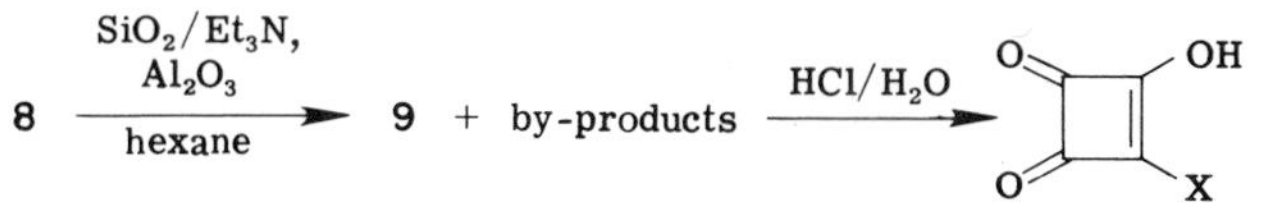

(10)
(10a. $X = C_6H_5$)

This synthesis is described by Bellus *et al.* [*13,24*]. Besides the derivatives of semisquaric acid (see Table II), the parent acid itself can be synthesized via this

Table II. Substitution Pattern on Compound **10** and Physical Properties[a]

X	mp (°C)[b]	Yield (%)
CH_3	160–161	83
$CH(CH_3)_2$	84–86	92
$C(CH_3)_3$[a]	121–122	93
$(CH_2)_5CH_3$	33–35	94
Cyclohexyl	105.5–107	83
Benzyl	155–156	92
$CH{=}CCl_2$	188	58
$CH_2CO_2CH_3$[c]	156–157	73
C_6H_5	207–208	65
$4\text{-}CH_3OC_6H_4$	220	86
$4\text{-}ClC_6H_4$	227–228	68

[a] From Bellus [*24*].
[b] Most compounds melt under decomposition.
[c] Under application of **3** instead of **6.**

sequence, which is based on Hoffmann's original route [*3*]. However, the newer synthesis profits from the convenient preparation of tetraalkoxyethenes according to the procedure of Scheeren *et al.* [*25*]. The application of **6** instead of **3** is of value because, with **6**, the oxetane derivative corresponding to **4** is formed only in negligible amounts during the reaction with ketenes.

2. *Semisquaric Acid by a Photochemical Route*

Semisquaric acid itself can be prepared by a photochemical route from the easily accessible starting material **11** [*26,27*]:

2 (11) —$h\nu$ (sens)→ (12) —H^+/H_2O→ 1

The reaction proceeds via head-to-tail [2+2] dimerization of **11** to **12**, which on hydrolysis gives semisquaric acid in good overall yield [*28*].

3. *N-Disubstituted Semisquaric Acid Amides*

A synthesis of N-disubstituted semisquarid acid amides **16** (Table III) [*29*] starting from the squaric acid bisamides **13** [*30*] was reported by Seitz *et al.* [*29*]:

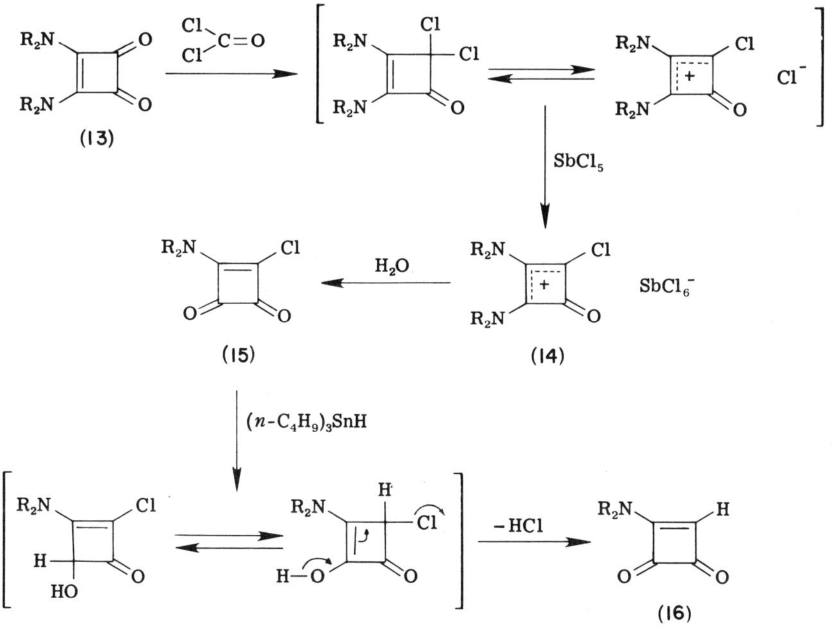

Intermediate compounds **14** and **15** may also be isolated [*31,32*].

B. CHEMICAL AND PHYSICAL PROPERTIES

For additional information, see Section I,B.

1. Acidity

The parent acid **1** and all related compounds **10** are strong vinylogous acids [*20*] (Table IV). The extraordinary acid strength of hydroxycyclobutenediones was attributed by Smutney, Caserio, and Roberts [*23*] to the contribution of 1,3-transannular bonding to the resonance hybrid, as indicated by canonical

Table III. Amides of Semisquaric Acid[a] **16**

R	R	mp (°C)	Yield (%)
CH_3	CH_3	83	50
$-(CH_2)_4-$		99	66
$-(CH_2)_5-$		77	42
$-(CH_2)_2-O-(CH_2)_2-$		141	51

[a] From Seitz *et al.* [*29*].

Table IV. Values of pK_a for Various Derivatives of Compound **10**

X	pK_a value
H	0.88 ± 0.03 [*6*]; 0.0 ± 0.05 [*13*]
CH_3	+0.20 ± 0.1 [*24*]
$C(CH_3)_3$	+0.28 ± 0.05 [*24*]
$CH{=}CCl_2$	−1.0 ± 0.05 [*24*]
C_6H_5	−0.22 ± 0.1 [*20*]
$4\text{-}ClC_6H_4$	−0.32 ± 0.05 [*24*]

structures D and E. Patton and West [*20*], however, argued that favorable dipolar

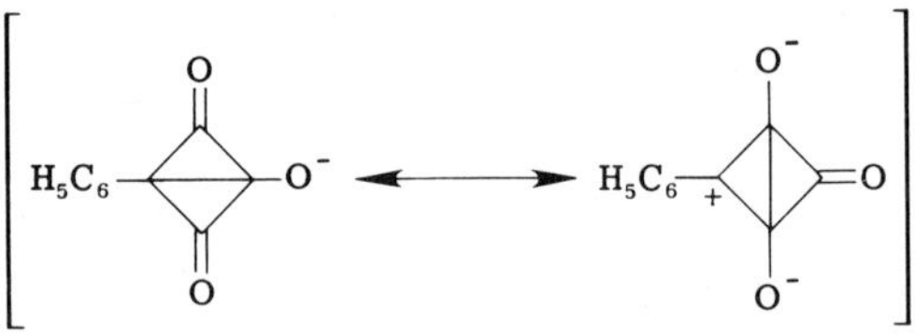

(D) (E)

interactions [*33*] between CO groups can account for most of the increased acid strengths of four-membered ring 1,3-dicarbonyl compounds. Furthermore, entropic effects [*34*] may also be important.

2. *Oxidation*

Bromination and dehydrobromination of **1** yield **17** (squaric acid monobromide), which upon hydrolysis gives squaric acid (**18**) in good yield [13].

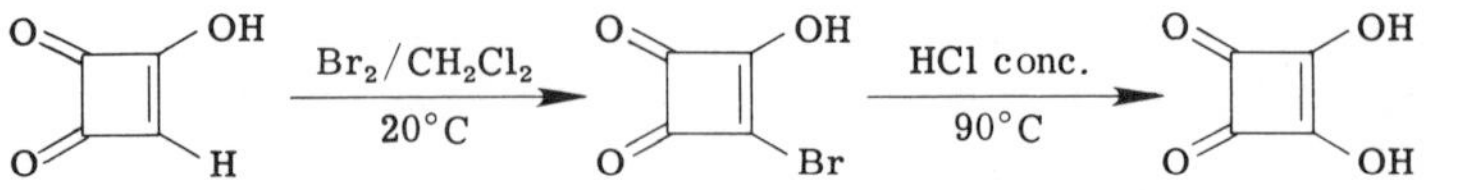

Br_2/CH_2Cl_2, 20°C; HCl conc., 90°C

(1) (17) (18)

3. *Simple Functional Derivatives*

The reactivity of alkyl- and arylhydroxycyclobutenediones (**10**) is similar to that of semisquaric acid itself. Self-catalyzed esterification or amidation of **10** leads to esters (**19**) and amides (**21**) [*6,18*] and treatment with phosgene/DMF gives stable chlorides (**20**) [*24*]. Esters, amides or thioesters (**22**) can be made from the chlorides [*35*]. The chemical behavior of compounds of the class **10** is guided by their electrophilicity and corresponds to that of squaric acid under similar conditions [*23,30,36*].

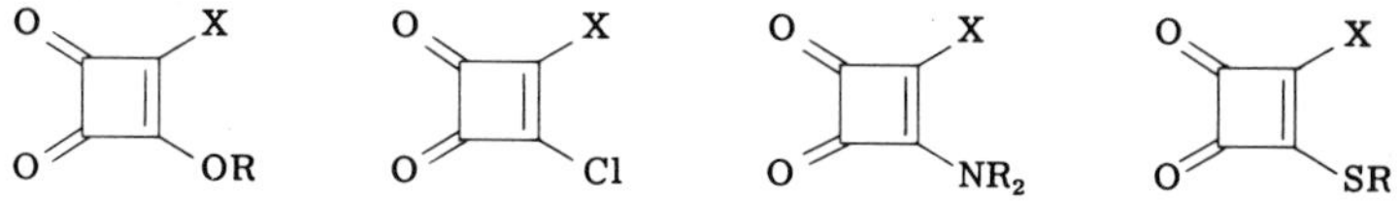

O X O X O X O X
OR Cl NR_2 SR
O O O O

(19) (20) (21) (22)

X = alkyl or aryl

4. Ammonium Salts and Amides

2-Hydroxy-1-phenylcyclobut-1-ene-3,4-dione (**10a**) can react with amines to give either 3-substituted condensation products or ammonium salts [*8*]. The course of the reaction depends on the basicity of the amines. Less basic (more nucleophilic) amines such as primary aromatic amines with **10a** yield the 3-arylamino-1-phenylcyclobutenediylium 2,4-diolates **23** [*37*]. Compounds of

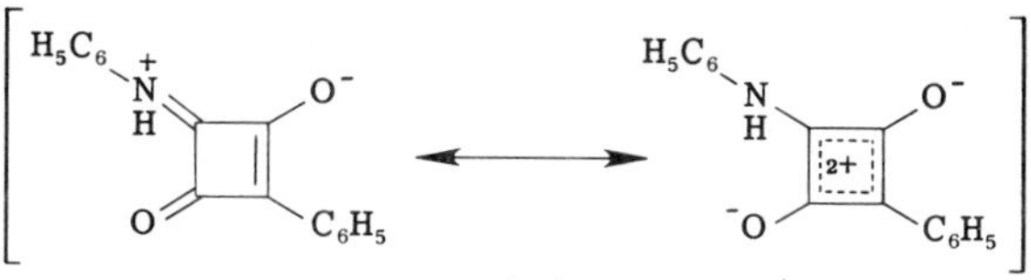

H_5C_6 N^+ H O^- O C_6H_5 ⟷ H_5C_6 N H O^- 2+ ^-O C_6H_5

(23)

type **23** are yellow- to orange-colored crystalline derivatives, closely related in their structure to the condensation products of squaric acid with activated pyrroles [*38*] and amines [*39,40*]. Aliphatic amines react with **10a** to give ammonium salts, but when these are refluxed in *n*-butanol the corresponding phenylcyclobutenediylium diolates (**23**) are formed. The latter are more weakly colored than their aromatic analogs.

Under similar conditions the more basic aliphatic amines primarily form ammonium salts with **10a;** for example, aziridine or NH_3 gives salts as yellow, crystalline materials. The electrophilicity of the parent compound **1** is obviously greater than that of **10a** since no salt formation is observed if **1** is treated with NH_3 or aniline at 0°C. Instead, amides are immediately formed [*6*]. The amide formation, however, corresponds to the acidity values of the vinylog acids given by Scharf *et al.* [*6*]. A synthesis of N,N-disubstituted amides (**16**) starting with squaric acid was reported by Seitz *et al.* [*29*] (see Section II,A).

5. Synthetic Thio Analogs

Nucleophilic substitution of 2-chloro-1-phenylcyclobut-1-ene-3,4-dione (**24**) by H_2S leads to the thio analog of phenylsemisquaric acid (**26**) [*21*]. Treatment

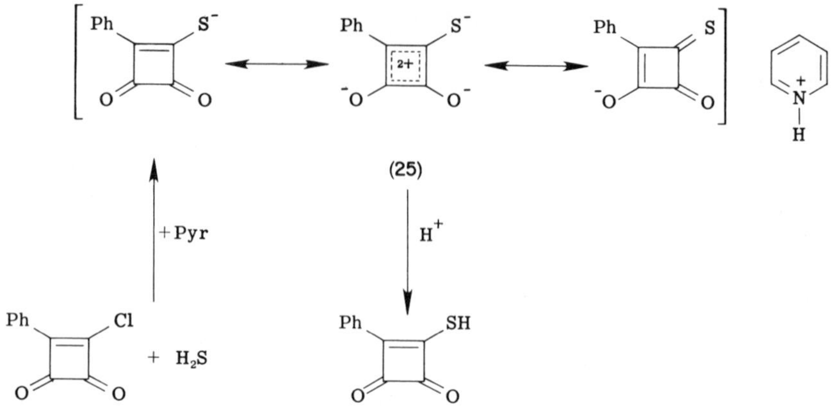

Ph S⁻ Ph S⁻ Ph S

(25)

+Pyr H⁺

Ph Cl Ph SH

+ H_2S

(24) (26)

of **24** with **26** yields the vinylogous thioanhydride **27**. Compound **26** behaves as

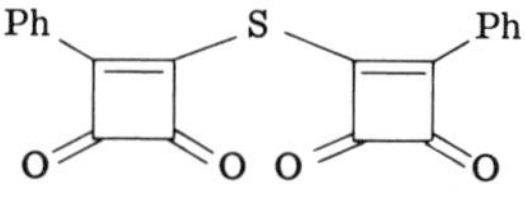

(27)

an acid, forming pyridinium salts (**25**) with pyridine and undergoing esterification with diazomethane [*21*]. Because **25** is electrophilic, it reacts with aromatic amines ($ArNH_2$) in THF at 0°C, yielding deeply colored condensation products **28**. The more basic aliphatic amines, however, react with **26** to give ammonium salts, demonstrating the acidity of the thio analog [*9*].

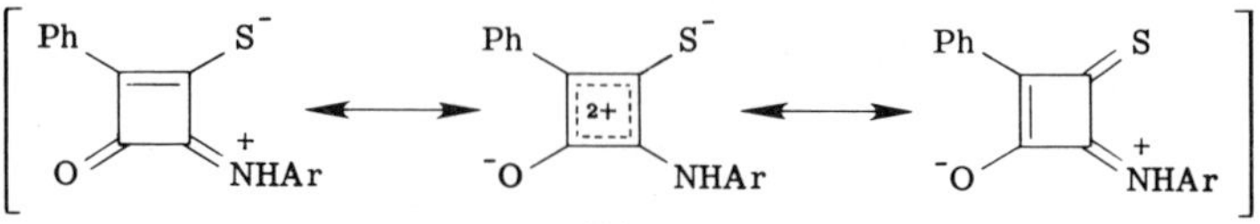

(28)

It may be concluded that within this class of vinylogous acids there is clear-cut complementary behavior between the acidity of the acid and the basicity of the amine, on one hand, and the electrophilicity of the acid and the nucleophilicity of the amine, on the other.

III. The Series of Semioxocarbons $CH(CO)_m^{(-)}M^{(+)}$

Parallel to the well-known class of oxocarbons of the type $(CO)_m^{(2-)}$ [*7*], such as deltic acid [*41–44*], squaric acid [*30,45*], croconic acid [*46,47*], and rhodizonic acid [46,47], there is a potential series of semioxocarbons $CH(CO)_m^-$ (**1, 29** [48], **30, 31, 32**). Investigation of this series intensified following the discovery of semisquarate ($m = 3$) as a natural product, as described in Section I. Of

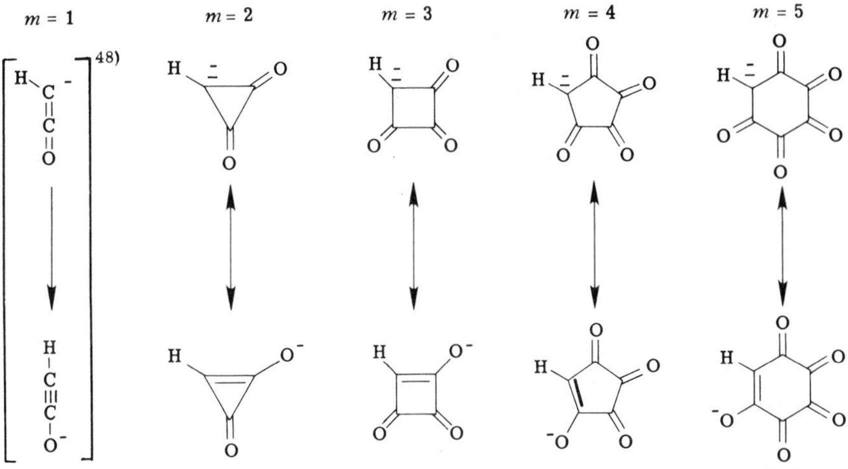

$m = 1$ $m = 2$ $m = 3$ $m = 4$ $m = 5$

(29) (30) (1) (31) (32)

the parent acids in this series only **1** ($m = 3$) is known, but the phenyl-substituted derivatives **33** ($m = 2$, 3, and 4) have been synthesized and investigated [20]. Spectroscopic [49] and X-ray analysis [20,50,51] indicate that these compounds

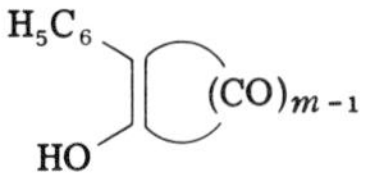

(33)

are planar and symmetric, implying complete π-electron delocalization. The stability of the corresponding anions is demonstrated by very low pK_a values of the parent hydroxides (**33**), which resemble those of inorganic rather than organic acids (Table V).

Table V. Dissociation Constants of Compounds **33** ($m = 2, 3, 4$)[a]

m	pK_a	Ref.
2	2.01 ± 0.03	20
	2.0 ± 0.5	51
3	−0.22 ± 0.1	20
	0.37 ± 0.04	23
4	1.64 ± 0.09	20
	1.75	52

[a] From Patton and West [20].

The most acidic compound in this series is the phenyl-substituted semisquaric acid **33** ($m = 3$) (see also **10a** and Tables II and IV). As mentioned in Section II,B, the reason for this strength is still uncertain [*20,23*].

The synthesis of **33** ($m = 2$) is achieved by hydrolysis of 2-phenyl-1,1,3-trichlorocyclopropene (**34**) [*53*]. The compound slowly decom-

(34) $\xrightarrow{H_2O/H^+}$ (33) $m = 2$

poses in aqueous solution. Phenylhydroxycyclopentenetrione (**33**), ($m = 4$) was synthesized according to Yamazaki *et al.* [*52*]. However, the chemical properties of these compounds, except for those of semisquaric acid, are not yet very well known.

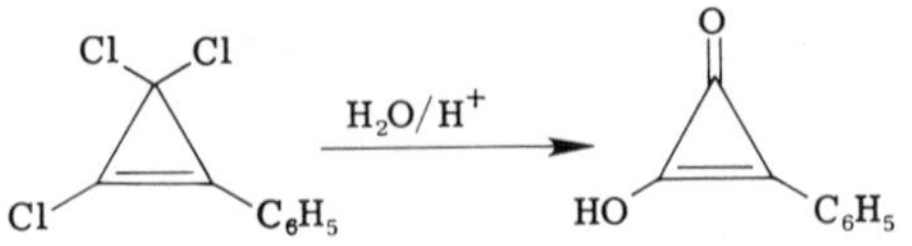

IV. Naturally Occurring Phenylog Substances

Plant natural products that inhibit the growth of other plants are called allelopathic agents and can be thought of as agents in chemical warfare that plants wage against each other [*54*]. Two of the best known allelopathic substances are juglone (**35**) [*55–57*] and lawsone (**36**) [*58*]. These compounds can be regarded

(35) (36)

as phenylogs or vinylogs of semisquaric acid. Walnut trees, in effect, synthesize **35** by air oxidation of the corresponding hydroquinone, which is produced in leaves and husks. The oxidation proceeds outside of the organism in nearly 100% yield, but the protosubstance is locked in the tissue structure in a harmless form [*59*]. Juglone is quite allelopathic to members of the heath family (e.g., azaleas, blueberries), tomatoes, and various other plants but seems to be ineffective against such plants as grasses and does not affect the growth behavior of corn. It will be most interesting to investigate whether the biological activity of **35** and **36** is related to that of the semisquaric acid derivatives. Clear-cut evidence on this point will be important for understanding biochemical mechanisms of allelopathy.

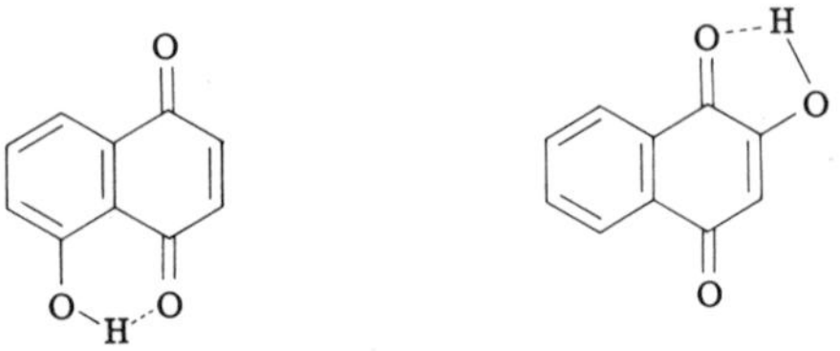

The characterization of juglone as well as its physical and spectroscopic properties have been reviewed [*60*]. Juglone (**35**) is an orange crystalline solid [*61*], mp 164°–165°C [*61,62*]. Lawsone (**36**) is a pale yellow crystalline solid, mp 192°C (dec.). Both juglone and lawsone act as acid–base indicators, giving a

35 $\xrightarrow[-H_2O]{NaOH}$ [⟷] Na^+

pink color

36 $\xrightarrow[-H_2O]{NaOH}$ [⟷] Na^+

red color

slight yellow color in acidic medium and pink and red colors, respectively, with aqueous alkalins [*61*]. The transition interval for juglone lies at pH 7.4–8.2, showing that it is a weak acid like typical phenols. However, lawsone has its transition from pH 2.6 to 3.4 and is therefore a moderately strong acid, somewhat resembling the oxocarbons.

References

1. R. J. Cole, J. W. Kirksey, H. G. Cutler, B. L. Doupnik, and J. C. Peckham, *Science* **179,** 1324 (1973).
2. J. P. Springer, J. Clardy, R. J. Cole, J. W. Kirksey, R. K. Hill, R. M. Carlson, and J. L. Isidor, *J. Am. Chem. Soc.* **96,** 2267 (1974).
3. R. W. Hoffmann, U. Bressel, J. Gehlhaus, and H. Häuser, *Chem. Ber.* **104,** 873 (1971).
4. W. Pinske, Dissertation, Technische Hochschule Aachen, Germany, 1972.
5. H.-D. Scharf, *Angew. Chem., Int. Ed. Engl.* **13,** 520 (1974).
6. H.-D. Scharf, H. Frauenrath, and W. Pinske, *Chem. Ber.* **111,** 168 (1978).
7. R. West and J. Niu, *in* "Nonbenzenoid Aromatics" (J. Snyder, ed.), Vol. 1, p. 312. Academic Press, New York, 1969.
8. W. Ried, A. H. Schmidt, G. Isenbruck, and F. Bätz, *Chem. Ber.* **105,** 325 (1972).
9. A. H. Schmidt, W. Ried, P. Pustoslemsek, and W. Schuckmann, *Angew. Chem., Int. Ed. Engl.* **14,** 823 (1975).
10. A. T. Blomquist and R. A. Vierling, *Tetrahedron Lett.* p. 655 (1961).
11. J. D. Park, S. Cohen, and J. R. Lacher, *J. Am. Chem. Soc.* **84,** 2919 (1962).
12. The point of decomposition reported in literature differs widely: 145–150°C[(3)]; 158°C[(13)] in spite of this most of the samples gave correct elementary analyses.

13. D. Bellus, H. Fischer, H. Greuter, and P. Martin, *Helv. Chim. Acta* **61,** 1784 (1978).
14. J. A. Lansden, R.J. Clarkson, W. C. Neely, R. J. Cole, and J. W. Kirksey, *J. Assoc. Off. Anal. Chem.* **57,** 1392 (1974).
15. H. Leismann, Institut für Organische Chemie der Techn. Hochschule Aachen, Prof.-Pirlet-Straße 1, D-5100 Aachen, private communication; H. Leismann and H. Frauenrath, to be published elsewhere.
16. Synthesized by the route described in Bellus *et al.* [*13*] (see also II.A.).
17. The Bayer AG, Leverkusen, Germany, very kindly provides us with testing results.
18. H. Fischer and D. Bellus, Offenlegungsschrift 26 16 756, Deutsches Patentamt from 28.10.76, Ciby-Geigy AG, Basel, 1976.
19. Lectured by Dr. W. F. O. Marasas, National Research Institute of Nutritive Diseases in Tygerberg, South-Africa, during the International Conference on Mycotoxins held in Munich, Germany, 14–15 August, 1978; see also *Selecta* **7,** 578 (1979).
20. E. Patton and R. West, *J. Am. Chem. Soc.* **95,** 8703 (1973).
21. A. H. Schmidt, W. Ried, P. Pustoslemsek, and H. Dietschmann, *Angew. Chem.* **84,** 110 (1972).
22. A. H. Schmidt, W. Ried, P. Pustoslemsek, and W. Schuckmann, *Angew. Chem.* **87,** 879 (1975).
23. E. J. Smutny, M. C. Caserio, and J. D. Roberts, *J. Am. Chem. Soc.* **82,** 1793 (1960).
24. D. Bellus, *J. Am. Chem. Soc.* **100,** 8026 (1978).
25. J. W. Scheeren, R. J. F. M. Staps, and R. J. F. Nivard, *Recl. Trav. Chim. Pays-Bas* **92,** 11 (1973).
26. J. J. Kucera and P. C. Carpenter, *J. Am. Chem. Soc.* **57,** 2346 (1935).
27. R. K. Summerbell and H. E. Lunk, *J. Am. Chem. Soc.* **79,** 4802 (1957).
28. H. Frauenrath and H.-D. Scharf, *Tetrahedron Lett.* (in press).
29. G. Seitz, H. Morck, R. Schmiedel, and R. Sutrisno, *Synthesis* p. 361 (1979).
30. G. Maahs and P. Hegenberg, *Angew. Chem. Int. Ed. Engl.* **5,** 888 (1966).
31. G. Seitz and H. Morck, *Chimia* **26,** 368 (1972).
32. H. Erhardt, S. Hünig, and H. Pütter, *Chem. Ber.* **110,** 2506 (1977).
33. R. B. Johns and A. B. Kriegler, *Aust. J. Chem.* **17,** 765 (1964).
34. L. M. Schwartz and L. O. Howard, *J. Phys. Chem.* **75,** 1798 (1971).
35. A. H. Schmidt and W. Ried, *Synthesis* p. 1 (1978).
36. S. Cohen and S. G. Cohen, *J. Am. Chem. Soc.* **88,** 1533 (1966).
37. W. Ried, W. Kunkel, and G. Isenbruck, *Chem. Ber.* **102,** 2688 (1969).
38. A. Treibs and K. Jakob, *Justus Liebigs Ann. Chem.* **699,** 153 (1966); **712,** 123 (1968); *Angew. Chem., Int. Ed. Engl.* **6,** 553 (1967).
39. H. E. Sprenger and W. Ziegenbein, *Angew. Chem., Int. Ed. Engl.* **5,** 894 (1966).
40. G. Manecke and J. Ganger, *Tetrahedron Lett.* p. 3509 (1967); p. 1339 (1968); *Chem. Ber.* **103,** 2696 (1970).
41. E. V. Dehmlow, *Tetrahedron Lett.* p. 1271 (1972).
42. D. Eggerding and R. West, *J. Am. Chem. Soc.* **97,** 207 (1975).
43. D. Eggerding and R. West, *J. Am. Chem. Soc.* **98,** 3641 (1976).
44. M. A. Pericás and F. Serratosa, *Tetrahedron Lett.* p. 4437 (1977).
45. R. West and J. Niu, *in* "The Chemistry of the Carbonyl Group" (J. Zabicky, ed.), p. ■. Wiley (Interscience), New York, 1970.
46. K. Yamada, M. Mizuno, and Y. Hirata, *Bull. Chem. Soc. Jpn.* **31,** 543 (1962).
47. J. F. Heller, *Justus Liebigs Ann. Chem.* **24,** 1 (1837).
48. T. B. Grindley, K. F. Johnson, A. R. Katritzky, H. J. Keogh, C. Thirkettle, R. C. T. Brownlee, J. A. Munday, and R. D. Topsom, *J. Chem. Soc., Perkin Trans. 2* p. 276 (1974); M. J. Leleu, *Cah. Notes Doc.* **82,** 127 (1976); A. N. Hayhurst and D. B. Kittelson, *Combust. Flame* **31,** 37 (1978).

49. M. Ito and R. West, *J. Am. Chem. Soc.* **85,** 2580 (1963).
50. W. M. Macintyre and M. S. Werkema, *J. Chem. Phys.* **42,** 3563 (1964); N. C. Baenzinger and J. J. Hegenbarth, *J. Am. Chem. Soc.* **86,** 3250 (1964).
51. D. G. Farnum, J. S. Chickes, and P. E. Thurston, *J. Am. Chem. Soc.* **88,** 3075 (1966).
52. T. Yamazaki, T. Oohama, T. Doiuchi, and T. Takizawa, *Chem. Pharm. Bull.* **20,** 238 (1972).
53. J. S. Chickos, E. Patton, and R. West, *J. Org. Chem.* **67,** 3112 (1976).
54. H. R. Bode, *Planta* **51,** 440 (1958).
55. L. F. Fieser and J. T. Dunn, *J. Am. Chem. Soc.* **59,** 1016 (1937).
56. A. Bernthsen and A. Semper, *Ber. Dtsch. Chem. Ges.* **20,** 934 (1887).
57. R. Willstätter and A. S. Wheeler, *Ber. Dtsch. Chem. Ges.* **47,** 2796 (1914).
58. L. F. Fieser, *J. Am. Chem. Soc.* **48,** 2922 (1926).
59. R. G. Jesaitis and A. Krantz, *J. Chem. Educ.* **49,** 436 (1971).
60. R. W. Hanson, *J. Chem. Educ.* **53,** 400 (1976).
61. K. C. Joshi, P. Singh, and G. Singh, *Z. Naturforsch., Teil B* **32,** 890 (1977).
62. The reported melting point is spread over a wide range [*59,61*]. According to Hanson [*60*] the range is markedly dependent on the starting temperature and rate of heating.

7

Raman Spectra and Jahn–Teller Effects of Oxocarbon Dianions

Mitsuo Ito, Koji Kaya, and Machiko Takahashi

I. Introduction 121
II. Raman Intensity and Electronic States 122
III. Electronic Absorption Spectra of Oxocarbon Ions 127
IV. Raman Spectra of $C_nO_n^{2-}$ 128
V. Jahn–Teller Effect Suggested by Raman Intensities 133
VI. Jahn–Teller Deformation of Ions 135
VII. Excitation Profile and Jahn–Teller Effect 138
VIII. Conclusion 139
References 140

I. Introduction

In an early study of the vibrational spectra of $C_4O_4^{2-}$ and $C_5O_5^{2-}$, Ito and West [1] noticed unusual features in their Raman spectra. The number and frequencies of the observed Raman lines agreed perfectly with the expected high degree of symmetry of these ions, but the intensities appeared to be anomalous. Raman intensities of totally symmetric vibrations are in general much stronger than those of nontotally symmetric vibrations. However, some Raman lines of the nontotally symmetric vibrations of the oxocarbon ions were found to be very strong, whereas some totally symmetric Raman lines were extremely weak. In this chapter we consider the implications of these unusual Raman intensities, which have remained unexplained since 1963.

Vibrational Raman spectroscopy is generally understood as a means for studying vibrational states of a molecule in its ground state. This is true for Raman frequencies, but Raman intensities are related to various electronic excited states of a molecule as well as the ground state. Correlation of Raman intensities with

OXOCARBONS

ISBN 0-12-744580-3

electronic excited states is readily understood from the fact that Raman intensities greatly increase when the exciting incident light frequency approaches the electronic absorption region of the molecule. The intensity enhancement of Raman lines due to the change of excitation wavelength is called the resonance Raman effect. Therefore, the nature of the various electronic excited states of a molecule should be reflected in the Raman intensities, and information about electronic excited states is to be obtained from them.

Oxocarbon ions are very symmetric. Because of their high degree of symmetry, these anions have many degenerate electronic excited states, in which a Jahn–Teller effect is expected. The result of this effect is that the degenerate excited-state ions have a lower degree symmetry due to dynamic interaction between electron and nuclear motions. Jahn–Teller effects are often studied from the vibrational structure of electronic absorption and emission spectra. However, experimental verification of the Jahn–Teller effect is very difficult, especially for species such as oxocarbon ions that show no fine structure in their spectra. Raman intensities may provide useful information for these molecules or ions because Raman spectra always give sharp lines, so that the vibrational coordinates of the Jahn–Teller distortion are well defined.

In this chapter we briefly describe the correlation between Raman intensities and the nature of electronic excited-state molecules and then explain how the Jahn–Teller effect of the degenerate electronic excited states of oxocarbons is reflected in their Raman intensities. Geometric distortion of the degenerate excited states of these ions is discussed on the basis of experimental results.

II. Raman Intensity and Electronic States

The quantum mechanical theory of Raman scattering was first given by Kramers and Heisenberg in 1925. According to the theory, the intensity of a vibrational Raman line due to the transition from the ith vibrational state of the electronic ground state g to the jth vibrational state of the same ground state is given by the following:

$$I_{gi,gj} = \frac{2^7\pi^5}{3^2c^4} I_0(\nu_0 \pm \nu_{gi,gj})^4 \sum_{\rho\sigma} |(\alpha_{\rho\sigma})_{gi,gj}|^2 \quad (1)$$

where I_0 and ν_0 are the intensity and frequency, respectively, of the incident light, and $\nu_{gi,gj}$ is the Raman frequency; $(\alpha_{\rho\sigma})_{gi,gj}$ is the $\rho\sigma$ component of the molecular polarizability tensor associated with the transition from the gi state to the gj state and is expressed by

$$(\alpha_{\rho\sigma})_{gi,gj} = \sum_e \sum_v \left[\frac{\langle gi|R_\rho|ev\rangle\langle ev|R_\sigma|gj\rangle}{E_{ev,gi} - E_0} + \frac{\langle gi|R_\sigma|ev\rangle\langle ev|R_\rho|gj\rangle}{E_{ev,gj} + E_0} \right] \quad (2)$$

Here, $\langle gi|R_\rho|ev\rangle$ is the ρ component of the electronic transition moment between the gi and ev states (e and v represent the electronic excited state and vibrational state associated with e, respectively). In order to have nonzero α, that is, to have Raman intensity, both transition moments $\langle gi|R_\rho|ev\rangle$ and $\langle ev|R_\sigma|gj\rangle$ should be simultaneously nonzero. The existence of such ev states is essential for Raman scattering, and they are called intermediate or virtual states. Therefore, the intensity of a vibrational Raman line involves information about the electronic excited states of a molecule. Equation (2) also tells us that, when the exciting light energy E_0 ($= h\nu_0$) approaches $E_{ev,gi}$, the first term of Eq. (2) greatly increases, leading to enormous intensity enhancement of the Raman line. This is the origin of the resonance Raman effect.

To the extent that the Born–Oppenheimer approximation, that is, the assumption of electron–nuclear separability, is valid, the wavefunctions can be separated into electronic and vibrational parts, giving

$$\begin{aligned} |gi\rangle &= |g\rangle \| i\rangle \\ |gj\rangle &= |g\rangle \| j\rangle \\ |ev\rangle &= |e\rangle \| v\rangle \end{aligned} \tag{3}$$

where $|g\rangle$ and $|e\rangle$ are wavefunctions of electronic ground and excited states, $\|i\rangle$ and $\|j\rangle$ are vibrational wavefunctions of initial and final vibrational states in the electronic ground state, and $\|v\rangle$ is a vibrational wavefunction of the excited state e. By substituting Eq. (3) into Eq. (2) and by expanding the electronic wavefunctions in terms of normal coordinates Q_a around the equilibrium nuclear configuration of the molecule, Albrecht [*2,3*] obtained the following expression for the polarizability:

$$(\alpha_{\rho\sigma})_{gi,gj} = A + B \tag{4}$$

Here, A and B are complicated equations. However, under several reasonable assumptions, the A term is simplified and given by the following [*4–6*]:

$$A = \sum_e \frac{2(E_{eg}^2 + E_0^2)}{(E_{eg}^2 - E_0^2)^2} \langle g^0|R_\rho|e^0\rangle\langle e^0|R_\sigma|g^0\rangle \hbar\omega_a \left[\frac{\mu\omega_a}{2h}\right]^{\frac{1}{2}} |\Delta_a| \tag{5}$$

where $|g^0\rangle$ and $|e^0\rangle$ are electronic ground- and excited-state wavefunctions at the equilibrium nuclear configuration, ω_a the vibrational frequency of the normal mode Q_a, and Δ_a the displacement of the origin of the normal coordinate Q_a in the e state relative to that in the g state (see Fig. 1). It is known from experience that the Raman intensity of the totally symmetric vibration is mainly contributed from the A term. Therefore, the intensity of a totally symmetric Raman line is derived from symmetry preserving geometric change in the excited state relative to the equilibrium structure of the ground-state molecule. Among many excited states, any state the geometric structure of which is greatly different from

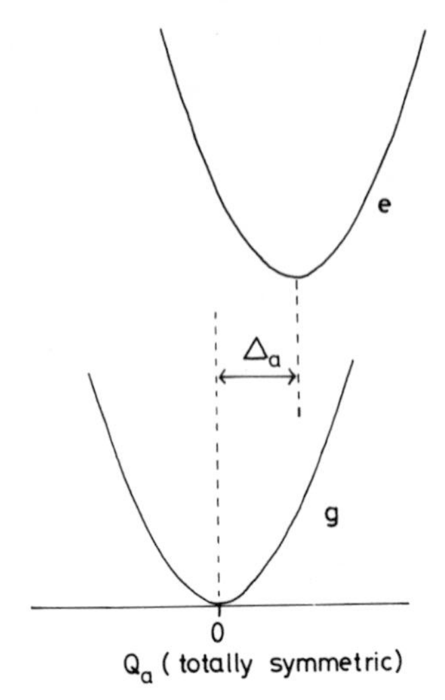

Fig. 1. Potentials of ground (g) and excited (e) states along normal coordinate of a totally symmetric vibration.

that of the ground state will make a large contribution to the intensity of a totally symmetric Raman line.

On the other hand, the B term is given by Eq. (6) [2]:

$$B = -\sum_{e}\sum_{s\neq e}\sum_{a}(\langle g^0|R_\rho|e^0\rangle\langle e^0|h_a|s^0\rangle\langle s^0|R_\sigma|g^0\rangle +$$

$$\langle g^0|R_\sigma|e^0\rangle\langle e^0|h_a|s^0\rangle\langle s^0|R_\rho|g^0\rangle)\frac{(E^0_{eg}E^0_{sg} + E^2_0)\langle i\|Q_a\|j\rangle}{[(E^0_{eg})^2 - E^2_0][(E^0_{sg})^2 - E^2_0]} \quad (6)$$

$$h_a = (\delta H/\delta Q_a)_0$$

In this equation, $\langle e^0|h_a|s^0\rangle$ represents vibronic coupling of two electronic excited states e^0 and s^0 through the normal vibration Q_a. By requirements of symmetry, the Raman intensity of a nontotally symmetric vibration comes exclusively from this B term. An important effect of the vibronic coupling is distortion of the vibrational potential of the relevant electronic state. Figure 2 shows the potentials of e and s states along the normal coordinate of a nontotally symmetric vibration with and without vibronic coupling between them. Vibronic coupling induces flattening and sharpening of the potential for lower and higher electronic states, respectively. Therefore, such distortion of the vibrational potential is reflected in the Raman intensity of nontotally symmetric vibrations [7].

The complexity of Raman intensity arises from the fact that all the electronic excited states can contribute to the intensity, as shown by the summation of Eqs.

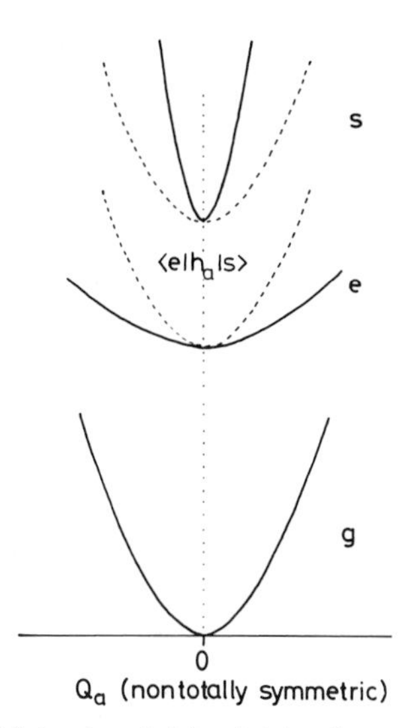

Fig. 2. Potentials of ground (g) and excited (e, s) states along normal coordinate of a nontotally symmetric vibration. Dashed and solid curves are the potentials without and with vibronic coupling, respectively.

(5) and (6). However, if we choose the exciting light energy E_0 very close to E_{eg} or E_{sg}, one term in the summation becomes predominant because of the energy denominator, and the other terms become negligible. Therefore, by adjusting the exciting light to meet this condition, one can obtain information about geometric structure and vibronic coupling of a specific electronic excited-state molecule. In the extreme case of $E_0 = E_{eg}$ or E_{sg} (the "rigorous resonance case"), Raman intensity becomes infinite according to Eqs. (5) and (6). Actually, it becomes extremely strong but still finite. For the rigorous resonance case, Eq. (2) should be slightly modified by introducing the damping factors of the intermediate state. Furthermore, Eqs. (5) and (6) are no longer valid because the approximations used in their derivation are not appropriate in the rigorous resonance case. Nevertheless, Eqs. (5) and (6) are known to be applicable for the case in which $E_{eg} - E_0 > 3000\ \text{cm}^{-1}$ (the "preresonance case").

As explained earlier, the Jahn–Teller effect (strong vibronic coupling within a degenerate electronic state) is expected for species of very high symmetry such as the oxocarbon ions. Jahn–Teller coupling may be regarded as an extreme case in which the e and s states in Eq. (6) and Fig. 2 are degenerate. Figure 3 illustrates the Jahn–Teller effect of the doubly degenerate electronic state of $C_4O_4^{2-}$ on the vibrational potential of a nontotally symmetric vibration. At the equilibrium configuration of the ground-state molecule, the electronic excited state is degenerate. However, degeneracy is removed at the displaced position, resulting in splitting of the potential. When vibronic interaction is small, we have in a first

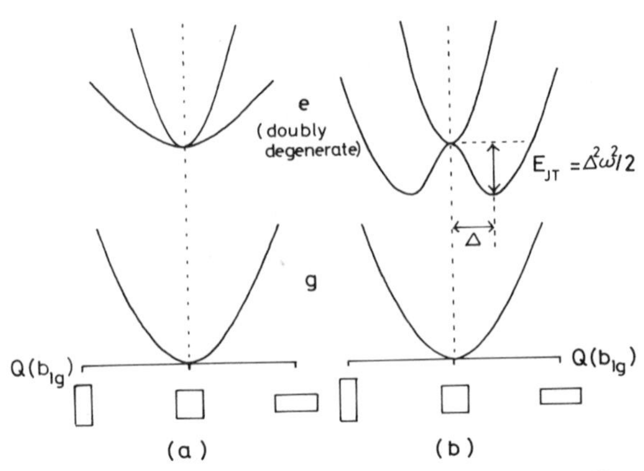

Fig. 3. Potentials of (a) weak Jahn–Teller coupling and (b) strong Jahn–Teller coupling for a b_{1g} C—C stretching vibration of a symmetric four-membered ring.

approximation simply two parabolic potentials, both of which have common minima at the equilibrium position, as shown in Fig. 3a. However, when vibronic interaction is large, one of the potentials is subject to a great deformation such that the potential minima occur at displaced positions and another potential becomes sharp, as shown in Fig. 3b. The two potentials cross each other at the symmetric position. Thus, in such circumstances, the most stable configuration of the molecule is not the symmetric one, but one of reduced symmetry given by the minima of the lower potential. A nontotally symmetric change of the geometric configuration of the molecule is called Jahn–Teller distortion and is measured by Δ (see Fig. 3b). The energy difference between the potentials at the symmetric position and the minima is called Jahn–Teller stabilization energy, which is given by $E_{JT} = \Delta^2 \omega^2/2$.

The effect of Jahn–Teller interaction on Raman intensity is not simple. However, as far as the shift of the potential minimum is concerned, the situation is very similar to the case of the totally symmetric vibration shown in Fig. 1. The difference is that the normal coordinate along which the distortion occurs is the nontotally symmetric one for the Jahn–Teller system and the totally symmetric one for the usual system. As expected from this similarity, theory [8] shows that the Raman intensity of a Jahn–Teller active nontotally symmetric vibration behaves like a totally symmetric Raman line, and Eq. (5) may be approximately used for a Jahn–Teller active nontotally symmetric Raman line in the preresonant condition. It is anticipated, therefore, that the Raman line of a nontotally symmetric vibration whose normal mode is closely related to the Jahn–Teller distortion will gain large intensity and its intensity will change with the change of exciting light frequency according to the energy factor given by Eq. (5).

The Raman intensity of a Jahn–Teller system in the rigorous resonance condition depends in a complicated way on the strength of vibronic coupling, the vibrational frequency of the Jahn–Teller active vibration, and the relative position of the exciting light with respect to the electronic absorption. However, it still provides very useful information about the Jahn–Teller distortion, as will be described in Section VII.

III. Electronic Absorption Spectra of Oxocarbon Ions

Electronic absorption spectra of $C_4O_4^{2-}$, $C_5O_5^{2-}$, and $C_6O_6^{2-}$ are shown in Fig. 4. There is only one strong absorption in the visible or ultraviolet region for each ion, and it shifts progressively toward longer wavelength with increasing ring size. It is also seen from the figure that the spectral shapes of the absorptions are very similar for all three anions: a broad peak at longer wavelength and a weak shoulder at shorter wavelength, with frequency separation of about 2000 cm^{-1} for each ion.

The electronic states of oxocarbon ions were calculated by Sakamoto and I'Haya [9] by a Pariser–Parr type of SCF–LCAO–MO method. Their calculated results are shown in Fig. 5, in which solid and broken lines indicate the states allowed and forbidden, respectively, for electronic transition from the ground state. It is seen from the figure that there are only two allowed excited states below 8 eV for each ion. The lowest allowed state is located in the visible or near-ultraviolet region and the other in the far ultraviolet. This is consistent with the experimental results reported by Michl and West in Chapter 5. Thus, the electronic absorptions of oxocarbon ions in the visible or near-ultraviolet region must

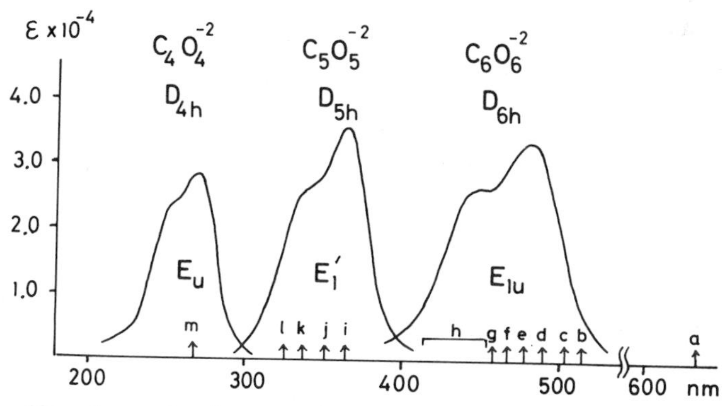

Fig. 4. Absorption spectra of aqueous solutions of $C_4O_4^{2-}$, $C_5O_5^{2-}$, and $C_6O_6^{2-}$. The arrows are laser lines used for Raman measurements: (a) 632.8 nm, He–Ne; (b) 514.5 nm, Ar; (c) 501.7 nm, Ar; (d) 488.0 nm, Ar; (e) 476.5 nm, Ar; (f) 465.8 nm, Ar; (g) 457.9 nm, Ar; (h) 450–420 nm, tunable dye; (i) 363.8 nm, Ar; (j) 351.1 nm, Ar; (k) 337.1 nm, N_2; (l) 325.0 nm, He–Cd; (m) 266.2 nm, fourth harmonics of YAG.

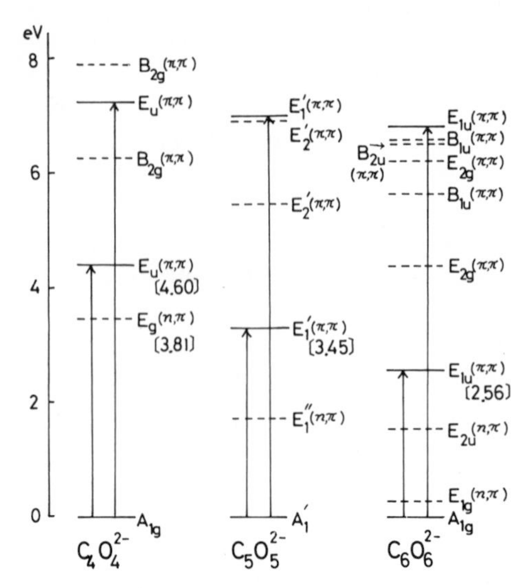

Fig. 5. Calculated energies of low-lying singlet electronic states of oxocarbon ions [*9*]. Solid lines and arrows represent optically allowed excited states and transitions. The figures in brackets are observed energies.

be due to transitions $E_u(\pi,\pi^*) \leftarrow A_{1g}$, $E'(\pi,\pi^*) \leftarrow A_1'$, and $E_{1u}(\pi,\pi^*) \leftarrow A_{1g}$ for $C_4O_4^{2-}$, $C_5O_5^{2-}$, and $C_6O_6^{2-}$, respectively. Since the lowest allowed excited state is the degenerate π,π^* state for all the ions, the occurrence of a Jahn–Teller effect is expected.

The existence of the Jahn–Teller effect is usually demonstrated by analysis of the vibrational fine structure of the absorption spectrum. However, the long-wavelength absorptions of oxocarbon ions are very broad and show only two broad peaks. Nevertheless, the observed absorption band shapes strongly suggest a Jahn–Teller effect. Muramatsu *et al.* calculated quantum mechanically the absorption band shapes of $C_4O_4^{2-}$ [*10*], $C_5O_5^{2-}$ [*11*], and $C_6O_6^{2-}$ [*12*] and found that the observed band shapes can be reproduced very well with suitable values of the strength of Jahn–Teller interaction. It is very probable, therefore, that the Jahn–Teller effect occurs in the lowest degenerate π,π^* state of the oxocarbon ion. However, the absorption spectrum does not reveal which vibrations are effective in the Jahn–Teller effect. This kind of information is supplied from the study of Raman intensities, which is described in the following sections.

IV. Raman Spectra of $C_nO_n^{2-}$

As will be understood from Eq. (2), Raman intensities depend on the energy difference between the exciting light and the electronic states serving as intermediate states in Raman scattering. The intermediate electronic state should be one to which the electronic transition from the ground state is dipole-allowed. Since such transitions usually give strong absorption, the Raman intensities depend on the relative energies of exciting light and the strong absorption. For oxocarbon ions, the lowest degenerate π,π^* state giving strong absorption in the visible or ultraviolet region can serve as an intermediate electronic state in Raman scattering. Therefore, the Raman spectra are characterized by the relative position of the exciting light with respect to the absorption. For the sake of convenience, we will call the spectra off-resonance, preresonance, or rigorous resonance depending on whether the exciting light is far from, near, or within the absorption, respectively.

Figures 6, 7, and 8 show the Raman spectra of aqueous solutions of $C_4O_4^{2-}$, $C_5O_5^{2-}$, and $C_6O_6^{2-}$, respectively, at various excitation wavelengths [*8,11,13*]. The relative positions of the exciting laser light with respect to the electronic absorption are seen in Fig. 4. The most characteristic feature of the spectra that is common to all the ions is the strong appearance of Raman lines of nontotally symmetric vibrations in off-resonance and preresonance conditions. They are $\nu_{10}(b_{2g})$(647 cm^{-1}) and $\nu_5(b_{1g})$(1123 cm^{-1}) for $C_4O_4^{2-}$, $\nu_{11}(e_2')$(555 cm^{-1}) for $C_5O_5^{2-}$, and $\nu_{16}(e_{2g})$(1252 cm^{-1}), $\nu_{17}(e_{2g})$(436 cm^{-1}), and $\nu_{18}(e_{2g})$(346 cm^{-1}) for $C_6O_6^{2-}$. Normal vibrations of the ions are summarized in Tables I, II, and III.

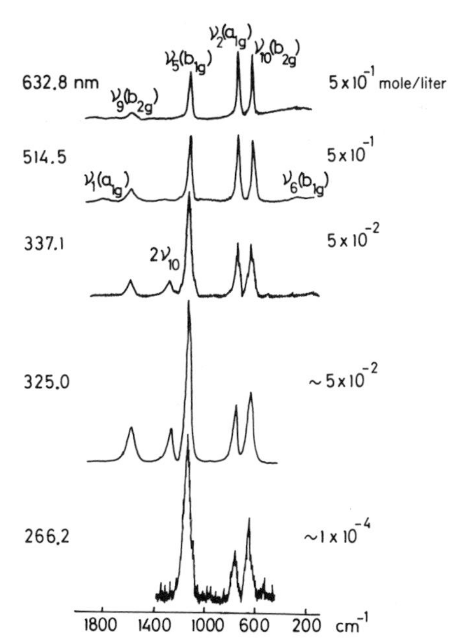

Fig. 6. Raman spectra of aqueous solutions of $C_4O_4^{2-}$ at various excitation wavelengths.

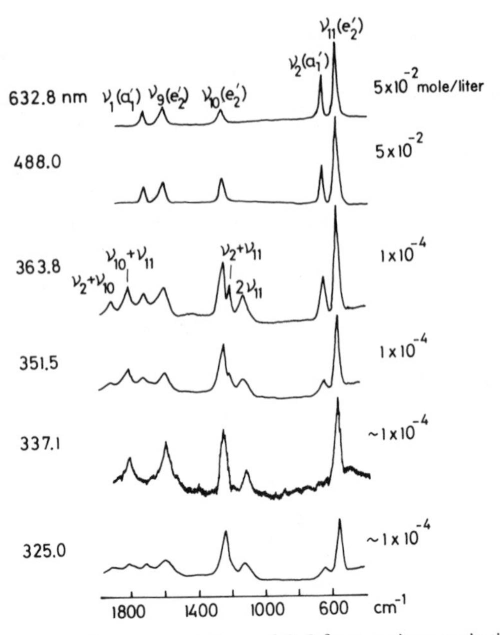

Fig. 7. Raman spectra of aqueous solutions of $C_5O_5^{2-}$ at various excitation wavelengths.

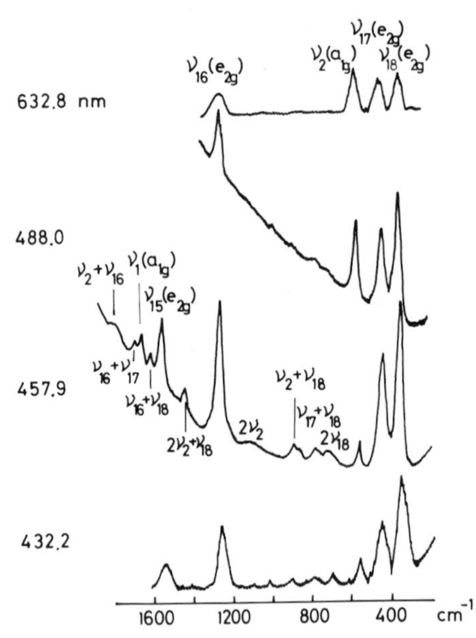

Fig. 8. Raman spectra of aqueous solutions of $C_6O_6^{2-}$ at various excitation wavelengths. Strong background observed in 488.0 nm excitation is due to fluorescence of impurity.

Table I. Normal Vibrations of $C_4O_4^{2-}$ [a]

Symmetry species (D_{4h})	Selection rule	Number of normal vibration	Observed frequency (cm^{-1}) Raman	IR	Vibrational mode
a_{1g}	Raman (p)	ν_1	1794(w)		Symmetric CO stretching
		ν_2	723(s)		Ring breathing
a_{2g}	Inactive	ν_3			In-plane CO bending
	Raman (inv. p)				
a_{2u}	IR	ν_4		259(s)	Out-of-plane CO bending
b_{1g}	Raman (dp)	ν_5	1123(vs)		CC stretching
		ν_6	294(w)		In-plane CO bending
b_{1u}	Inactive	ν_7			Out-of-plane CO bending
		ν_8			Ring twisting
b_{2g}	Raman (dp)	ν_9	1593(s)		CO stretching
		ν_{10}	647(s)		Ring bending
e_g	Raman (dp)	ν_{11}	662(vw)[b]		Out-of-plane CO bending
e_u	IR	ν_{12}		1530(vs)	CO stretching
		ν_{13}		1090(s)	CC stretching
		ν_{14}		350(m)	In-plane CO bending

[a] From Ito and West [1].
[b] Observed in the solid state only.

Table II. Normal Vibrations of $C_5O_5^{2-}$ [a]

Symmetry species (D_{5h})	Selection rule	Number of normal vibration	Observed frequency (cm^{-1}) Raman	IR	Vibrational mode
a_1'	Raman (p)	ν_1	1718(w)		Symmetric CO stretching
		ν_2	637(m)		Ring breathing
a_2'	Inactive	ν_3			In-plane CO bending
	Raman (inv. p)				
a_2''	IR	ν_4		248(s)	Out-of-plane CO bending
e_1'	IR	ν_5		1570(vs)	CO stretching
		ν_6		1100(w)	CC stretching
		ν_7		374(m)	In-plane CO bending
e_1''	Raman (dp)	ν_8			Out-of-plane CO bending
e_2'	Raman (dp)	ν_9	1591(s)		CO stretching
		ν_{10}	1243(m)		CC stretching
		ν_{11}	555(s)		Ring bending
		ν_{12}			In-plane CO bending
e_2''	Inactive	ν_{13}			Ring twisting
		ν_{14}			Out-of-plane CO bending

[a] From Ito and West [1].

Table III. Normal Vibrations of $C_6O_6^{2-}$ [a]

Symmetry species (D_{6h})	Selection rule	Number of normal vibration	Frequency (cm^{-1}) Observed Raman[b]	IR	Calculated	Vibrational mode
a_{1g}	Raman (p)	ν_1	1669		1594	Symmetric CO stretching
		ν_2	553		580	Ring breathing
a_{2g}	Inactive Raman (inv. p)	ν_3			854	In-plane CO bending
a_{2u}	IR	ν_4		235	—	Out-of-plane CO bending
b_{1u}	Inactive	ν_5			1551	CO stretching
		ν_6			489	Ring bending
b_{2g}	Inactive	ν_7			—	Out-of-plane CO bending
		ν_8			—	Ring twisting
b_{2u}	Inactive	ν_9			1320	CC stretching
		ν_{10}			451	In-plane CO bending
e_{1g}	Raman (dp)	ν_{11}			—	Out-of-plane CO bending
e_{1u}	IR	ν_{12}		1449	1587	CO stretching
		ν_{13}		1051	1031	CC stretching
		ν_{14}		386	340	In-plane CO bending
e_{2g}	Raman (dp)	ν_{15}	1546		1562	CO stretching
		ν_{16}	1252		1222	CC stretching
		ν_{17}	436		420	In-plane CO bending
		ν_{18}	346		339	Ring bending
e_{2u}	Inactive	ν_{19}			—	Out-of-plane CO bending
		ν_{20}			—	Ring twisting

[a] From Takahashi *et al.* [8].
[b] R. T. Bailey, *J. Chem. Soc.* (*B*), 627 (1971).

These nontotally symmetric Raman lines show great intensity enhancement when the exciting light approaches the absorption. In the rigorous resonance condition, the Raman intensities increase to 10^2–10^3 times the intensities in the off-resonance condition. Moreover, the spectral features change sensitively with the exciting wavelength. For example, for $C_5O_5^{2-}$ (Fig. 7), with the excitation at 363.8 nm, which almost coincides with the longer-wavelength peak of the absorption, several overtones and combinations involving $\nu_{11}(e_2')$, $\nu_{10}(e_2')$, and $\nu_2(a_1')$ are observed in addition to the fundamental lines. On the other hand, with the excitation at 337.1 nm, which coincides with the shorter-wavelength shoulder of the absorption, the spectrum consists entirely of the fundamentals $\nu_{11}(e_2')$, $\nu_{10}(e_2')$, and $\nu_9(e_2')$ and their overtones and combinations, and the totally symmetric fundamentals and combinations involving them are apparently missing from the spectrum.

A similar situation is seen for $C_6O_6^{2-}$, as shown in Fig. 8. The whole absorp-

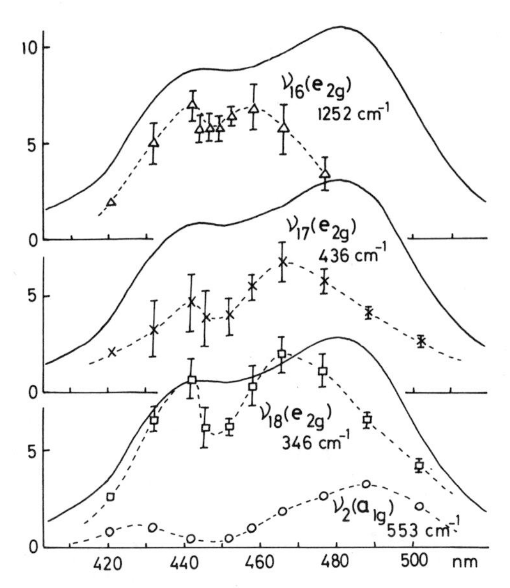

Fig. 9. Excitation profiles of Raman intensities of $\nu_2(a_{1g})$, $\nu_{18}(e_{2g})$, $\nu_{17}(e_{2g})$, and $\nu_{16}(e_{2g})$ of $C_6O_6^{2-}$. Solid curves represent the electronic absorption. Vertical bars indicate experimental error.

tion region of rhodizonate ion can be covered with the lasing lines of an Ar ion laser and a tunable dye laser. Therefore, we could obtain the excitation profiles—Raman intensity as a function of exciting wavelength within the absorption—of the individual Raman lines. The results are shown in Fig. 9 for ν_2, ν_{16}, ν_{17}, and ν_{18}. It is seen from the figure that the intensity of the totally symmetric fundamental $\nu_2(a_{1g})$ is the weakest of the four lines in the whole spectral absorption region. The excitation profiles of ν_2 and ν_{18} definitely exhibit two maxima, whereas the existence of two maxima is not so clear for ν_{16} and ν_{17}. It is noted that the position of the longer-wavelength maxima shifts to shorter wavelength roughly in order of the vibrational frequency for the e_{2g} Raman lines, while the maximum for the $\nu_2(a_{1g})$ is located at the longest-wavelength position. The implication of these observations is discussed in Section VII.

V. Jahn–Teller Effect Suggested by Raman Intensities

The most characteristic features of the Raman spectra of $C_nO_n^{2-}$ are the strong intensities of the nontotally symmetric vibrations in the off-resonance condition and their great intensity enhancement in the preresonance and rigorous resonance conditions. As described in Section II, the Raman intensities of nontotally symmetric vibrations arise from vibronic coupling in electronic excited states. Referring to the low-lying electronic states of $C_nO_n^{2-}$ shown in Fig. 5, the electronic excited states to which the transitions from the ground state are symmetry-allowed are two E_u (π, π^*), two E'_1 (π, π^*), and two E_{1u} (π, π^*) for $C_4O_4^{2-}$, $C_5O_5^{2-}$, and $C_6O_6^{2-}$, respectively. Therefore, the Raman intensities of the nontotally symmetric vibrations arise from the following two possibilities: (a) the vibronic coupling between the two different degenerate states, (b) the vibronic coupling within the lowest degenerate state, that is, Jahn–Teller coupling.

In case (a), the active vibrations coupling the two different degenerate states should be of symmetry species derived from the direct product of the two same degenerate species, which are listed in the second column of Table IV. As seen from the table, we may expect Raman inactive vibrations belonging to a_{2g}, a'_2, and a_{2g} of $C_4O_4^{2-}$, $C_5O_5^{2-}$, and $C_6O_6^{2-}$, respectively, in the preresonance condition. Such inactive Raman lines were actually found to appear in the preresonance Raman spectra of cytochrome *c* as the result of the vibronic coupling between the two different degenerate electronic states [*14*]. It was also found that these inactive Raman lines exhibit inverse polarization (depolarization ratio of the Raman line is greater than ¾). A careful search for these Raman lines [ν_3 (a_{2g}), ν_3 (a'_2), and ν_3 (a_{2g}) for $C_4O_4^{2-}$, $C_5O_5^{2-}$, and $C_6O_6^{2-}$, respectively] in the expected frequency regions shows no indication of their occurrence. According to the calculated results for electronic states of $C_nO_n^{2-}$ shown in Fig. 5, the upper degenerate state is located more than 3 eV higher in energy than the lowest degenerate state. Effective vibronic coupling is almost impossible between two degenerate states having such a large energy separation. For these reasons, case (a) is unlikely.

Let us examine case (b), that is, Jahn–Teller coupling in the lowest degenerate state. The symmetry species of vibrations effective in the Jahn–Teller coupling should be of the symmetry species derived from the symmetric product of the same degenerate species, which are given in the third column of Table IV. In the fourth column, the nontotally symmetric vibrations showing great intensity enhancement in the preresonance and rigorous resonance conditions are listed. The symmetry species coincide perfectly with those expected from Jahn–Teller coupling. It is suggested, therefore, that the unusual intensities of the nontotally symmetric Raman lines of $C_nO_n^{2-}$ are due to a Jahn–Teller effect in the degenerate π,π^* state of each ion.

Table IV. Vibrations of $C_4O_4^{2-}$, $C_5O_5^{2-}$, and $C_6O_6^{2-}$

Ion	Direct product	Symmetric product	Anomalous Raman lines
$C_4O_4^{2-}$	a_{1g}, a_{2g}, b_{1g} b_{2g}	a_{1g}, b_{1g}, b_{2g}	ν_5 (b_{1g}), ν_{10} (b_{2g})
$C_5O_5^{2-}$	a'_1, a'_2, e'_2	a'_1, e'_2	ν_{11} (e'_2)
$C_6O_6^{2-}$	a_{1g}, a_{2g}, e_{2g}	a_{1g}, e_{2g}	ν_{16} (e_{2g}), ν_{17} (e_{2g}), ν_{18} (e_{2g})

VI. Jahn–Teller Deformation of Ions

The next question to consider is, Which vibrations among the many of possible symmetry species are effective in Jahn–Teller coupling? As mentioned in Section II, the Raman intensities of Jahn–Teller active nontotally symmetric vibrations behave like those of totally symmetric vibrations in the preresonance condition, and Eq. (5) given for the totally symmetric Raman line can also be applied to these nontotally symmetric Raman lines. The summation over e in Eq. (5) can be divided into two parts for the preresonance condition; one is the term contributed from the lowest degenerate π,π^* state and the other the term from all the allowed states other than the lowest state. Since the allowed states other than the lowest one are located at very high energies and $E_{eg} \gg E_0$ for these states, the latter term can be assumed to be independent of the excitation wavelength. Then, the intensities of the Jahn–Teller active vibrations can be given by

$$I = |xF_A + y|^2 \qquad F_A = \frac{2(E_{eg}^2 + E_0^2)}{(E_{eg}^2 - E_0^2)^2} \tag{7}$$

where xF_A is the term contributed from the lowest degenerate π,π^* state and y the nonresonant part arising from the higher electronic states. In Fig. 10, the square roots of the observed intensities of $\nu_5(b_{1g})$, $\nu_{10}(b_{2g})$, and $\nu_2(a_{1g})$ of $C_4O_4^{2-}$ are plotted against F_A for six excitation wavelengths in the preresonance region. The plots give straight lines for all three Raman lines, supporting the above assumption. A similar linear relationship was also found for $\nu_1(a_1')$, $\nu_2(a_2')$, $\nu_{10}(e_2')$, and $\nu_{11}(e_2')$ of $C_5O_5^{2-}$. The values of x and y obtained from the straight lines are summarized in Table V. The values of x and y for other Raman lines are

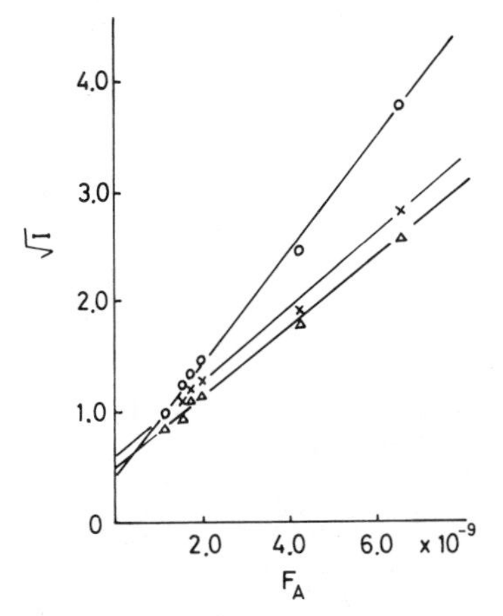

Fig. 10. Plots of the square root of observed Raman intensities against F_A for $C_4O_4^{2-}$. Key: X, $\nu_2(a_{1g})$; ○, $\nu_5(b_{1g})$; △, $\nu_{10}(b_{2g})$.

Table V. Values of x and y of the Raman Lines of $C_5O_5^{2-}$ and $C_6O_6^{2-}$ [a]

Raman line	Raman frequency (cm^{-1})	x		y
$C_4O_4^{2-}$				
ν_2 (a_{1g})	723	3.4×10^8	(0.23)	0.60
ν_5 (b_{1g})	1123	5.1×10^8	(1.92)	0.42
ν_{10} (b_{2g})	647	3.2×10^8	(1.00)	0.49
$C_5O_5^{2-}$				
ν_1 (a_1')	1718	0.5×10^8	(0.66)	0.60
ν_2 (a_2')	637	1.6×10^8	(0.32)	0.54
ν_{10} (e_2')	1243	1.8×10^8	(0.96)	0.28
ν_{11} (e_2')	555	2.9×10^8	(1.00)	0.37

[a] The values in parentheses are calculated relative values of vibronic coupling strength.

very small compared with those given in the table. For $C_6O_6^{2-}$, the corresponding values were not obtained because of the lack of experimental points in the preresonance region.

It is seen from comparison between Eqs. (5) and (7) that x is proportional to Δ, which is the displacement of the origin of the normal coordinates in the electronic excited state relative to that in the ground state for the totally symmetric vibrations and the Jahn–Teller distortion in the degenerate excited state for the nontotally symmetric vibrations. In $C_4O_4^{2-}$, x is largest for $\nu_5(b_{1g})$. This suggests a great deformation of the excited state along this normal coordinate. Since $\nu_5(b_{1g})$ is the C—C stretching vibration as shown in Fig. 11, a great deformation of the ring from square to rectangular shape occurs in the excited state. A fairly large x value for $\nu_{10}(b_{2g})$ suggests that considerable deformation of the ring to a rhombus form also occurs. In $C_5O_5^{2-}$, x is largest for $\nu_{11}(e_2')$, which is a ring-bending vibration, as shown in Fig. 11. This indicates that the lowest degenerate excited state is greatly distorted along this normal coordinate.

In order to confirm the above conclusions, we carried out calculations of Jahn–Teller coupling in the lowest degenerate π,π^* states of the oxocarbon ions [*8*]. The electronic wavefunctions used were of the SCF–LCAO–MO type obtained by Sakamoto and I'Haya [*9*]. The normal modes were calculated by using the same force field as that used by Ito and West [*1*] for $C_4O_4^{2-}$ and $C_5O_5^{2-}$, and the force field given by Takahashi *et al.* [8] for $C_6O_6^{2-}$. The normal modes of several Jahn–Teller active vibrations are shown in Fig. 11. Since the details of the calculations have already been reported [*8*], only the results will be described here. A theoretical quantity that can be compared with the experimental value x is the vibronic coupling strength [*8*]. The relative values of calculated vibronic coupling strength are given in parentheses in Table V. As seen from the table, for $C_4O_4^{2-}$ the calculated value is largest for $\nu_5(b_{1g})$, in agreement with the largest x

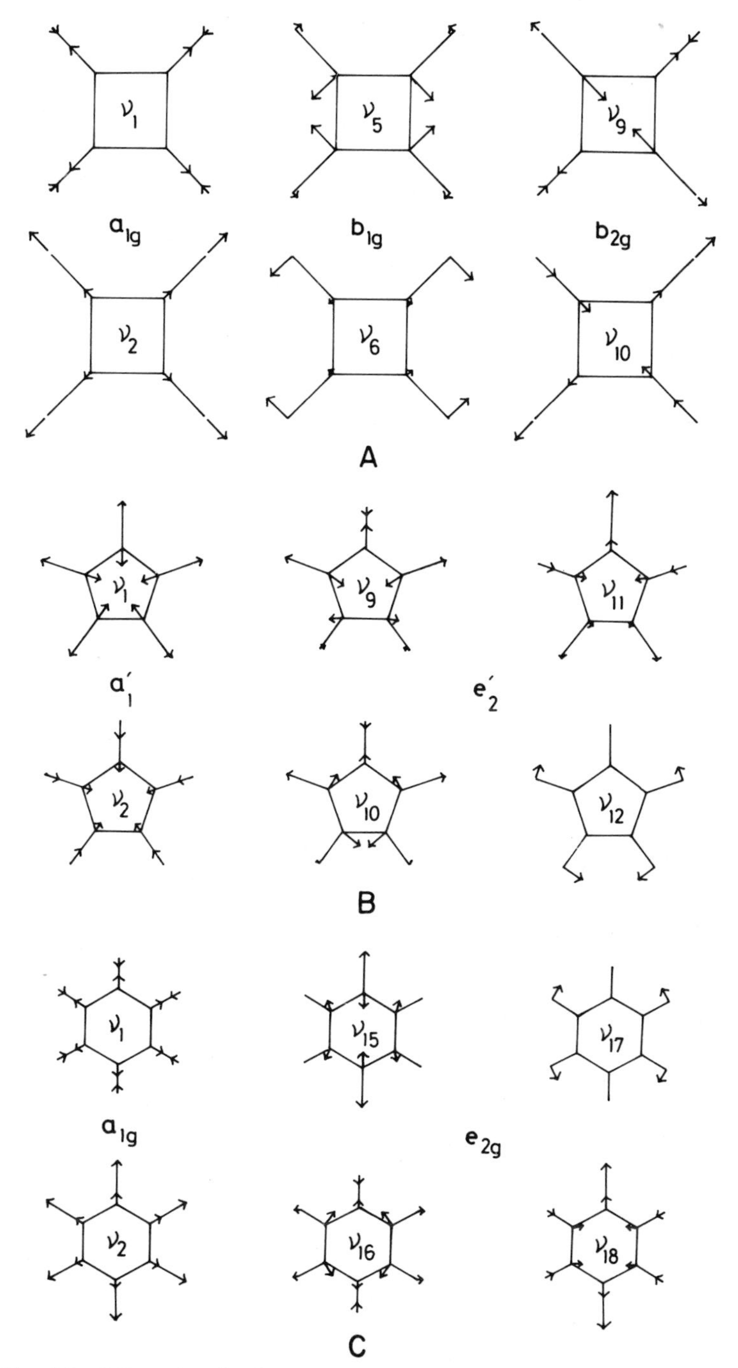

Fig. 11. Jahn–Teller active vibrational modes of (A) $C_4O_4^{2-}$, (B) $C_5O_5^{2-}$, and (C) $C_6O_6^{2-}$. For degenerate vibrations, only one component of each mode is given.

value for this vibration. In the case of $C_5O_5^{2-}$, the largest calculated value is for $\nu_{11}(e_2')$, again in agreement with the experimental result. There also seems to be a qualitative parallel relation between the orders of the magnitude of x and the calculated strength, except for ν_1. Because of many assumptions involved in the calculations, quantitative agreement with the experimental results is not expected at present. However, the qualitative agreement described above seems to show that the observed x values actually reflect the Jahn–Teller distortion in the electronic excited state.

One of the characteristics of the Raman spectra of oxocarbon ions is strong Raman intensities of the nontotally symmetric vibrations even in the off-resonance condition. This is shown by large y values for the nontotally symmetric vibrations, which are comparable with the y values of the totally symmetric vibrations, as seen from Table V. As described above, y represents contributions of higher allowed excited states other than the lowest degenerate π,π^* state. Because of the high degree of symmetry of oxocarbon ions, all the allowed π,π^* states are restricted to degenerate states, for which we can also expect Jahn–Teller coupling. Actually, the second allowed π,π^* excited state is predicted from our calculation to have Jahn–Teller deformation similar to that of the lowest excited state. Then all higher degenerate states can contribute to the Raman intensities of Jahn–Teller active nontotally symmetric vibrations. This results in large y values, leading to anomalously large Raman intensities of the nontotally symmetric vibrations even in the off-resonance condition.

VII. Excitation Profile and Jahn–Teller Effect

As described in Section IV, we obtained the excitation profiles of the individual Raman lines of $C_6O_6^{2-}$ for the whole absorption region. Let us consider how the observed excitation profiles can be explained with the Jahn–Teller effect in the electronic excited state. If the Jahn–Teller effect exists, the degeneracy of the electronic state E_{1u} is removed and the nuclear potential splits into two branches, as illustrated in Fig. 3. The normal coordinates that are responsible for the splitting are e_{2g} vibrations. The potential of the E_{1u} state is expected to be greatly distorted along the e_{2g} normal coordinates for which the vibronic coupling is large. For such a normal coordinate, we have favorable Franck–Condon overlapping between the vibrational states of the E_{1u} state and the vibrationless state of the electronic ground state. Then, the Raman line of this particular e_{2g} vibration gains its intensity from the Franck–Condon overlap by the same mechanism as that for a totally symmetric Raman line. When the wavelength of exciting light is continuously changed from longer to shorter wavelengths, it coincides in succession with the individual vibronic levels of the E_{1u} state. The totally symmetric Raman line is expected to show an intensity enhancement at the vibrationless level in the excited state. On the other hand, the Raman lines of the e_{2g} vibrations, the potentials of which are greatly distorted, will have enhanced intensities when the excitation frequency is resonant with the vibronic level involving the e_{2g} vibrations. This gives a qualitative explanation for the observed fact that the maximum of the excitation profile is located at the longest wavelength for a $\nu_2(a_{1g})$ Raman line and moves progressively to shorter wavelength with increasing frequency of the e_{2g} vibration.

The existence of two maxima in the excitation profile is readily explained by the potential characteristic of the Jahn–Teller effect. One of the maxima occurs when the excitation frequency coincides with the vibronic level of the lower potential where the Franck–Condon overlap becomes large. Another maximum is expected at the position where the Franck–Condon factor for the vibronic level in the upper potential is large. In a strong Jahn–Teller coupling case, theory [*15*] predicts two distinct maxima when the coupling constant of the Jahn–Teller active nontotally symmetric mode is larger than that of the totally symmetric mode, whereas two maxima are smeared out as the coupling of the nontotally symmetric vibration becomes weaker than that of the totally symmetric vibration. On the basis of this prediction, we can conclude that $\nu_{18}(e_{2g})$, the excitation profile of which exhibits two distinct maxima, must have a larger Jahn–Teller coupling constant than $\nu_{16}(e_{2g})$ and $\nu_{17}(e_{2g})$.

Calculation [*8*] shows that, among the three e_{2g} vibrations ν_{16}, ν_{17}, and ν_{18}, vibration ν_{18} has the largest value of the coupling constant, which is also larger than that of ν_2 (a_{1g}). Consequently, we observed two distinct maxima in the excitation profile for $\nu_{18}(e_{2g})$. The calculated Jahn–Teller coupling constant for $\nu_{15}(e_{2g})$ at 1546 cm^{-1} is comparable to that of $\nu_{18}(e_{2g})$. We expect, therefore, two distinct maxima also for the $\nu_{15}(e_{2g})$. Unfortunately, however, we could not confirm this because of the appearance of strong impurity fluorescence in this frequency region.

VIII. Conclusion

In the present work, the vibronic coupling within the degenerate excited states of three oxocarbon ions was studied by means of resonance Raman spectra and calculations of the Jahn–Teller interactions. Good agreement was found between the relative vibronic coupling strengths obtained from experiment and calculation. On the basis of these results, we are convinced of the occurrence of the Jahn–Teller effect in the lowest allowed electronic excited states of the three oxocarbon ions. The most effective vibrations that give rise to the Jahn–Teller deformation in the excited state are $\nu_5(b_{1g})$ at 1123 cm^{-1} of the C—C stretching mode and $\nu_{10}(b_{2g})$ at 647 cm^{-1} of the ring-bending mode for $C_4O_4^{2-}$, $\nu_{11}(e_2')$ at 555 cm^{-1} of the ring-bending mode for $C_5O_5^{2-}$, and $\nu_{18}(e_{2g})$ at 346 cm^{-1} of the ring-bending mode for $C_6O_6^{2-}$, respectively. The symmetric structures of these ions will be distorted along these normal coordinates in their excited states.

In the study of the Jahn–Teller effect, band shape analysis of the absorption spectrum is frequently useful. However, the absorption spectrum is often very broad, having no fine structure. In these cases, it is impossible to obtain information about the effective mode for the Jahn–Teller interaction from the absorption spectrum only. On the other hand, the resonance Raman spectra under preresonance and rigorous resonance conditions and their proper analysis inform us of the relative values of the Jahn–Teller coupling strength for all Jahn–Teller active vibrations. It is concluded, therefore, that resonance Raman spectroscopy is a promising means of studying the Jahn–Teller effect.

References

1. M. Ito and R. West, *J. Am. Chem. Soc.* **85,** 2580 (1963).
2. J. Tang and A. C. Albrecht, *in* "Raman Spectroscopy" (H. A. Szymanski ed.), Vol. 2, p. 33. Plenum, New York, 1970.
3. A. C. Albrecht, *J. Chem. Phys.* **34,** 1476 (1961).
4. C. Ting, *Spectrochim. Acta, Part A* **24,** 1177 (1966).
5. F. A. Savin, *Opt. Spectrosc. (USSR)* **19,** 308 (1965).
6. W. L. Peticolas, L. Nafie, P. Stein, and B. Franconi, *J. Chem. Phys.* **52,** 1576 (1970).
7. M. Ito, *in* "Proceeedings of the Fifth International Conference on Raman Spectroscopy" (E. D. Schmid, ed.), p. 267, and the references therein. Hans Ferdinand Schulz Verlag, Freiburg, 1976.
8. M. Takahashi, K. Kaya, and M. Ito, Chem. Phys. **35,** 293 (1978).
9. K. Sakamoto and Y. J. I'Haya, *J. Am. Chem. Soc.* **92,** 2636 (1970).
10. S. Muramatsu and N. Sakamoto, *Chem. Phys. Lett.* **39,** 273 (1976).
11. S. Muramatsu, N. Nasu, M. Takahashi, and K. Kaya, *Chem. Phys. Lett.* **50,** 284 (1977).
12. S. Muramatsu and K. Nasu, *J. Phys. Soc. Jpn.* **46,** 189 (1979).
13. M. Iijima, Y. Udagawa, K. Kaya, and M. Ito, *Chem. Phys.* **9,** 229 (1975).
14. T. G. Spiro and T. C. Streakas, *Proc. Natl. Acad. Sci. U.S.A.* **69,** 2622 (1972).
15. V. Hizhnyakov and I. Tehver, *in* "Proceedings of the Second International Conference on Light Scattering in Solid" (M. Balkanski, ed.), p. 57. Flammarion, Paris, 1971.

The Structural Phase Transition and Dielectric Properties of Squaric Acid

Jens Feder

I. Introduction . . . 141
II. Structure Determinations . . . 143
III. Raman and Infrared Spectra . . . 147
IV. NMR Spectra . . . 149
V. Birefringence, Dielectric, and Elastic Properties . . . 152
VI. Theory . . . 160
VII. Conclusion . . . 165
References . . . 166

I. Introduction

Cohen, Lacher, and Park first synthesized squaric acid ($C_4O_4H_2$; 3,4-dihydroxy-3-cyclobutene-1,2-dione) in 1959 [*1*,*2*]. The structure of the planar molecule is shown in Fig. 1. Squaric acid is a strongly hydrogen-bonded and remarkably stable solid with a decomposition point of about 293°C and a solubility of only about 3% by weight in water at room temperature. Squaric acid forms clear crystals and is an unusually strong dibasic acid with $pK_1 = 0.52$ [*3*] and $pK_2 = 3.48$ [*4*].

Much of the early work on squaric acid concentrated on properties of the dianion $C_4O_4^{2-}$. Cohen *et al.* [*1*] suggested that the dianion would have much resonance stabilization since the four oxygens should be equivalent through resonance in the dianion. The square planar structure (D_{4h} symmetry) was first established by vibrations spectroscopy [*5*]. The discovery of squaric acid and

OXOCARBONS

ISBN 0-12-744580-3

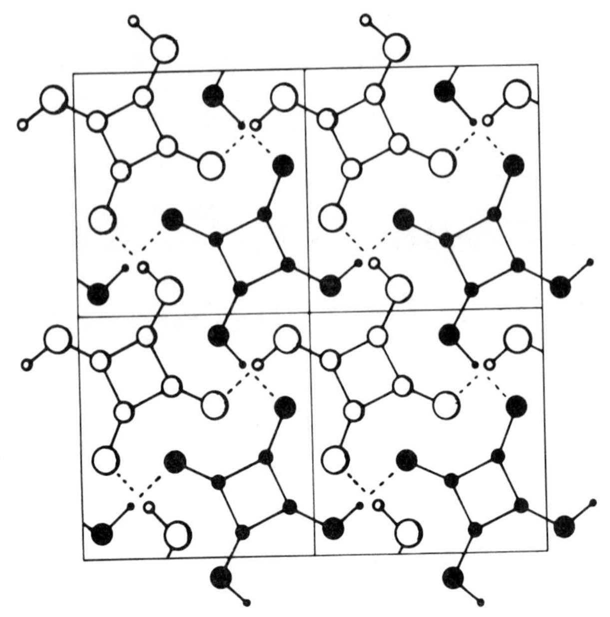

Fig. 1. The structure of squaric acid at room temperature. The unique axis is vertical. Two molecular layers are shown. Big circles, oxygen; intermediate circles, carbon; small circles, hydrogen. The hydrogen bonds are indicated by dashed lines. After Semmingsen [*11*].

squarate ion led to the identification of the oxocarbon dianions as a new aromatic series [*5–9*]. More recently, Semmingsen determined the structure of squaric acid by X-ray diffraction [*10,11*], and his structure is shown in Fig. 1.

In 1974 it was discovered that squaric acid undergoes a structural phase transition at 100°C, which increases to 243°C for the deuterated compound [*12*]. The low temperature phase has a most unusual structure with ordered hydrogen bonds, which can be described as two-dimensional ferroelectric with the polarization in the plane of the molecular layers. Neighboring layers have polarization in the opposite direction, and weak coupling between layers results in a three-dimensional antiferroelectric crystal structure [*13*]. This unique structure makes squaric acid a valuable model for hydrogen-bonded ferroelectrics and has led to the recent extensive physical studies of this compound.

Much of the recent work on squaric acid has concentrated on its structure and bonding as functions of temperature. It should be stressed that the various experimental techniques used measure different aspects of the crystal properties. Neutron and X-ray diffraction techniques give information on the average structure and also the long-range order parameter. These results are reviewed in Section II. Raman and infrared absorbtion spectra give information on molecular and atomic vibrations and reveal the local symmetry of the squaric acid molecule, as discussed in Section III. Nuclear magnetic resonance probes the local environment of the nuclei under study, and thus the order parameter and its fluctuations are measured locally. Our discussion of recent high-resolution measurements can be found in Section IV. Fluctuations of the order parameter are measured by birefringence, dielectric, and ultrasonic methods. These results are considered in Section V, where we also present our new high-resolution dielectric measurements on squaric acid. In Section VI, the theoretical situation is reviewed and the critical behavior at the phase transition is discussed.

II. Structural Determinations

The first crystallographic investigation relating to the squarate anion was the X-ray structure determination of potassium squarate monohydrate by Macintyre and Werkema [*13a*]. They concluded that the squarate anion had D_{4h} symmetry, with mean bond lengths given by C—O = 1.259 Å and C—C = 1.457 Å. These bond lengths are the same as the averages of the corresponding bond lengths in the squaric acid molecule (see Table I).

The first X-ray structure determination of squaric acid was done by Semmingsen [*10,11*], and more recently he has investigated squaric acid by neutron diffraction [*14*]. The high-temperature phase has been investigated by neutron diffraction [*15*], and an independent X-ray and neutron structure determination of squaric acid has been published [*16*]. The deuterated squaric acid structure has also been determined [*17*]. Some of the crystal data of squaric acid are compiled in Table I.

Squaric acid is monoclinic at room temperature and has the space group $P2_1/m$, with two molecules per unit cell. The $C_4O_4H_2$ molecule is planar and is expected to have an electric dipole moment in the plane. As seen in Fig. 1, the structure consists of C_4O_4 groups linked by hydrogen bonds to four neighboring C_4O_4 groups to form a planar quadratic lattice. The molecules in one layer are related to molecules in the layers above and below by a screw axis, producing a pseudo-body-centered tetragonal structure.

In the high-temperature phase the structure is body-centered tetragonal with space group $I4/m$. The hydrogens are "disordered" about the inversion center at the midpoint of the O···H···O hydrogen bond [*16*].

In the low-temperature phase only two hydrogens are close to any C_4O_4 unit. In Fig. 2, the vector sum of the hydrogen displacements from the symmetry center is shown at the center of each molecule. This total averaged displacement is expected to be proportional to the electric dipole moment of the molecule. The direction of the polarization in the plane is not fixed by symmetry, and it may be any direction in the plane. Squaric acid consists of planar ferroelectric layers that are stacked so that the crystal as a whole is antiferroelectric. The wave vector

Table I. Crystal Data of Squaric Acid

Parameter	$C_4O_4D_2$, 19°C[a]	Average	$C_4O_4H_2$, 25°C[b]	Average	$C_4O_4H_2$, 121°C[c]
a (Å)	6.152		6.143		6.137
b (Å)	5.269		5.286		5.337
c (Å)	6.165		6.148		6.137
β	89.92°		89.96°		90°
ρ (gm/cm³)			1.90		1.88
Bond length (Å)					
C-1—O-1	1.295		1.289		
C-2—O-2	1.295	1.260	1.287	1.258	1.256
C-3—O-3	1.226		1.227		
C-4—O-4	1.225		1.230		
C-1—C-4	1.464		1.464		
C-2—C-3	1.462	1.460	1.461	1.460	1.457
C-1—C-2	1.407		1.414		
C-3—C-4	1.506		1.500		
O-1···O-3	2.5731	2.574	2.553	2.554	2.548
O-2···O-4	2.5751		2.554		
O-3···H-1	1.56	1.550	1.524	1.521	1.536
O-4···H-2	1.54		1.517		
O-1—H-1	1.01	1.025	1.030	1.034	1.014
O-2—H-2	1.04		1.037		

[a] T_c = 243°C; monoclinic; $P2_1/m$, Z = 2. Semmingsen [14].
[b] T_c = 100.2°C; monoclinic, $P2_1/m$, Z = 2. Semmingsen *et al.* [15].
[c] Tetragonal; $I4/m$, Z = 2. Hollander *et al.* [16].

associated with the phase transition is $q_z = (\overline{1/2}, 1/2, 1/2)$—the wave vector of the z point of the Brillouin zone of the high-temperature body-centered structure.

Considering the crystal data in Table I, it is clear that, although the crystal is monoclinic at room temperature, the β angle deviates from 90° by only about 0.05°, and the crystals are pseudotetragonal. In fact, Weissenberg photographs of generally twinned crystals show Laue symmetry $4/m$ [*17*]. The twinned crystals consist of domains with domain walls that form an angle of about 36° with the crystallographic axes in the basal plane. Domains are easily identified in the polarizing microscope. The twins represent exchanges of a and c axes [*12, 18*].

Comparing $C_4O_4D_2$ and $C_4O_4H_2$, one sees a small but significant change of the lattice constants $\delta(a, b, c) = (0.009, -0.017, 0.013)$ Å, which may be compared to the change of $(-0.006, 0.051, -0.011)$ Å obtained by increasing the temperature to 121°C. The C—O and C—C bonds, on the other

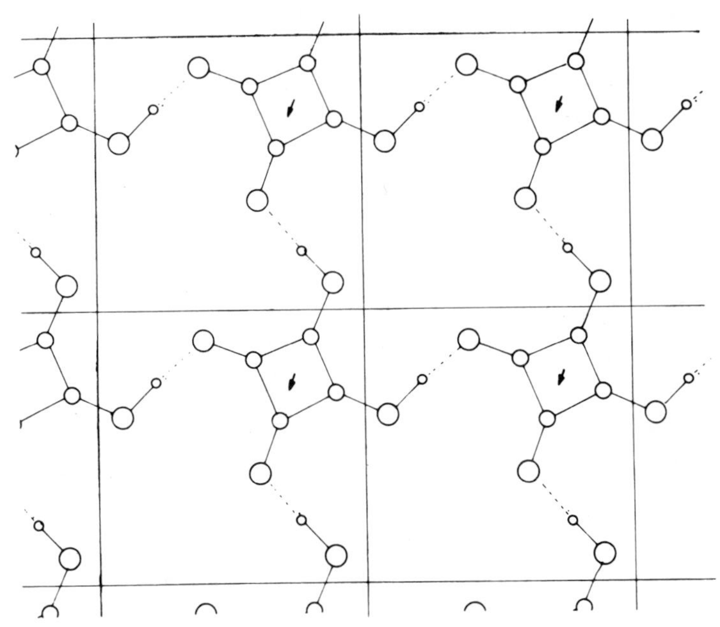

Fig. 2. One layer of squaric acid molecules, with the vector sums of the two hydrogen displacements indicated at the center of the C_4O_4 unit to which they are closest.

hand, hardly change with temperature, and they show no significant isotope effect.

In contrast, there is a clear isotope effect and temperature dependence of the hydrogen bond. Going to the deuterated crystal at room temperature changes the linear O—H—O bond length from 2.554 to 2.574 Å, a change of 0.020 Å. This change is somewhat larger than the change in the lattice constants a and c. Comparing the O···O and O—H bond lengths with those found in other hydrogen-bonded solids [*19*], as shown in Fig. 3, one sees that the hydrogen bond is a relatively short and asymmetric bond that falls nicely on the empirical curve fitted to the earlier data on other systems.

In a neutron diffraction investigation, Samuelsen and Semmingsen [*20,21*] studied the temperature dependence of three superlattice reflections in squaric acid (Fig. 4). They confirmed the continuous nature of the phase transition discovered by birefringence investigations and found that the Bragg intensities of superlattice reflections have the form

$$I(h,k,l) = C(h,k,l)[(T_c - T)/T_c]^{2\beta} \tag{1}$$

with

$$\beta = 0.137 \pm 0.010 \tag{2}$$

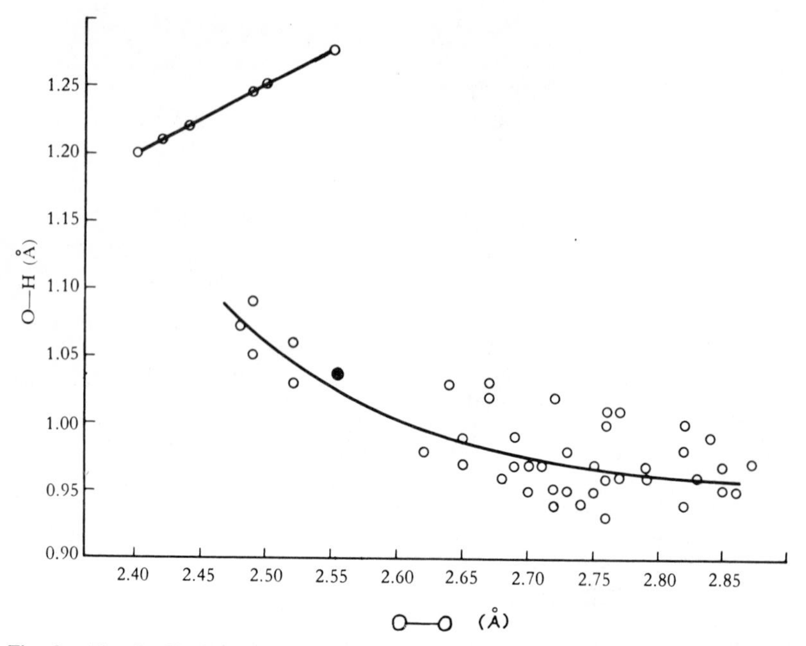

Fig. 3. The O—H distance as a function of the O···O distance as determined by neutron diffraction for a number of compounds containing O—H···O hydrogen bonds (open circles) after Hamilton and Ibers [*19*]. The curved line represents the best least-squares fit to the points, but the deviation from the line is significant in many cases. The closed circle represents the result for squaric acid at room temperature.

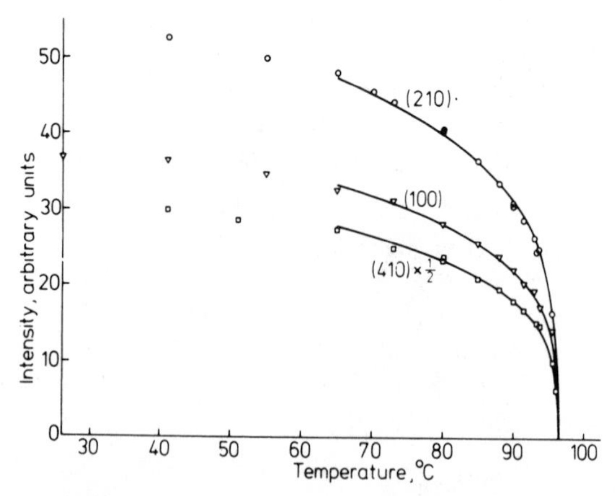

Fig. 4. Temperature dependence of the intensity at superlattice reflections in squaric acid obtained by neutron diffraction. From Samuelsen and Semmingsen [*20,21*].

Here β is the critical index of the order parameter, or the sublattice polarization. Such a low value of the critical index is characteristic of a two-dimensional phase transition. The observed value of β is very close to the theoretical value $\beta = 1/8$ of the two-dimensional Ising model. On the basis of thermal diffuse scattering, Samuelsen and Semmingsen concluded that the individual molecules retain their low-temperature shape even at temperatures well above the transition temperature. The correlation length of ordered molecules, i.e., the "dimension" of low-temperature domains or clusters, in the high-temperature phase is of the order of 25 lattice constants in the plane, whereas the planes are correlated only over three to five layers.

On the basis of these structure determinations, one concludes that the phase transition in squaric acid is essentially two-dimensional with weak correlation between planes. Although the phase transition is continuous and the high-temperature phase is tetragonal, clusters of the low-temperature phase persist as correlated domains in the high-temperature phase. The crystal has tetragonal symmetry only when averaged over volumes much bigger than the clusters and/or over sufficiently long times.

III. Raman and Infrared Spectra

The first infrared spectra of squaric acid were taken by Cohen *et al.* [*1*], who found broad absorption bands at 2330, 1820, and 1640 cm^{-1}, which they ascribed to hydrogen bonding, carbonyl groups, and C=C bonds, respectively, Baglin and Rose [*22*] collected both Raman and infrared spectra of squaric acid, but their assignment of vibrational modes was incomplete since the crystal structure and the phase transition of squaric acid were unknown at the time. Nevertheless, they correctly concluded that squaric acid has an asymmetric hydrogen bond and estimated the O···O distance to be 2.49 ± 0.05 Å, in agreement with later structure determinations. Nakashima and Balkanski [*23*] studied the Raman spectra of squaric acid both above and below the phase transition, and their spectra are shown in Fig. 5.

The modes below 300 cm^{-1} are external modes in which the molecule moves as an almost rigid unit. The region above 800 cm^{-1} represents motion of hydrogen and/or carbon and oxygen atoms. The intermediate region 300–800 cm^{-1} was ascribed to internal molecular vibrations.

The most striking temperature dependence is that of the 149 cm^{-1} band, which splits into three bands at 77 K. The 153 and 158 cm^{-1} pair is thought to be the external rotary B_g modes, with the rotary axis in the basal plane, whereas the 164 cm^{-1} line was attributed to the translational B_g mode in which the C_4O_4 unit moves normal to the basal plane. The doublet at 89 and 90 cm^{-1} was taken to

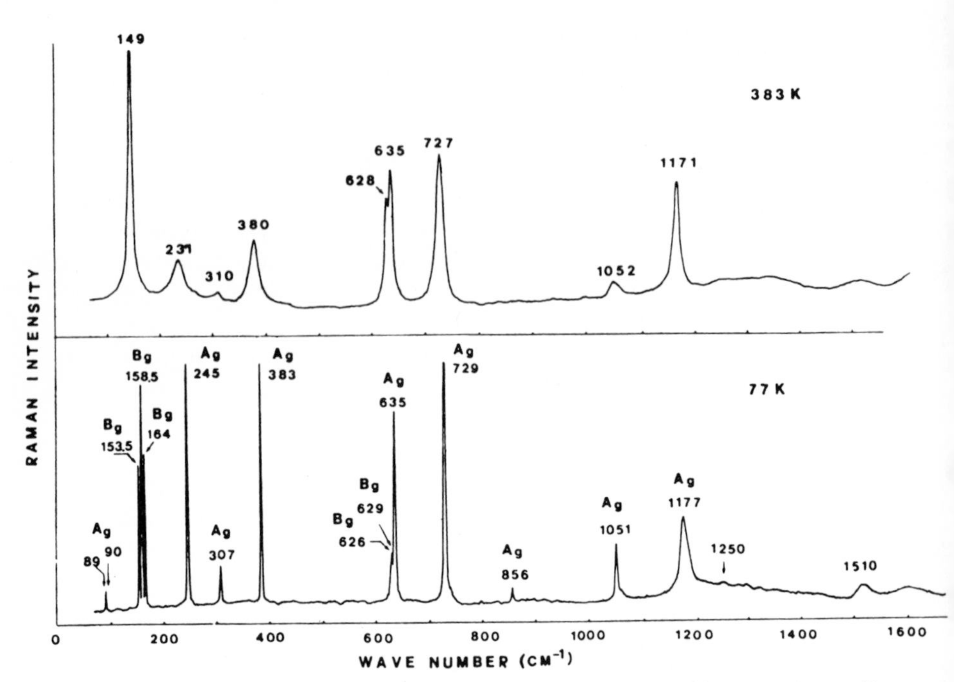

Fig. 5. Unpolarized Raman spectra below and above the transition temperature. From Nakashima and Balkanski [*23*].

represent the shear, rigid, layerlike modes of B_g symmetry. None of the observed modes exhibited phonon softening at the phase transition.

Bougeard and Novak [*24*] reexamined the Raman and infrared spectra of squaric acid and deuteriated squaric acid. They reassigned the C=O bending mode so that the Raman bands at 306 and 379 cm^{-1} belong to A_g vibrations and the infrared counterparts at 307 and 383 cm^{-1} correspond to the B_u species. The OH stretching fundamental is assigned to the 1320 cm^{-1} infrared absorption band. This OH band changes very little upon deuteriation, with $\nu OH/\nu OD = 0.96$. Comparing the high- and low-temperature spectra, they concluded that all the intermolecular vibrations of the low-temperature phase persist in the high-temperature spectra at similar frequencies. This led to the conclusion that the molecular symmetry of $C_4O_4H_2$ is practically unchanged by the phase transition. This was also pointed out by Thackeray *et al.* [*25*].

Samuelsen, Fjaer, and Semmingsen [*26*] made a detailed study of the temperature variation of the 83 cm^{-1} Raman Line. The integrated intensity of this line as a function of temperature is shown in Fig. 6. The curve fitted to the data has the form

$$I \sim [(T_c - T)/T_c]^{2\beta} \tag{3}$$

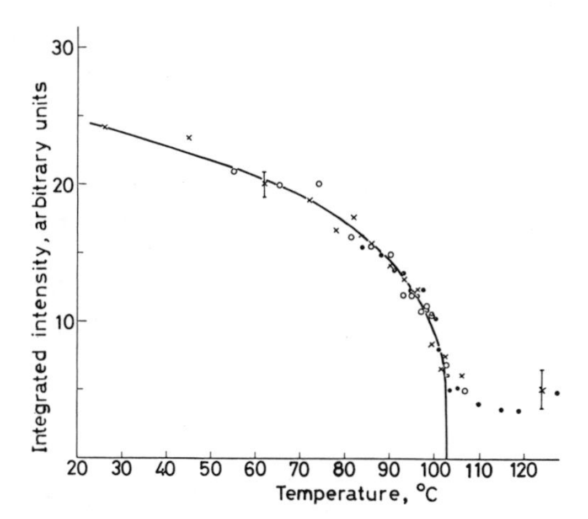

Fig. 6. Temperature dependence of the integrated intensity of the 83 cm^{-1} line in the Raman spectrum of squaric acid. From Samuelsen, Fjaer, and Semmingsen [*26*].

where for β the experimental value 0.137 [*20,21*] of the order parameter critical index has been used. These authors argued that the integrated intensity in fact should behave as the order parameter square, a point that is nicely brought out in Fig. 6. The fact that the 83 cm^{-1} line persists into the high-temperature phase is taken as evidence for the nontetragonal symmetry of the squaric acid molecule in the high-temperature phase.

In conclusion, one may say that spectroscopic investigations have not revealed any soft mode associated with the phase transition. The spectroscopic evidence is that the molecules retain their low-temperature symmetry above the transition temperature, and the temperature dependence of the order parameter is reflected only in the intensity variations of various lines.

IV. NMR Spectra

The proton chemical shift tensor in squaric acid was determined by Suwelack *et al.* [*27*] using multiple-pulse NMR. The chemical shift tensor $\boldsymbol{\sigma}$ is defined in terms of the effective field at the nuclear site H' and the applied field H_0 by the relation $H' = (1 - \boldsymbol{\sigma}) \cdot H_0$. The principal values of the shift tensor are $\sigma_{11} = -26.5$, $\sigma_{22} = -20.2$, and $\sigma_{33} = 1.0$ ppm with respect to tetramethylsilane. The most shielded axis (3 axis) is parallel with the hydrogen bond, and the least shielded axis (1 axis) is perpendicular to the molecular plane. No change was found in the chemical shift tensor in going to the high-temperature phase. Therefore, there is no change in the electronic environment of the hydrogen bond at the phase transition.

Investigations of the ^{13}C chemical shift tensor in squaric acid were done by Becker, Suwelack, and Mehring [*28*] as a function of temperature. As seen in Fig. 7, the chemical shifts for the four carbon atoms on the squaric acid molecule change with temperature so that the line pair splittings decrease, and only two lines are found above T_c. Since the resonance lines have constant line width, ~60 Hz, and amplitude, these authors concluded that the molecular wavefunction can be represented as a superposition of two pseudo spin states representing the hydrogen positions. This leads to the conclusion that the line splitting is proportional to the molecular dipole movement; thus, the line splitting is proportional to the order parameter and

$$\Delta\sigma = \left(\frac{T_c - T}{T_c}\right)^{\beta} \Delta\sigma_0 \tag{4}$$

From the observed line splittings, they found $\beta = 0.14$, consistent with the β value combined by neutron diffraction [*20,21*].

Recently, Mehring and Suwelack [*29*] published high-resolution ^{13}C-NMR spectra of squaric acid (Fig. 8). In contrast with the earlier work they found that the line splittings do not vanish above the transition temperature. Instead there remains a splitting of about 54% of the room-temperature value even above T_c.

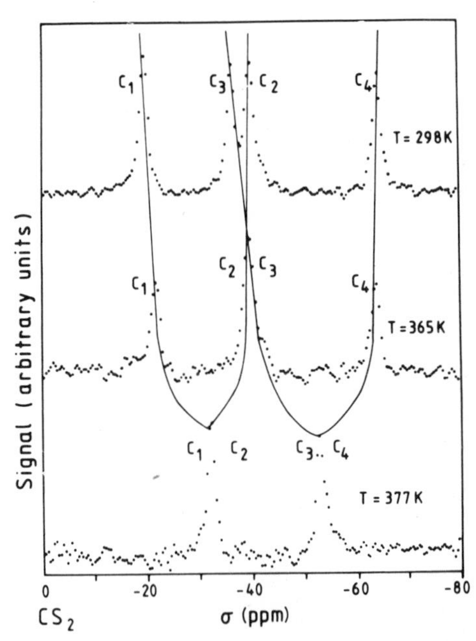

Fig. 7. ^{13}C-NMR spectra of squaric acid at three different temperatures for an arbitrary orientation of the magnetic field. The four spectral lines corresponding to four different carbon nuclei in one molecule are labeled C_1, C_2, C_3, and C_4. From Suwelack, Becker, and Mehring [*27*].

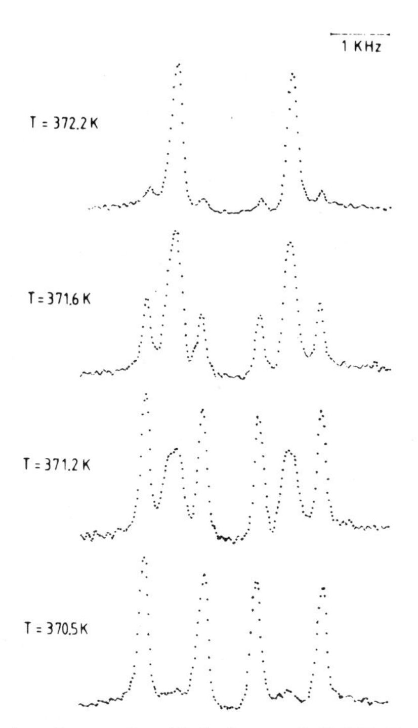

Fig. 8. ^{13}C-NMR spectra of a squaric acid single crystal (0.03% Cr) near the phase transition with arbitrary orientation of the magnetic field, for temperatures below and above the transition temperature. From Mehring and Suwelack [*29*].

There is a transfer in intensity from the doublet characteristic of the low-temperature structure to a central line in the middle of each pair. This central line is taken to correspond to molecules in the high-temperature phase. The transfer of intensity from the doublet to the central line occurs over a narrow temperature range, as seen in Fig. 9. The transition region is interpreted as representing the coexistence of clusters of molecules in the high-temperature phase in the low-temperature structure below T_c, and vice versa above T_c. It should be noted that this interpretation assumes an unspecified form of the squaric acid molecule in the high-temperature phase, in disagreement with the conclusion based on the neutron and Raman scattering data that molecules retain their low-temperature shape even above T_c. There is agreement, however, in the sense that all the experimental evidence discussed so far indicates the presence of clusters with the low-temperature structure in the high-temperature phase. Let us also remark that, since NMR is an entirely local probe, only the local order parameter can be measured, and the strictly long-range macroscopic order parameter cannot be measured directly with local methods.

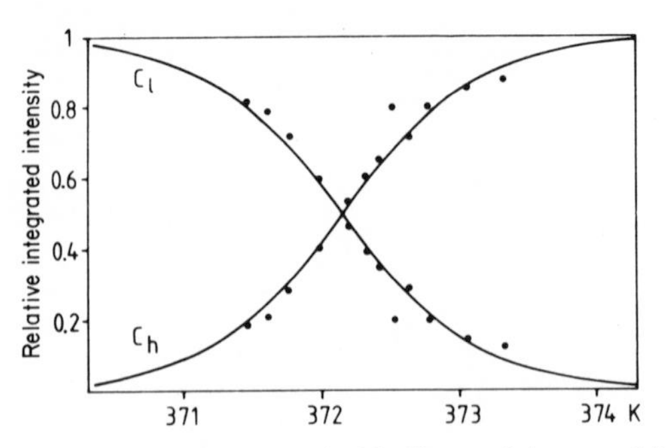

Fig. 9. Relative integrated intensity of the doublet lines and the central line in Fig. 8. The C_l corresponds to low-temperature clusters; C_h corresponds to high-temperature clusters. From Mehring and Suwelack [29].

A very interesting result obtained by Mehring and Suwelack is the variation of the transition temperature in squaric acid with impurity concentration. As little as 0.03% Cr decreases the transition temperature by 1.9°C, and the transition region increases in width.

V. Birefringence, Dielectric, and Elastic Properties

The first study of the phase transition in squaric acid [12] was done by measuring the birefringence obtained with light propagating perpendicular to the basal plane, and in Fig. 10 the observed birefringence in squaric acid is plotted as a function of reduced temperature $\tau = (T_c - T)/T_c$. Over more than two decades in τ, we find that the critical behavior of the birefringence is given by

$$\Delta n \sim \tau^{2\tilde{\beta}} \tag{5}$$

with the values

$$\tilde{\beta}_H = 0.17 \pm 0.01 \qquad \tilde{\beta}_D = 0.18 \pm 0.02 \tag{6}$$

for squaric acid and the deuteriated compounds, respectively.

Birefringence measures the anisotropies in the dielectric constant at optical frequencies. The dielectric tensor in the low-temperature phase is ϵ_{ij}, where $i, j = 1, 2, 3$ label the cartesian coordinates. The high-temperature dielectric tensor is ϵ_{ij}^0. The inverse dielectric tensor $\epsilon_{ij}^{-1} = (\boldsymbol{\epsilon}^{-1})_{ij}$ may be expanded in terms of the local order parameter $\boldsymbol{\eta}(r) = (\eta_1(r), \eta_2(r))$, which is proportional to the local dipole moment,

$$\epsilon_{ij}^{-1} = (\epsilon_{ij}^0)^{-1} + g_{ijkl}\langle \eta_k \eta_l \rangle \tag{7}$$

with summation over k and l. There the term linear in η vanishes by symmetry, and we neglect higher-order terms. Both $(\epsilon_{ij}^0)^{-1}$ and g_{ijkl} must satisfy the $4/m$

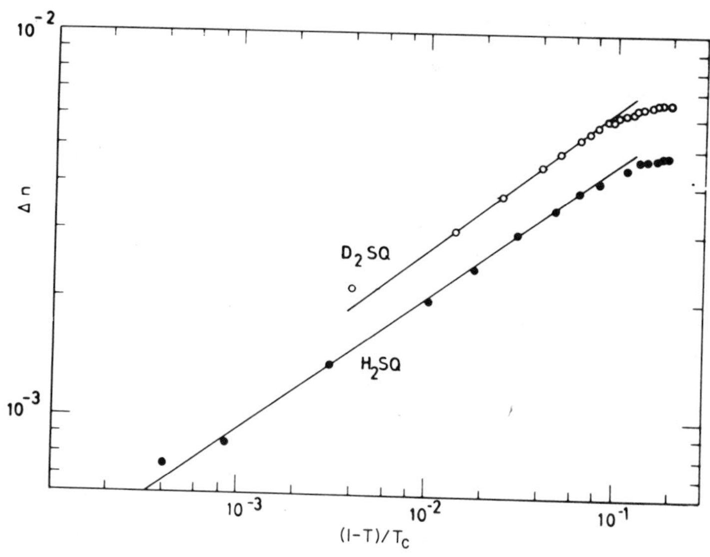

Fig. 10. Birefringence of squaric acid as a function of reduced temperature in the low-temperature phase. $\Delta n = \alpha[(1 - T)/T_c]^{2\tilde{\beta}}$; $T_{CH} = 98°C$; $\tilde{\beta}_H = 0.17 \pm 0.01$; $T_{CD} = 243°C$; $\tilde{\beta}_D = 0.18 \pm 0.02$. From Semmingsen and Feder [12].

point symmetry of the high-temperature phase, and consequently most of the tensor components vanish or are related by symmetry. On the other hand, ϵ_{ij}^{-1} and ϵ_{ij} have only the $2/m$ symmetry of the low-temperature monoclinic phase. Taking these symmetry restrictions into account, it follows that the birefringence measured by propagating light along the unique axis is given by

$$\Delta n \sim g_1\langle \eta_1^2 - \eta_2^2 \rangle + g_2\langle \eta_1 \eta_2 \rangle \tag{8}$$

so that by a convenient choice of axes one obtains $\Delta n \sim \langle \eta^2 \rangle$ [13]. Here g_1 and g_2 are coupling constants related to g_{ijkl}. The angular brackets represent spatial averages of the local order parameter over the volume of the crystal probed in the experiment, as discussed by Courtens [30]. The expression for birefringence [Eq. (8)] is consistent with the general form suggested by Gehring [31], but it neglects nonlocal terms of the form $\eta(0)\eta(r)$. Gehring [31] concluded that the birefringence should exhibit a critical behavior given by

$$\Delta n \sim \tau^{2\tilde{\beta}} \tag{9}$$

where the critical index $\tilde{\beta}$ is given by the relationship

$$2\tilde{\beta} = 2 - \alpha - \phi = 2\beta + \gamma - \phi \tag{10}$$

Here α, β, and γ are the critical indices of the specific heat, the order parameter, and the order paremeter susceptibility, respectively, whereas ϕ is the crossover exponent from Heisenberg to Ising behavior. There is however, an extra term of the form, $\langle \eta^2 \rangle \sim \tau^{2\beta}$ that should be included in Eq. (9), as discussed in Section VI. Thus, birefringence does not measure the order parameter critical index β directly, but also $\tilde{\beta}$, which in general is larger than β, since $\gamma - \phi$ is larger than zero.

The low-frequency dielectric tensor also couples to the order parameter, as given by Eq. (7), but instead of taking local values, $\boldsymbol{\epsilon}$ and $\langle \eta_k \eta_l \rangle$ are for wave-number q $\simeq$ 0, and frequency $\omega \simeq 0$. Preliminary dielectric measurements at 1 kHz were reported earlier [*13*], and a marked anomaly in the dielectric constant was found. In Fig. 11, we present new high-resolution 10-kHz dielectric constant measurements as functions of temperature for a (100) plate [(010) is the unique axis] of squaric acid [*32*]. The dielectric constant increases rapidly with temperature, and slightly above the transition temperature there is a maximum in the dielectric constant ϵ_{ac} in the basal plane. The maximal value varies somewhat from sample to sample but is generally found in the range 680–720. Similar values were obtained by Yasuda *et al.* [*33*] and Maier, Müller, and Petersson [*34*]. The observed maximal value is about twice the preliminary value reported [*13*], which was obtained in a geometry suitable for crystals that cleave easily but for which the proper geometric scale factor can only be estimated.

We argued earlier [*13*] that in analogy with antiferromagnets one expects the transition temperature to occur at the point of maximal slope of ϵ. In fact one expects [*13*] the specific heat to be given by $c \sim d(T\epsilon)/dT$, so that the critical part of the dielectric constant behaves as $c(T) - C_0 \sim \tau^{1-\alpha}$. In order to obtain the critical behavior of the dielectric constant ϵ_{ac}, we plot the reduced dielectric constant or capacitance versus the reduced temperature in the log–log

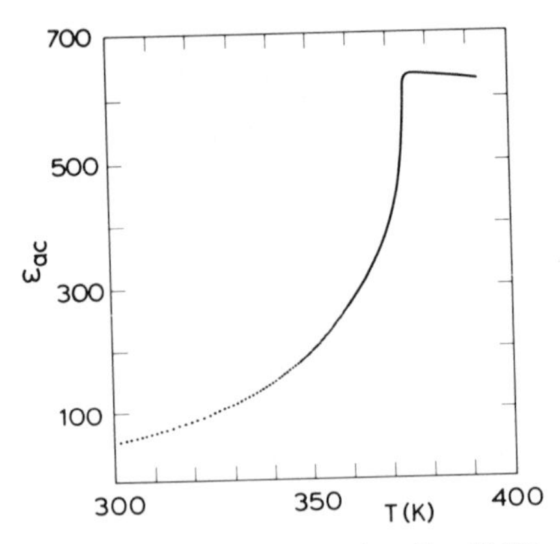

Fig. 11. Dielectric constant of a (100) plate of squaric acid at 10 kHz as a function of temperature. The sample is not a monodomain sample so the dielectric constant is some average of the dielectric constant in the *a* and *c* directions.

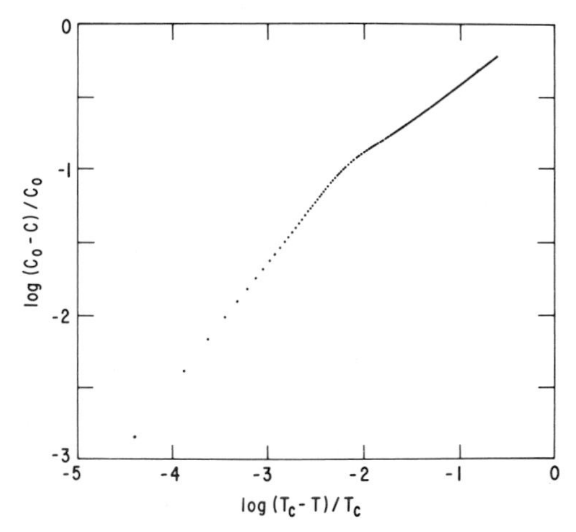

Fig. 12. Relative change of the dielectric constant of a (101) plate of squaric acid at 100 kHz as a function of reduced temperature in the low-temperature phase. The transition temperature T_c = 100.265°C has been chosen so that the slope near T_c is the same above and below T_c.

plot of Fig. 12 for a (101) plate. The detailed form of the curve in Fig. 12 near T_c depends strongly on the choice of T_c and C_0. These parameters have been chosen in such a way that the slopes of the curve just above and just below T_c coincide as expected from scaling theories. This choice of T_c (100.265°C for this sample) coincides with the point of maximum $d\epsilon/dT$.

Keeping these remarks in mind, one sees that there is a marked change in slope as T_c is approached. By a least-squares fit, we find that the effective critical index α is

$$\alpha = \alpha(2D) = 0.52 \qquad -2 \leqslant \log t \tag{11}$$

and in the critical region

$$\alpha = \alpha(3D) = 0.08 \qquad \log t \leqslant -2.5 \tag{12}$$

We interpret our result to be the first observation of crossover from two-dimensional to three-dimensional behavior in squaric acid.

We take $\alpha(3D)$ to represent the correct critical index valid in the critical region proper of the three-dimensional crystal, whereas $\alpha(2D)$ is the effective value of α farther away ($\tau \gtrsim 5 \times 10^{-3}$) from the critical point where there is little correlation between the planes, and the crystal behaves as if it were in the wider critical region of a two-dimensional phase transition.

As discussed in Section VI, squaric acid is expected to behave as a three-dimensional *XY* model in the critical region. The critical indices for this model are known, and one finds that

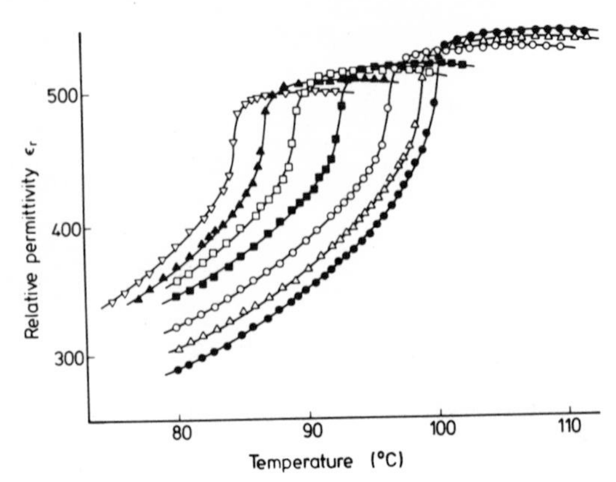

Fig. 13. Temperature dependence of the relative permittivity along the *a* (*c*) axis under various constant pressures. Key: ●, 0; △, 0.4; ○, 0.6; ■, 0.8; □, 1.0; ▲, 1.2; ▽, 1.4 kbar. From Yasuda *et al.* [33].

$$\alpha_{XY}(3D) = -0.08 \tag{13}$$

consistent with our observed $\alpha(3D)$. For other samples and frequencies we find $\alpha(3D)$ values in the range 0.12–0.08. These results will be reported in detail elsewhere [32].

Yasuda *et al.* [33] measured the pressure dependence of the dielectric constant in squaric acid (Fig. 13). The transition temperature decreases strongly with increasing pressure, $-dT_c/dp = 13°C/kbar$. They also measured the thermal expansion as a function of temperature and pressure. Their data in Fig. 14 show anomalies similar to those seen in the dielectric constant. This is to be expected since the strain tensor e_{ij} will couple to the order parameter just as the dielectric constant does, and one may simply replace ϵ_{ij} in Eq. (7) by e_{ij} and change the coupling coefficients accordingly.

The pressure dependence of the dielectric properties of squaric acid and deuterated squaric acid were investigated by Samara and Semmingsen [35]. For

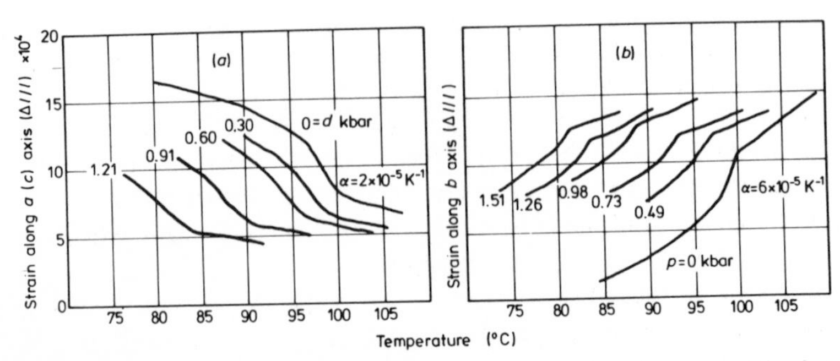

Fig. 14. The temperature dependence of the strain (a) along the *a* (*c*) axis and (b) along the *b* axis under various constant pressures. From Yasuda *et al.* [33].

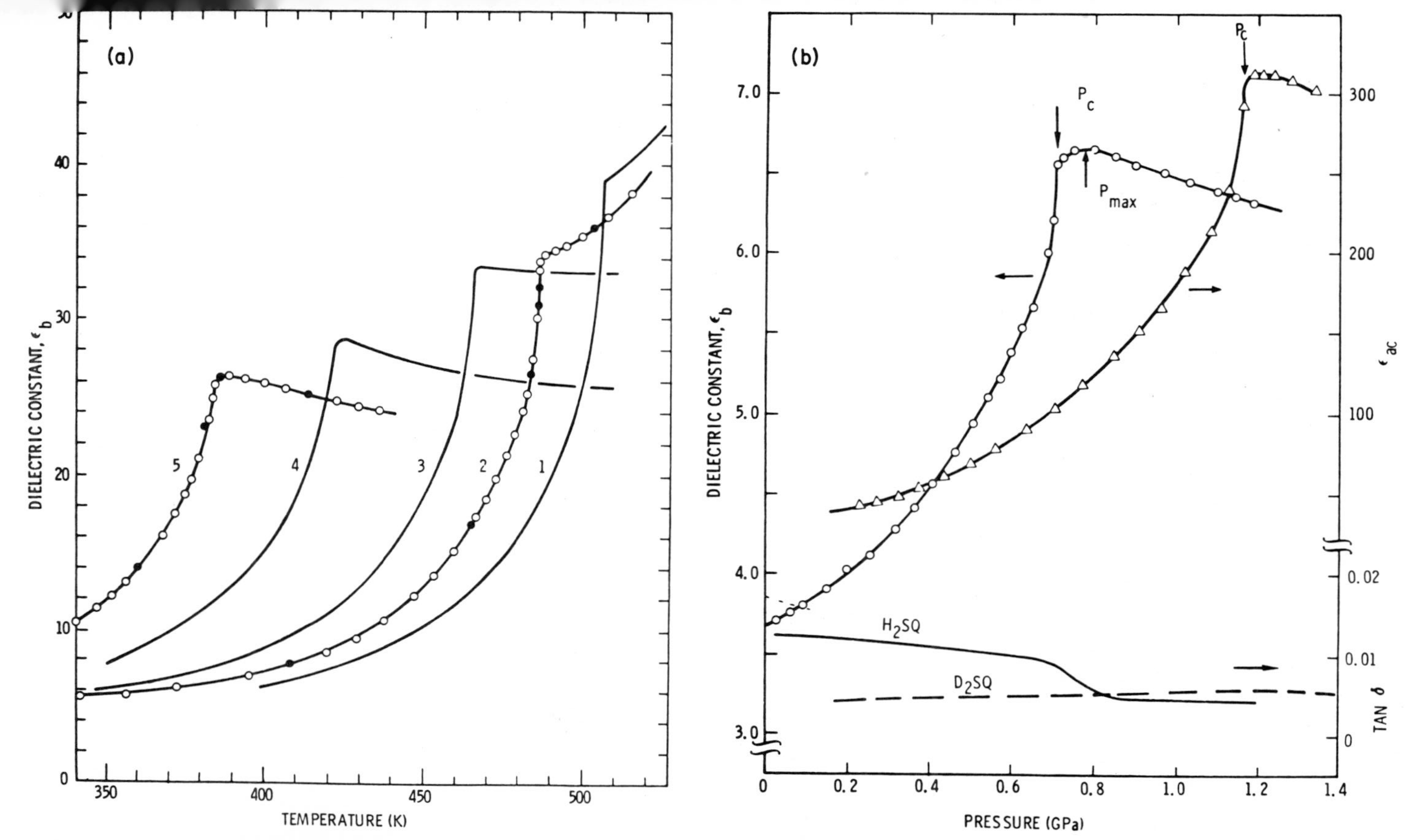

Fig. 15. (a) Temperature dependence of the dielectric constant ϵ_b along the unique axis of deuterated squaric acid. Key: ○, heating; ●, cooling; curve 1, 0.20 GPa; 2, 0.40 GPa; 3, 0.60 GPa; 4, 1.20 GPa; 5, 180 GPa. (b) Pressure dependence of the dielectric constant ϵ_b for squaric acid and deuterated squaric acid. The dielectric loss given as tan δ is also given as a function of pressure. Key: —○—, H_2SQ, 300 K, 10 kHz; —△—, D_2SQ, 422 K, 100 kHz. From Samara and Semmingsen [*35*].

Fig. 15a, the dielectric constant ϵ_b, measured perpendicular to the basal plane, is shown as a function of temperature for deuterated squaric acid. In Fig. 15b, ϵ_e as a function of pressure exhibits the phase transition at constant temperature both for squaric acid and the deuteriated compound. As seen in Fig. 16, the transition temperature decreases dramatically with pressure, with initial slopes of 106 and 103 K/GPa for $C_4O_4H_2$ and $D_2C_2O_4$, respectively. The remarkable feature of this result is that there is no isotope effect on dT_c/dp. Samara and Semmingsen argued that this means that proton tunneling is unimportant and that the large isotope effect in the transition temperature can be understood in terms of simple bond length considerations. Thus, from the observed compressibility [*33*] and the pressure dependence of T_c, one finds a pressure that reduces the O—H···O bond length by 0.020 Å from the observed value in $D_2C_4O_4$ to the value in $C_4O_4H_2$ (see Table I) and also reduces the transition temperature from $T_{c,D}$ to $T_{c,H}$.

Maier *et al.* [*34*] studied the frequency dependence of the dielectric constant of squaric acid as a function of temperature. They were able to deduce the dc conductivity from the data, and Fig. 17 shows that above T_c the conductivity has an activation energy of 1.0 eV, whereas below T_c the conductivity varies with the order parameter and has an average activation energy of 1.3 eV. They

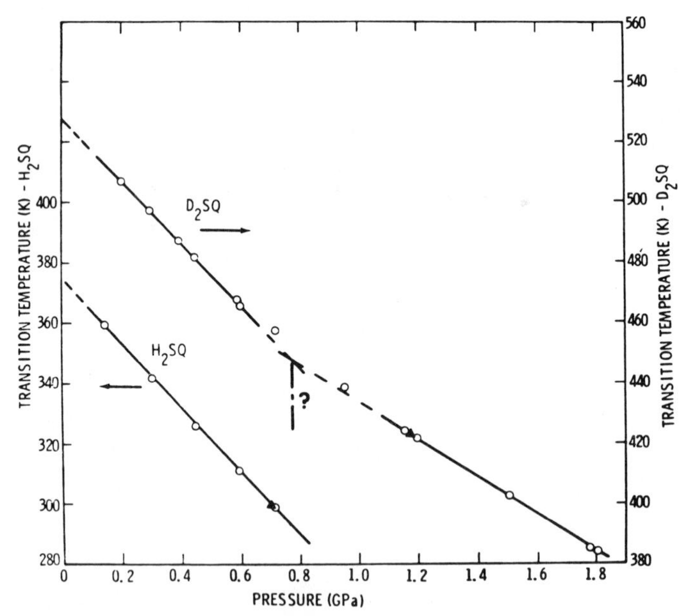

Fig. 16. Pressure dependence of the transition temperature of squaric acid and deuterated squaric acid. Key: ○, from $\epsilon(T)$ isobars; ▲, from $\epsilon(P)$ isotherm. From Samara and Semmingsen [*35*].

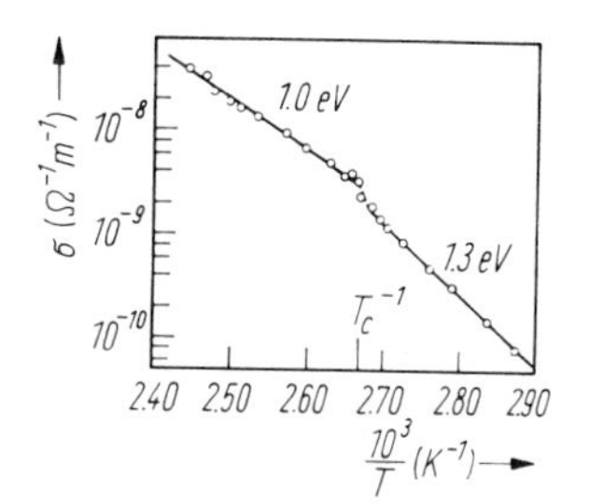

Fig. 17. Arrhenius plot of the dc conductivity of squaric acid. From Maier, Müller, and Petersson [*34*].

interpreted this activation energy as representing diffusional motion, and at 100°C the relaxational frequency is 0.7 Hz. By analyzing their dielectric constants near T_c, they found a critical exponent of $\tilde{\beta} = 0.20$, consistent with our $\tilde{\beta}$(2D).

Rehwald [*36,37*] measured the elastic constants of squaric acid as a function of temperature. As seen in Fig. 18, only C'_{11} and C'_{66} exhibit a strong anomaly near T_c. Rehwald and Vonlanthen [*37*] suggested that the stiffness function $(C_{11} - C'_{12})/2$, C'_{66}, and C'_{16} have a critical behavior given by

$$\Delta C_b(t) \sim t^{2\tilde{\beta}} \tag{14}$$

and they found by fitting the observed temperature dependence that $\tilde{\beta} = 0.27 + 0.03$, which is consistent with our $\tilde{\beta}$(2D).

In conclusion we may say that birefringence, dielectric, and elastic experiments all show anomalies in the critical region controlled by α, β, and the fluctuation exponent $\tilde{\beta}$. High-resolution dielectric measurements show that α changes

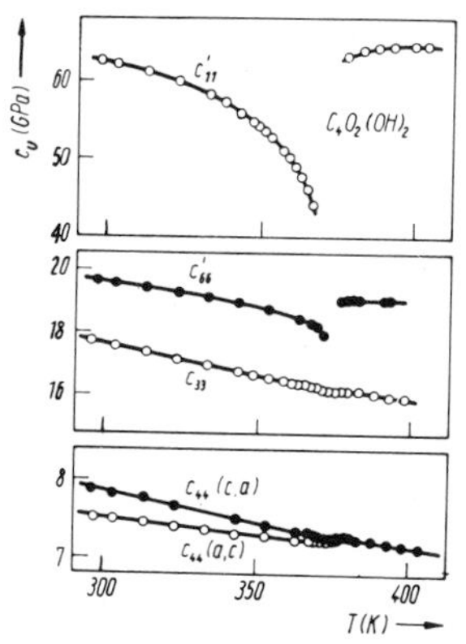

Fig. 18. Temperature dependence of the elastic functions (notation for tetragonal crystals is used). From Rehwald [*36*].

from its two-dimensional value of about 0.52 to a value of 0.08 in the three-dimensional regime.

VI. Theory

The theoretical understanding of the phase transition in squaric acid has been evolving since the first observations of the transition. Clearly, the hydrogen bonds are instrumental in the phase transition as evidenced by the large isotope effect, and most theoretical models that have been proposed concentrate on the positions and motions of the hydrogens in the structure.

It has become clear through scaling, and later renormalization group theories of phase transitions that critical behavior is controlled by the dimensionality of the space and the symmetry of the order parameter.

The experimental evidence is that in squaric acid there is little interaction between layers and the phase transition is essentially two-dimensional, in the sense that the correlations are long range in the plane but short range between planes. Very near the transition point, one expects to see a crossover to three-dimensional behavior of the type we have found in the dielectric constant.

The fundamental step in any theory of a phase transition is the identification of the order parameter describing the essential physical quantity that changes by going through the phase transition. Some order parameters, such as the magnetization for a ferromagnetic phase transition or the polarization for a ferroelectric transition, are well known and easily understood. Other cases, such as the order parameter associated with the superconducting phase transition or the superfluid transition in helium, do not lend themselves to an intuitive interpretation and can be fully understood only within the quantum theory of superfluid transitions.

For squaric acid the order parameter is reasonably simple and can be described as the polarization in the molecular planes caused by the hydrogen displacements. The connection between the two components of a polarization in the plane and the displacement of hydrogen along linear bonds can best be appreciated by constructing a model Hamiltonian for squaric acid in terms of localized normal coordinates.

There are two hydrogens per C_4O_4 unit, and thus we may describe the hydrogen positions by giving their displacement $U_k(l)$ from the center of the kth ($k = 1, 2$) bond of the lth C_4O_4 unit. The model Hamiltonian for the hydrogen, then, is the sum of the kinetic and potential energies,

$$H_0 = \frac{1}{2} \sum_{kl} \left\{ \frac{1}{2m} p_k(l)^2 + V[U_k(l)] \right\} \tag{15}$$

where m is the hydrogen mass and $p_k(l)$ the momentum of the hydrogen with coordinate $U_k(l)$. The potential energy for a hydrogen may for definiteness be taken to have the form

$$V[U_k(l)] = \frac{1}{4} AU_k(l)^2 + \frac{1}{4} BU_k(l)^4 \tag{16}$$

with the model parameters chosen to satisfy $A < 0$, $B > 0$; the potential V has the characteristic double-well form.

The Hamiltonian H_0 describes a set of uncoupled anharmonic oscillators. In order to obtain a model Hamiltonian useful in describing the phase transition it is necessary to introduce coupling between the hydrogens. This coupling must have the proper symmetry and can best be introduced using local normal coordinates [*13*]. We have used local normal coordinates $\mathbf{R}(l) = (R_1(l),R_2(l))$, as shown in Fig. 19. If we let $\mathbf{R}(l) = (\delta,0)$, the hydrogens surrounding the lth C_4O_4 unit are displaced an amount proportional to δ in the directions shown in Fig. 19a, whereas $\mathbf{R}(l) = (0,\delta)$ produces displacements as shown in Fig. 19b. Since the hydrogens labeled 1, 2, 3, and 4 are shared between neighboring molecules, the actual hydrogen displacements are given by

$$\begin{aligned}
\mathbf{U}_1(l) &= [(R_1 + R_2)_l + (R_1 + R_2)_{l+n_1}] \quad (11)\\
\mathbf{U}_2(l) &= [(R_1 - R_2)_l + (R_1 - R_2)_{l+n_2}] \quad (11)\\
\mathbf{U}_3(l) &= [(R_1 + R_2)_l + (R_1 + R_2)_{l+n_3}] \quad (11)\\
\mathbf{U}_4(l) &= [(R_1 - R_2)_l + (R_1 - R_2)_{l+n_4}] \quad (11)
\end{aligned} \tag{17}$$

where $\mathbf{n}_1 = (10) = -\mathbf{n}_3$; $\mathbf{n}_2 = (01) = -\mathbf{n}_4$ are the nearest-neighbour vectors in the plane in units of the lattice constant a.

In terms of the local normal coordinates $\mathbf{R}(l)$, the coupling between hydrogens, takes the simple form

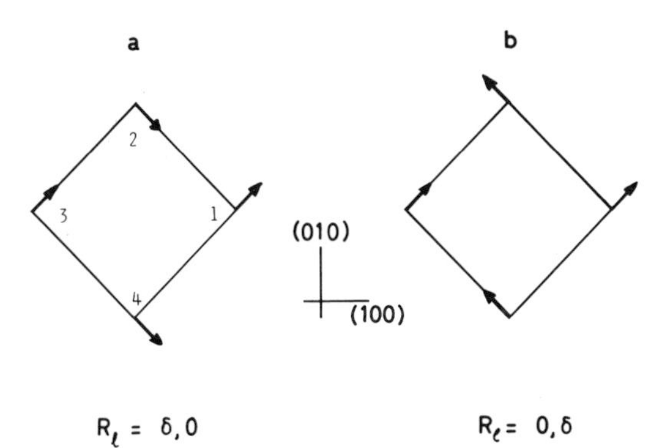

Fig. 19. Hydrogen displacements in terms of the local normal mode coordinate $R(l)$. The corners of the square lie on hydrogen bond centers, whereas the center of the square coincides with the molecular center. From Feder [*13*].

$$H_{\text{int}} = \frac{1}{2} \sum_{ll'} \mathbf{R}(l) \cdot \mathbf{C}'(l - l') \cdot \mathbf{R}(l') \tag{18}$$

The dynamic matrix $\mathbf{C}'$ couples hydrogens not only belonging to different C_4O_4 units in the plane, but also between planes.

If now H_0 is also rewritten in terms of the local normal coordinates using the relation given in Eq. (17), the model Hamiltonian $H = H_0 + H_{\text{int}}$ may be written in the form

$$\begin{aligned} H = \frac{1}{2} \sum_{l} P(l)^2 &+ \frac{1}{2} \sum_{ll'} \mathbf{R}(l) \cdot \mathbf{C}(l - l') \cdot \mathbf{R}(l') \\ &+ U \sum_{l} [R_1^4(l) + R_2^4(l)] \\ &+ V \sum_{l} R_1^2(l) R_2^2(l) \\ &+ W \sum_{l} R_1(l) R_2(l) [R_1^2(l) - R_2^2(l)] \end{aligned} \tag{19}$$

In rewriting H_0 in terms of the R's, we obtain a number of quadratic terms that have been collected in the new dynamic matrix $\mathbf{C}$, which also contains the $\mathbf{C}'$ from Eq. (18). The anharmonic terms in Eq. (19) are related to the B parameter in Eq. (16). We have, however, neglected nonlocal anharmonic terms such as $R_1^2(l) R_2^2(l')$ since it is known that these terms are irrelevant [*47*]. The dynamic matrix $\mathbf{C}$ contains only a few independent model parameters since it must satisfy the $I4/m$ symmetry. The layer anisotropy of squaric acid is contained in $\mathbf{C}$. By symmetry there are no cubic terms, and by limiting the expansion to single ion terms there are only three quartic terms. Also the W term may be eliminated by applying an appropriate rotation in the R_1,R_2 plane.

Since the C_4O_4 unit is not rigid, there are both nuclear and electronic displacements associated with changes in hydrogen positions. Any deformation that belongs to the E representation of the C_4 point group will couple linearly to the hydrogen displacement order parameter $\mathbf{R}$, and thus the order parameter proper $\boldsymbol{\eta}$ is a linear combination of such displacements. The molecular polarization $\mathbf{p}$, for instance, lies in the basal plane and transforms as $\mathbf{R}$; therefore, $\mathbf{p}$ is an aspect of the order parameter and the planes are ferroelectric. A Hamiltonian for squaric acid using $\boldsymbol{\eta}$ instead of $\mathbf{R}$ has the form given in Eq. (19), and the critical behavior may be derived from the Hamiltonian given.

Renormalization group theory leads to the conclusion that as the critical point is approached the quartic anisotropies become irrelevant and the critical behavior of Eq. (19) is that of a two-component three-dimensional Heisenberg model. The phase transition in squaric acid is therefore that of a three-dimensional XY model. For this model the critical indices are

$$\beta = 0.36 \qquad \gamma = 1.30 \qquad \phi = 1.16 \tag{20}$$

We considered the effect of applying an electric field to the crystal (13) using Landau theory and found that the transition temperature should decrease as

$$T_c(E^2) - T_c(0) \sim -E^2 \tag{21}$$

Mukamel [*38,39*] considered the phase diagram for a Hamiltonian of the form (19) with the presence of fields and found that there is a multicritical point at $E = 0$, and that the transition temperature should increase as

$$T_c(E^2) - T_c(0) \sim E^{2/\phi} \tag{22}$$

when ϕ is the crossover exponent.

In our experiments we have seen only the decrease in T_c predicted by Landau theory. We have also been unable to apply fields strong enough to reach the tricritical point.

The critical behavior of the birefringence and related properties may easily be derived by scaling arguments. If a perturbation H', given by

$$H' = g_1 \sum_l [R_1^2(l) - R_2^2(l)] + g_2 \sum_l R_1(l)R_2(l) \tag{23}$$

is applied to the Hamiltonian (19), the free energy will become a function of g_1 and g_2. Using scaling the form must be as follows:

$$\begin{aligned} \frac{1}{N} F(\tau, g_1, g_2) &= \tau^{2-\alpha} \mathscr{F}(g_1\tau^{-\phi_1}, g_2\tau^{-\phi_1}) \\ &\quad + g_1(\langle R_1(l)\rangle^2 - \langle R_2(l)\rangle^2) + g_2\langle R_1(l)\rangle\langle R_2(l)\rangle \end{aligned} \tag{24}$$

where α is the specific heat index, and $\phi_1, \phi_2 = \phi$ is the crossover exponent for *XY* to Ising behavior. From Eq. (24) one finds that the fluctuations have critical behaviors given by

$$\langle R_1^2(l) - R_2^2(l)\rangle = \frac{1}{N}\frac{\partial F}{\partial g_1}\bigg|_0 = \tau^{2-\alpha-\phi}\mathscr{F}_0' + (\langle R_1(l)\rangle^2 - \langle R_2(l)\rangle^2) \tag{25}$$

Here N is the number of unit cells in the crystal. The critical behavior of the birefringence is then obtained by using (20,21) for $\langle \eta_1^2 - \eta_2^2\rangle$ and $\langle \eta_1\eta_2\rangle$ in Eq. (8). Note that the last terms of Eqs. (25 and 26), have a critical behaviour given by $\tau^{2\beta}$, which dominates at T_c.

The crossover from two- to three-dimensional behavior has to our knowledge not been discussed for the Hamiltonian (15). The renormalization group theory gives quantitative information at the critical point, but away from the critical point phase diagrams are more easily discussed using Landau theory (see Feder [*13*] and Okada [*40*]).

Further insight into the squaric acid phase transition may be gained by exhibiting the relation between the vertex models and the hydrogen configuration in squaric acid. By projecting the hydrogen displacements from the bond centers onto the *a* and *c* axes, one obtains a configuration as shown in Fig. 20. A squaric acid molecule is located at each vertex, and there are 16 possible vertex configurations, ranging from C_4O_4 to $C_4O_4H_4$ groups. By assigning energies to the various vertices one may calculate the free energy exactly in special cases. As discussed elsewhere [*13*], we found that with the simplest assignment of energies squaric acid is an example of the so-called IF model that unfortunately has a transition temperature only at $T_c = 0$.

$$\langle R_1(l)R_2(l)\rangle = \frac{1}{N}\frac{\partial F}{\partial g_2}\bigg|_0 = \tau^{2-\alpha-\phi}\,\mathcal{F}_0' + \langle R_1(l)\rangle\langle R_2(l)\rangle \tag{26}$$

Abraham and Heilmann [*41*] also discussed the applicability of vertex models to squaric acid. They concluded that vertex models are inappropriate and suggested that the inclusion of electric dipole forces between hydrogen bonds would explain the phase transition in squaric acid.

Zinenko [*42*] has given a description in which hydrogen positions in the double well are given by pseudospins. Tunneling is taken into account by a transverse field. The interaction between layers is treated in the mean field approximation, whereas the strong in-plane hydrogen interactions are treated by

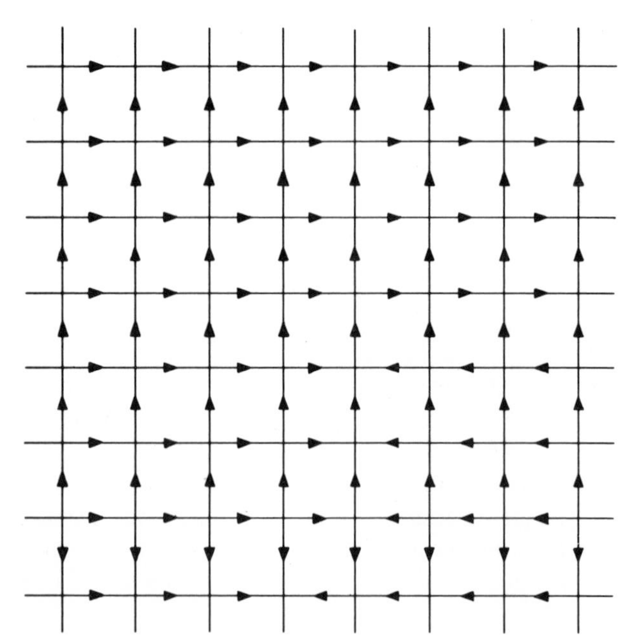

Fig. 20. Hydrogen displacements in squaric acid projected onto the *a* and *c* axes. A squaric acid molecule is centered at each vertex. The configuration shown is almost completely ordered.

assigning energies to configurations in a way similar to the vertex models. Zinenko found that this model gives a second-order phase transition both with and without the tunneling term.

Pumpernik *et al.* [43] studied the squaric acid phase transition by a molecular dynamics calculation. They treated the 4C ring as rigid and assigned harmonic potentials to C—O bending and stretching modes, whereas a rather realistic double-well potential was used for the O—H···O bond. Model parameters were chosen so that experimental data such as infrared frequencies, equilibrium distances, and dissociation energies were closely reproduced. The equations of motion for a set of 81 molecules were then solved numerically for several temperatures.

These authors found that at low temperatures the ordered phase remains stable. At high temperatures, however, the positions of the protons on the O- - -O line are almost uniformly distributed so that the mean O—H distance is at half the O- - -O distance. This indicates that the phase transition is displacive rather than of the order–disorder type. This result, however, is inconsistent with the neutron diffraction structure determination above T_c [16].

VII. Conclusion

Squaric acid represents an extremely interesting model system for the study of hydrogen bonding. The phase transition is continuous and can be described by a two-component order parameter in the basal plane proportional to the sublattice polarization. We have found that very near the transition temperature the dielectric constant has an anomaly described by the three-dimensional *XY* model. Away from the transition temperature, at a reduced temperature of about $5\ 10^{-3}$, there is a crossover to two-dimensional behavior with a critical exponent for the order parameter of 0.14, very close to the two-dimensional Ising value.

The experimental evidence shows that the squaric acid molecule retains its low-temperature structure even about T_c and that clusters of the low-temperature phase exist even above T_c. This is clearly seen in the high-resolution NMR results. We believe that such domains of the low-temperature phase above the transition temperature is a general feature of second-order phase transitions (see our discussion in Riste *et al.* [44] and Feder [45,46]). These domains will be observed by any local probe that does not average over times long enough to measure only true time averages.

Acknowledgment

It is a great pleasure to thank the General Electric Corporate Research and Development Center for the hospitality that was extended to me during my stay as a visiting scientist. Innumerable discussions with Ivar Giaever and Charles P. Bean made my stay enjoyable and also inspired me to write this review.

References

1. S. Cohen, J. R. Lacher, and J. D. Park, *J. Am. Chem. Soc.* **81,** 3480 (1959).
2. J. D. Park, S. Cohen, and J. R. Lacher, *J. Am. Chem. Soc.* **84,** 2919 (1962).
3. L. M. Schwartz and L. O. Howard, *J. Phys. Chem.* **75,** 1798 (1971).
4. L. M. Schwartz and L. O. Howard, *J. Phys. Chem.* **74,** 4374 (1970).
5. M. Ito and R. West, *J. Am. Chem. Soc.* **85,** 2580 (1963).
6. R. West and D. L. Powell, *J. Am. Chem. Soc.* **85,** 2577 (1963).
7. R. West and D. Powell, *J. Am. Chem. Soc.* **84,** 1324 (1962).
8. R. West, H. Y. Niu, D. L. Powell, and M. V. Evans, *J. Am. Chem. Soc.* **82,** 6204 (1960).
9. R. West and J. Niu. *In* "The Chemistry of the Carbonyl Group" (J. Zabicky, ed.) Vol. 2, p. 241. Interscience, New York (1970).
10. D. Semmingsen, *Tetratredron Lett.* No. 11, p. 807 (1973).
11. D. Semmingsen, *Acta Chem. Scand.* **27,** 3961 (1973).
12. D. Semmingsen and J. Feder, *Solid State Commun.* **15,** 1369 (1974).
13. J. Feder, *Ferroelectrics* **12,** 71 (1976).
13a. W. M. MacIntyre and M. S. Werkema, *J. Chem. Phys.* **40,** 3563 (1964).
14. D. Semmingsen, F. J. Hollander, and T. F. Koetzle, *J. Chem. Phys.* **66,** 4405 (1977).
15. F. J. Hollander, D. Semmingsen, and T. F. Koetzle, *J. Chem. Phys.* **67,** 4825 (1977).
16. Y. Wang, G. D. Stucky, and J. M. Williams, *J. Chem. Soc., Perkin Trans. 2* p. 35 (1974).
17. D. Semmingsen, *Acta Chem. Scand., Ser. A* **29,** 470 (1975).
18. I. Suzuki and K. Okada, *Solid State Commun.* **29,** 759 (1979).
19. W. C. Hamilton and J. A. Ibers, "Hydrogen Bonding in Solids," p. 53. Benjamin, New York, 1968.
20. E. J. Samuelsen and D. Semmingsen, *Solid State Commun.* **17,** 217 (1975).
21. E. Samuelsen and D. Semmingsen, *J. Phys. Chem. Solids* **38,** 1275 (1977).
22. F. G. Baglin and C. B. Rose, *Spectrochim. Acta, Part A* **26,** 2293 (1970).
23. S. Nakashima and M. Balkanski, *Solid State Commun.* **19,** 1225 (1976).
24. D. Bougeard and A. Novak, *Solid State Commun.* **27,** 453 (1978).
25. D. P. C. Thackeray, R. Shirley, and B. C. Stace, *Solid State Commun.* **25,** 1039 (1978).
26. E. J. Samuelsen, E. Fjaer, and D. Semmingsen, *J. Phys. C* **12,** 2007 (1979).
27. D. Suwelack, J. Becker, and M. Mehring, *Solid State Commun.* **22,** 597 (1977).
28. J. D. Becker, D. Suwelack, and M. Mehring, *Solid State Commun.* **25,** 1145 (1978).
29. M. Mehring and D. Suwelack, *Phys. Rev. Lett.* **42,** 317 (1979).
30. E. Courtens, *Phys. Rev. Lett.* **29,** 1380 (1972).
31. G. A. Gehring, *J. Phys. C* **10,** 531 (1977).
32. J. Feder, and R. Nystøl, to be published.
33. N. Yasuda, K. Sumi, H. Shimizu, S. Fujimoto, K. Okada, I. Suzuki, H. Sugie, K. Yoshino, and Y. Inuishi, *J. Phys. C* **11,** L229 (1978).
34. H. D. Maier, D. Müller, and J. Petersson, *Phys. Status Solidi B* **89,** 587 (1978).
35. G. A. Samara and D. Semmingsen, *J. Chem. Phys.* **71,** 1401 (1979).
36. W. Rehwald, *in* "Proceedings of the Third International Conference or Lattice Dynamics" (M. Balkanski, ed.), p. 644. Flammarion, Paris, 1978.
37. W. Rehwald and A. Vonlanthen, *Phys. Status Solidi B* **90,** 61 (1978).
38. D. Mukamel, *Phys. Rev. B* **14,** 1303 (1976).

39. M. Kerszberg and D. Mukamel, *Phys. Rev. B* **18,** 6283 (1978).
40. K. Okada, *J. Phys. Soc. Jpn.* **27,** 420 (1969).
41. D. B. Abraham and O. J. Heilmann, *J. Phys. C* **9,** L393 (1976).
42. V. I. Zinenko, *Phys. Status Solidi B* **78,** 721 (1976).
43. D. Pumpernik, I. Mendaš, B. Borštnik, and A. Ažman, *J. Phys. Chem. Solids* **40,** 463 (1979).
44. T. Riste, E. J. Samuelsen, K. Otnes, and J. Feder, *Solid State Commun.* **9,** 1455 (1971).
45. J. Feder, *Solid State Commun.* **9,** 2021 (1971).
46. J. Feder, *in* ''Anharmonic Lattices, Structural Transitions and Melting'' (T. Riste, ed.), p. 113. Noordhoff Int., Leiden, 1974.

Syntheses of Highly Oxidized Cyclobutanes via [2+2] Cycloaddition Reactions of Ketenes

Daniel Belluš

I. Introduction 169
II. Moniliformin 173
III. Alkylmoniliformins 175
IV. Arylmoniliformins 179
V. Squaric Acid 180
References 183

I. Introduction

The first, pioneering synthesis of four-membered monocyclic oxocarbons [*1*] took advantage of the strong propensity of polyhalogenated, especially di-, tri-, and tetrafluorinated, ethylenes to enter [2+2] cycloaddition reactions. Squaric acid (**1**) was first prepared by Cohen, Lacher, and Park [*2*] in 1959 by the aqueous hydrolysis of halogenated cyclobutene derivatives that originated from the 1,2-dichloro-2,2-difluoroethylene [2+2] cyclodimer. Similarly, Smutny, Caserio, and Roberts [*3*] employed the products of thermal cycloaddition of

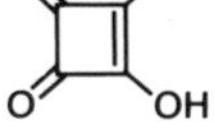

(1) 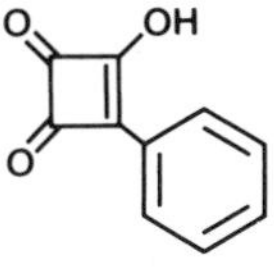 (2) 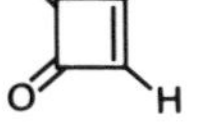 (3)

OXOCARBONS

ISBN 0-12-744580-3

trifluorochloroethylene as well as tetrafluoroethylene to phenylacetylene as precursors for the first monohydroxycyclobutenedione (**2**). In the following years, a number of substituted derivatives of **2** were prepared by this method (for recent reviews, see Ried and Schmidt [*4,4a*]. Nearly 20 years later, the parent hydroxycyclobutenedione (**3**) was also made by a similar route, employing the [2+2] cycloadduct of trifluorochloroethylene and 1,1-dichloroethylene as the precursor, followed by its dehydrochlorination and acid hydrolysis [*5*]. However, from the preparative viewpoint, especially as far as the synthesis of the derivatives of **3** substituted in the free position 4 are concerned, the "fluoroolefin" route possesses some shortcomings: Substituted acetylenes are not readily available; alkyl-substituted acetylenes do not react; the reaction has to be carried out under pressure; and the hydrolysis requires concentrated sulfuric acid. The interest in derivatives of **3**, which was triggered by the discovery of moniliformin (see below), called, therefore, for a new, general synthesis of four-membered cyclic oxocarbons.

In the course of a screening program for toxigenic products of the fungus *Fusarium moniliforme,* a microbial toxin was isolated by Cole *et al.* [*6*]. A year later, in 1974, X-ray analysis revealed that the mycotoxin in question, named by Cole *et al.* moniliformin,*,† possessed the surprisingly simple structure **3** in the form of its sodium salt [*7*]. Interesting plant growth-regulating and phytotoxic effects were reported [*6*]. Later, we found [*8,8a*] that the esters of squaric acid (**1**) show much higher desiccant activity than moniliformin or the sodium salt of **1**. This activity and the simple carbocyclic structures, which we expected would present no environmental problems, were reasons to look for an economically realistic synthesis of four-membered oxocarbons.

The obvious way to form a suitable four-membered ring was by means of ketenes: The ketenes were available in great variety and might be generated from the readily accessible acyl chlorides, and it could be envisaged that neither high temperatures nor autoclave reactions would be necessary. Furthermore, the pronounced ability of ketenes to undergo [2+2] cycloaddition reactions was well documented [*9*].

The feasibility of this synthetic strategy was well supported by three previous syntheses of **1** and **3** via a "ketene" route. Maahs found in 1964 [*10*] that perchlorovinylketene (**6**), generated either thermally at 160°–180°C from 1-ethoxypentachlorobuta-1,3-diene (**4**) or by HCl elimination from 2,3,4,4-tetrachloro-3-butenoic acid chloride (**5**) at 50°C, easily cyclized to tet-

*The name "moniliformin" is consequently used in the following text for the parent acid, 3-hydroxycyclobut-3-ene-1,2-dione (**3**). 4-Aryl derivatives of **3** are called arylmoniliformins; 4-alkyl derivatives are thus alkylmoniliformins.

†Compound **3** is a strong vinylogous acid with $pK_a = 0.0 \pm 0.05$ [*5*]. It is the first known "deoxy" member of a family of cyclic oxocarbons with the common formula $(CO)_nH_2$, since neither "deoxydeltic," "deoxycroconic," nor "deoxy rhodizonic" acids are known so far.

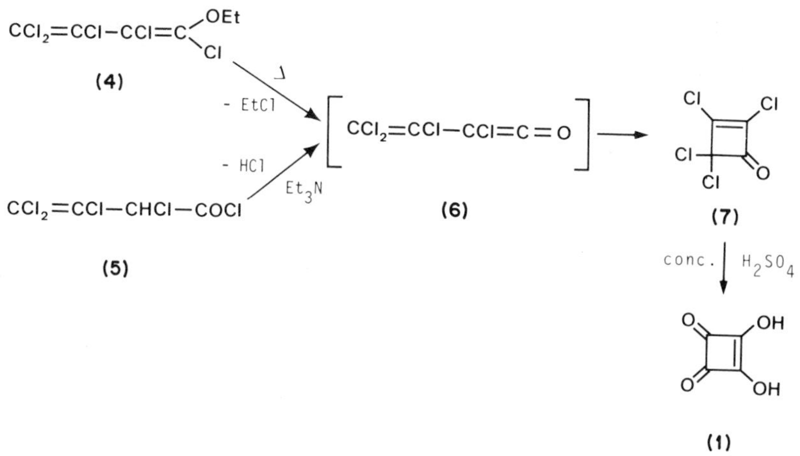

$CCl_2{=}CCl{-}CCl{=}C(OEt)Cl$ (4) —Δ, − EtCl→ [$CCl_2{=}CCl{-}CCl{=}C{=}O$] (6) → (7) —conc. H_2SO_4→ (1)

$CCl_2{=}CCl{-}CHCl{-}COCl$ (5) —Et_3N, − HCl→ (6)

Scheme 1

rachlorocyclobutenone (**7**). With concentrated sulfuric acid, **7** underwent at 110°C a hydrolysis to squaric acid (Scheme 1). This butadiene→cyclobutene type of cyclization remains apparently restricted to the reaction **6**→**7**. For example, repeated attempts to isolate 2,4,4-trichlorocyclobut-2-enone, a promising precursor of **3**, after triethylamine-induced HCl elimination from 2,4,4-trichloro-3-butenoic acid chloride failed [*11*], although the intermediate existence of chloro(2,2-dichlorovinyl)ketene was unequivocally established by chemical trapping [*12*].

Springer *et al.* [*7*] accomplished a short moniliformin synthesis via a ketene route to provide material for structural comparison with the isolated natural product (Scheme 2). The [2+2] cycloaddition of **8** with **9** formed *in situ* from CCl_2HCOCl and triethylamine afforded **10** (20% yield [*13*]), which could be hydrolyzed by aqueous HCl to **3**. Disadvantages of this synthesis are the low yields as well as the price and the instability of **8**.

The first synthesis of moniliformin (**3**), however, was reported by Hoffmann *et al.* [*14*]. It is a rather complicated one and, in its original form (Scheme 3), certainly not suitable for large-scale preparation of **3**. The crucial intermediate in this synthesis is tetramethoxyethylene (**12**), formed by dimerization of dimethoxycarbene, which in turn is extruded thermolytically from the norbor-

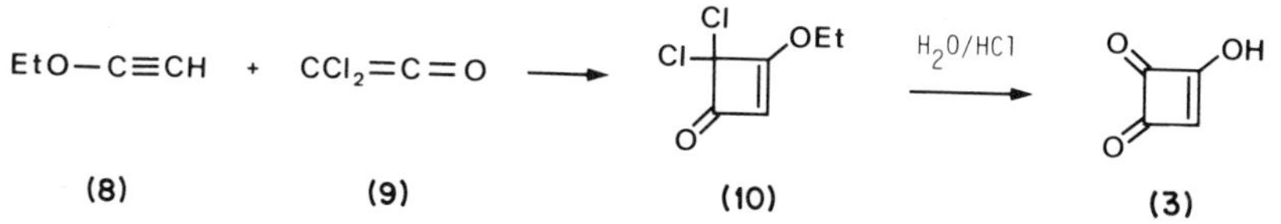

$EtO{-}C{\equiv}CH$ (8) + $CCl_2{=}C{=}O$ (9) → (10) —H_2O/HCl→ (3)

Scheme 2

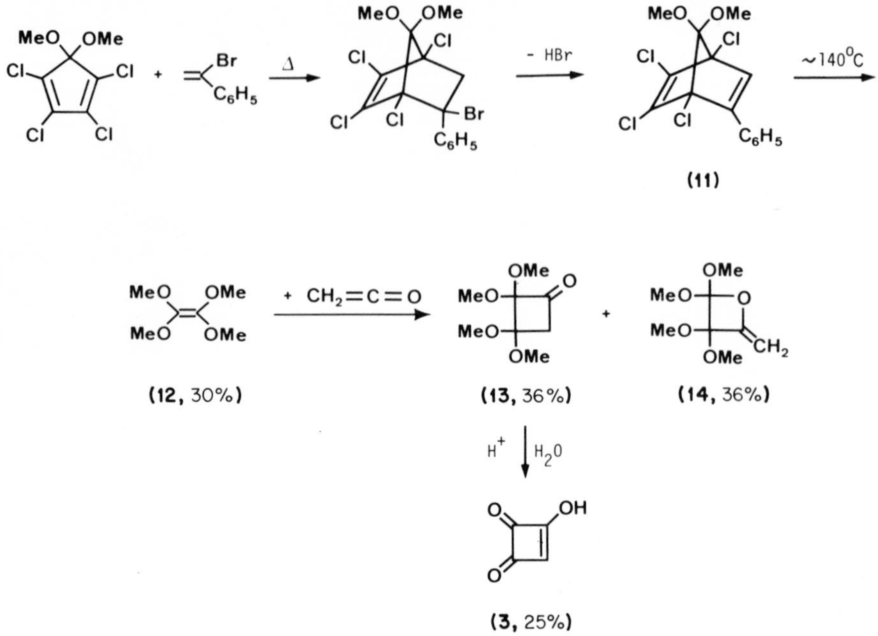

MeO OMe Cl Cl Cl Cl + Br C_6H_5 Δ → MeO OMe Cl Cl Cl Cl Br C_6H_5 – HBr → MeO OMe Cl Cl Cl Cl C_6H_5 (**11**) ~140°C →

MeO OMe MeO OMe (**12**, 30%) + $CH_2{=}C{=}O$ → MeO MeO OMe OMe O (**13**, 36%) + MeO MeO OMe OMe O CH_2 (**14**, 36%)

H^+ H_2O ↓

O O OH (**3**, 25%)

Scheme 3

nadiene derivative **11**. For the [2+2] cycloaddition of **12** to the C=C double bond of gaseous ketene as well as for the hydrolysis to moniliformin (**3**), only low yields were reported.

Fortunately, an important breakthrough in the synthesis of tetraalkoxyethylenes was achieved by Scheeren *et al.* [*15*] in 1973 (Scheme 4). The mixed orthoester of formic acid (**17**) can be made from alkyl orthoformate (**15**) and 4-chlorophenol (**16**) simply by heating and shifting the equilibrium by removing the volatile alcohol. Scheeren showed in his elegant work that treatment of mixed orthoester **17** with sodium hydride in glyme afforded the corresponding tetraalkoxyethylenes in yields of up to 50%. With some improvements in his original reaction conditions, e.g., by replacing the solvent glyme by tetraglyme

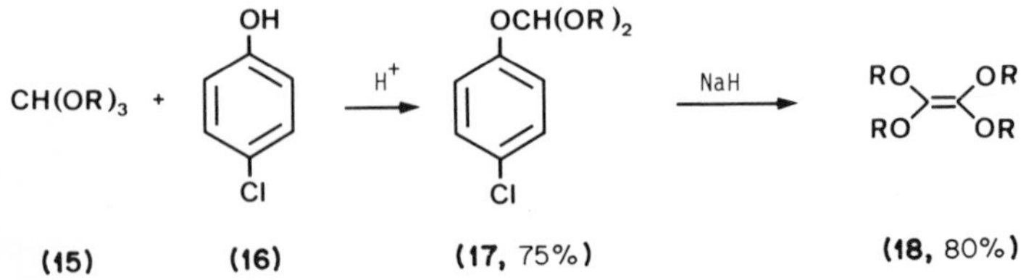

$CH(OR)_3$ (**15**) + OH Cl (**16**) H^+ → $OCH(OR)_2$ Cl (**17**, 75%) NaH → RO OR RO OR (**18**, 80%)

Scheme 4

in the preparation of tetramethoxyethylene (**18**, R = Me), the yields could be raised to 80–85% [*4,4a*].

This excellent access to the highly nucleophilic tetraalkoxyethylenes enables us to use them as the ideal ketenophiles for the general syntheses of four-membered cyclic oxocarbons, which are discussed in the following paragraphs.

II. Moniliformin

We reinvestigated in detail the last two steps of Hoffmann's synthesis of moniliformin (Scheme 3). The addition of gaseous ketene to tetraethoxyethylene (**19**) yielded crystalline, stable cyclobutanone (**21**) (Scheme 5) [*5*]. In contrast to the result using tetramethoxyethylene (**12**), no oxetane of type **14** was observed. The sole by-product isolated, ethyl 2,2-diethoxy-3-oxobutanoate (5.5% yield), indicated that the 1,4 dipole **20** is an intermediate in the [2+2] cycloaddition reaction of **18** with ketenes.

Surprisingly, the hydrolysis of **21** was not a trivial reaction. Under various experimental conditions large amounts of ring-opened products were formed. Finally, the hydrolysis with 18% HCl, in the presence of dioxane as phase mediator at 60°C, afforded **3** in 90% yield [*5*]. Owing to these high yields, the synthetic route shown in Scheme 5 represents a good practical-scale preparation of moniliformin (**3**).

It seemed feasible that so-called ketene equivalents [*16*] could replace gaseous ketene in this [2+2] cycloaddition reaction. Although 2-chloroacrylonitrile as well as 2-chloroacroyl chloride afforded [2+2] cycloadducts with **19** in 91 and 30% yield, respectively, only the latter adduct could be transformed to **3** in a low yield (5–10%) [*5*].

Cyclobutanes are similar to the chemical elements: They, too, can be found in eight oxidation levels (as well as in oxidation level zero, corresponding to cyclobutane itself). Thus, the question arises whether **3** could be prepared by oxidation or reduction of a cyclobutane in a neighboring oxidation level. This is, in fact, possible in a number of ways [*5*]. For illustration, Scheme 6 shows the synthesis of **3** accomplished by oxidation of a cyclobutane derivative of oxidation

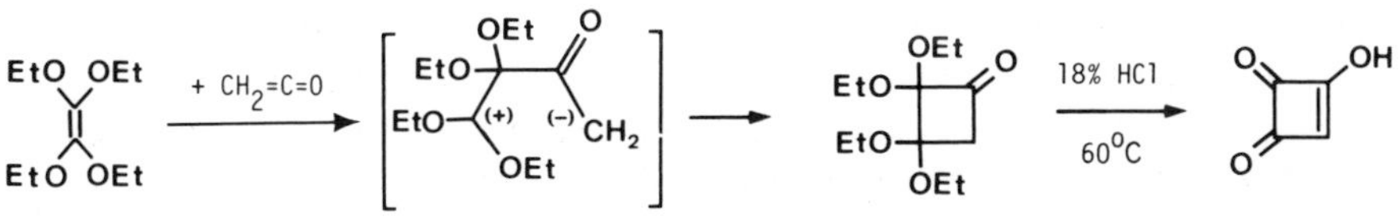

EtO OEt / EtO OEt (**19**) —+ CH_2=C=O→ [EtO, EtO, OEt, O, (+), (−), CH_2, OEt] (**20**) → EtO, EtO, OEt, O, OEt (**21**, 85%) —18% HCl, 60°C→ O, OH, O (**3**, 90%)

Scheme 5

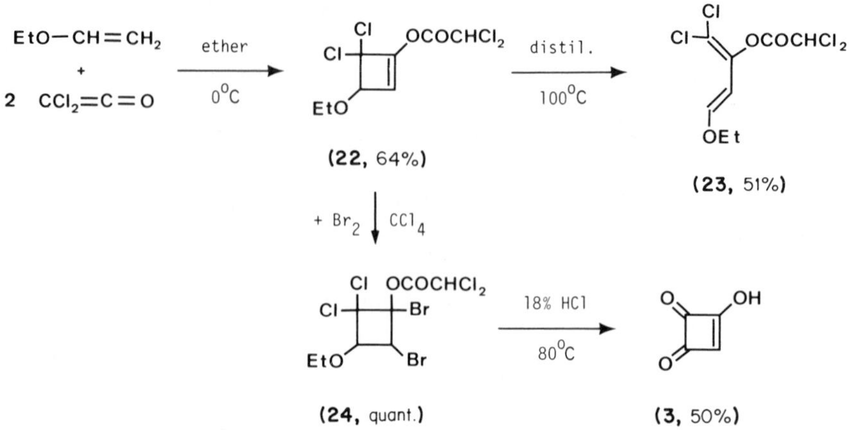

Scheme 6

level 5 and involving as a cyclobutane-forming step the [2+2] cycloaddition of ethyl vinyl ether with dichloroketene (formed *in situ* from dichloroacetyl chloride and Et_3N). This reaction afforded the enol ester **22**, which could be isolated at 0°C. Its bromination proceeded smoothly to give a mixture of stereoisomeric dibromides (**24**), now possessing the oxidation level 6. Because of their instability, they were directly hydrolyzed to **3**. The very simple reagents make this synthesis attractive.

The common feature of all three syntheses of moniliformin (**3**) presented so far was the construction of a proper cyclobutane derivative by a [2+2] cycloaddition of a suitable ketene with an alkoxyacetylene (Scheme 2) or an alkoxyethylene (Schemes 3, 5, and 6), e.g., of *two different* unsaturated compounds. For a number of reasons (e.g., different accessibility, stability, reactivity of starting materials) we would have preferred a synthesis of a moniliformin precursor by [2+2] cyclodimerization of *two identical* unsaturated compounds.

The dimerization of heterosubstituted ketenes (**25a–c**) appeared especially attractive since a number of ketenes are known to self-dimerize to cyclobutane-1,3-diones such as **26** (Scheme 7) [*9*]. Nothing was hitherto known about the cyclodimerization of **25a** and **25b**. There exists a negative [*17*] and a positive [*13*] report about dimerization of chloroketene (**25c**). In neither case, however, are experimental data or yields given.

In our experiments [*5*] we were not able to isolate any primary products of [2+2] cyclodimerization of **25c**, generated *in situ* from **27**, due to their instability. However, treatment of the crude reaction mixture at room temperature with 18% HCl afforded **3** in 1.6% yield! This microscopic yield was very reproducible but could not be improved by any optimization. In any case, it may be pointed out as a peculiarity unknown until now that every 1-liter bottle of inexpensive

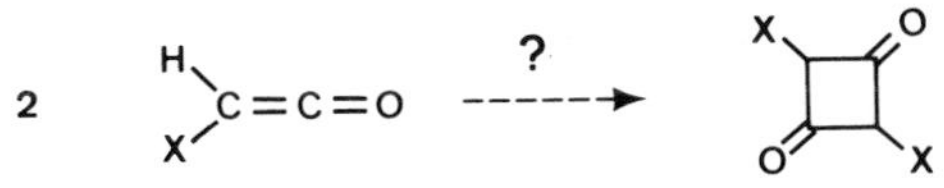

2 (25a, X = OMe) (25b, X = OCOMe) (25c, X = Cl) ? ⟶ (26)

$CH_2Cl{-}COCl$ + $N(CH_2CH_3)_3$ (27) ⟶ ether, 0^oC ⟶ [(28)] ⟶ 18% HCl, r.t. ⟶ (3, 1.6%)

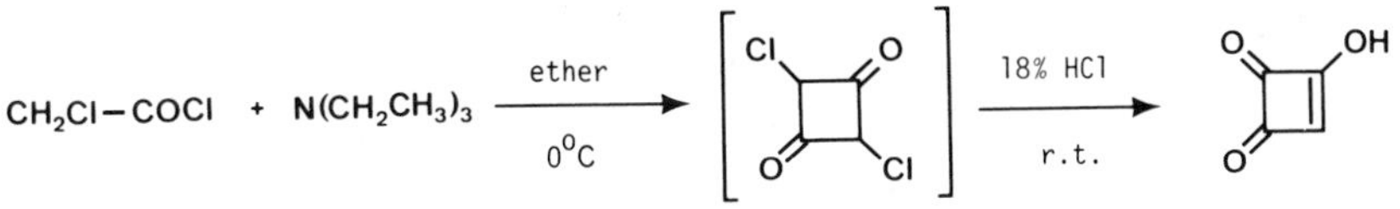

Scheme 7

chloroacetyl chloride (**27**) contains potentially about 10 gm of moniliformin (**3**).

Moreover, this is equally true for squaric acid (**1**), as the very simple treatment of **3** with 1 mole bromine in CH_2Cl_2, followed by evaporation of the solvent and hydrolysis of the monobromide of squaric acid (**29**) with concentrated HCl, affords **1** in 78% overall yield [*5*] (Scheme 8).

Thus, this quick and facile route may be the preferred one, if **3** or **1** is required only in small, e.g., spectroscopic, amounts.

Experiments with acetoxyketene (**25b**) and with methoxyketene (**25a**) provided no cyclobutane-1,3-dione (**26**). In the latter case, a derivative of hydroxytetronic acid was unexpectedly obtained as the only distillable product [*5*].

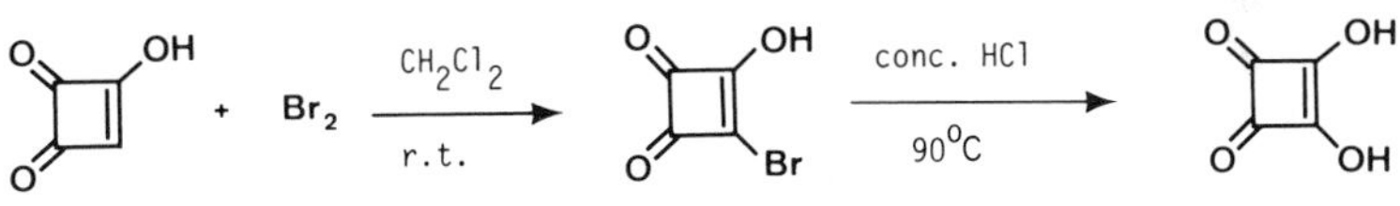

(3) + Br_2 ⟶ CH_2Cl_2, r.t. ⟶ (29, 89%) ⟶ conc. HCl, 90^oC ⟶ (1, 78%)

Scheme 8

III. Alkylmoniliformins

So far, only a few alkylmoniliformins were available by specific routes. The lowest member of this series of oxocarbons, methylmoniliformin (3-hydroxy-4-methylcyclobut-3-ene-1,2-dione) (**35**), was prepared first by Chickos [*18*] via addition of MeMgBr to diethyl squarate. Although no similar additions of Grignard reagents to dialkyl squarates were reported later, it is conceivable that this synthetic approach might have general validity for preparation of the title compounds. Some ethyl esters of oxo-substituted alkyl- and cycloalkylmoniliformins were obtained via addition of enamines [*19*] or β-dicarbonyl compounds [*20*] to diethyl squarate.

In spite of the clean, high-yield [2+2] cycloaddition of tetraethoxyethylene with gaseous, unsubstituted ketene (see Section II), it was not ensured *a priori* that methylketene, generated *in situ,* would give an analogous [2+2] cycloadduct with tetraalkoxyolefins, because Hoffmann *et al.* [*14*] had reported unsuccessful attempts to carry out [2+2] cycloaddition of dimethylketene and diphenylketene to tetramethoxyethylene. Actually, in the reaction of tetraethoxyethylene (**19**) with methylketene formed *in situ,* 2.6 equivalents of the methylketene precursor, propionyl chloride, were necessary to achieve complete conversion of **19**. The complex reaction mixture consisted of the expected [2+2] cycloadduct (**30**), the ester of its enol form with propionic acid (**31**), and the curious 1 : 4 adduct (**32**) of tetraethoxyethylene (**19**) and methylketene, accompanied by small amounts of by-product (methylketene dimer, **33**) (Scheme 9). The crude reaction mixture could be directly hydrolyzed to the desired methylmoniliformin (**35**) by 18% HCl at 60°C. This procedure took advantage of the pronounced tendency of **35** (as well as of all other alkyl- and arylmoniliformins, which will be mentioned in the following text) to crystallize, thus allowing

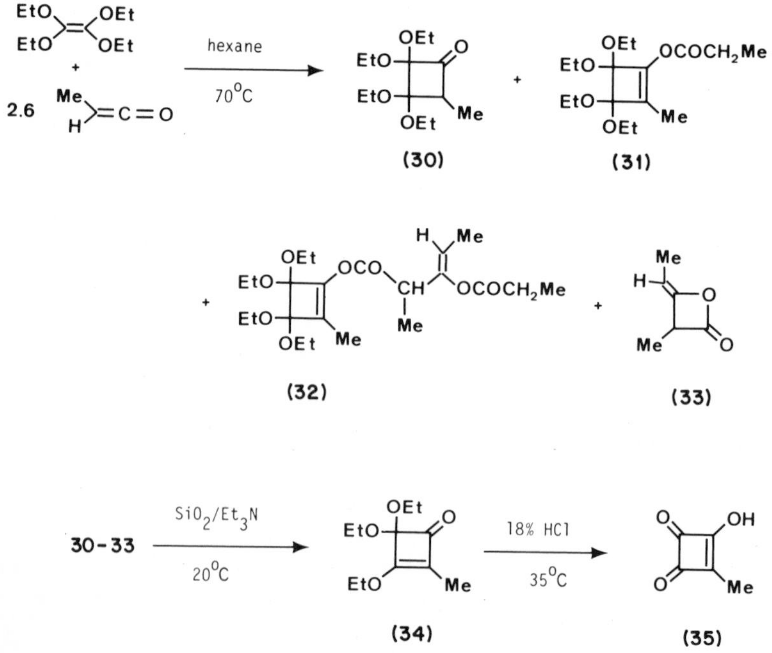

Scheme 9

removal of the oily by-products after hydrolysis by simple washing with hexane. The yield of **35** was about 50%. This method thus represents an extremely simple synthesis of **35** [*21,22*]. The acidic by-products (e.g., from **31**, **32**, or **33**), however, sometimes rendered the isolation of **35** in pure form difficult. To overcome such occasional isolation problems, we preferred to pretreat the hexane solution of the crude reaction mixture with a 10- to 15-fold weight excess of a silica gel–3% Et_3N system [*5*] or basic alumina [*23*] at room temperature. This procedure converted **30**, **31**, and **32** smoothly to "push–pull" cyclobutenone (**34**). Afterward, **34** was eluted by excess of hexane, all polar by-products thereby being retained. Compound **34** could now be hydrolyzed to **35** by 18% HCl at 35°C in 83% yield [*22*].

The procedure according to Scheme 9 is representative of the preparation of a number of alkyl-, cycloalkyl-, and substituted alkylmoniliformins as well as dichlorovinylmoniliformin [*21*]. However, the composition of the crude reaction mixture varied considerably depending on the ketene used. As a specific example, the [2+2] cycloaddition of tetramethoxyethylene to isopropylketene, which was formed *in situ* from 3-methylbutanoic acid chloride, afforded a nearly pure 1 : 1 adduct of type **30** ($CHMe_2$ instead of Me) in 92% yield [*21*]. *tert*-Butylketene behaved similarly. A possible explanation is that the mutual (Z) position of two bulky groups in enol esters of type **31** or **32** would be energetically very unfavorable, in spite of the large $C{=}C{-}C_{exo}$ valence angles of about 132°–133° in cyclobutene derivatives [*24*]. In the two latter cases, the pretreatment of the crude reaction mixture with a SiO_2–Et_3N system or basic alumina was essential, since the direct acid hydrolysis was accompanied to a large extent by acid-assisted heterolytic ring cleavage (for a possible mechanism see Belluš *et al.* [*5*]).

In Table I are summarized all the substituted moniliformins that we have prepared via the general procedure shown for **35** in Scheme 9. All moniliformins are strong vinylogous acids with very low pK_a values. Among them, 4-(2′,2′-difluoro-2′-chloroethyl)-3-hydroxycyclobut-3-ene-1,2-dione represents the strongest oxocarbon acid ($pK_a = -1.12 \pm 0.1$) measured so far [*25*]. The chemical reactivity of the substituted moniliformins presented in Table I is similar to that of moniliformin (**3**) [*8a*] and squaric acid (**1**) [*26*]: Self-catalyzed esterification leads to esters, and treatment with $(COCl)_2$ plus trace of DMF affords stable chlorides, from which amides and thio esters can be prepared in the usual way [*4a*]. In the ^{13}C-NMR spectrum of substituted moniliformins, only three ring carbons are detectable owing to rapid proton transfer, which cannot be suppressed either by steric hinderance (Table I, R = CMe_3) or by intramolecular hydrogen bonding (Table I, R = CH_2COOH) even at −95%C in acetone-d_6 [*21*].

We synthesized 1,1,2-triethoxy-2-trimethylsilyloxyethylene (**37**) in a simple one-pot reaction by Me_3SiCl trapping of the mesomeric anion generated from the readily accessible ester (**36**) by lithium 2,2,6,6-tetramethylpiperidide at −78°C

Table I. Yields and Properties of Moniliformins Prepared Via [2+2] Cycloaddition Reaction of Tetraalkoxyethylenes with R-Monosubstituted Ketenes[a]

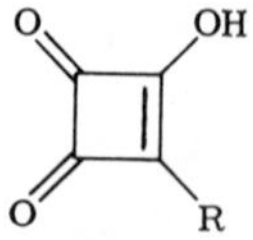

R	Overall yield (%)[b]	mp (°C)	pK_a[c]
H	75	158	0.0 ± 0.05
CH_3	47	160–161	0.20 ± 0.01
CH_2CH_3	52	67–68	—
$CH(CH_3)_2$	75	84–86	—
$CH(CH_3)_3$	77	121–122	0.28 ± 0.05
$(CH_2)_5CH_3$	73	33–35	—
Cyclohexyl	62	105.5–107	—
Benzyl	78	155–156	—
CH_2COOH	25	156–157	—
$CH{=}CCl_2$	36	188	−1.0 ± 0.05
CH_2CF_2Cl	40	138–139	−1.12 ± 0.1
C_6H_5	39	207–208	−0.22 ± 0.1
4-CH_3—C_6H_4	52	216	—
4-CH_3O—C_6H_4	67	220	—
4-Cl—C_6H_4	48	227–228	−0.32 ± 0.05
4-NO_2—C_6H_4	25	>160 (dec.)	−1.05 ± 0.1

[a] For general procedure, see Belluš [*21*].
[b] Overall yield based on tetraalkoxyethylene.
[c] For method of determination see Belluš *et al.* [*5*] and Patton and West [*25*].

[*27*]. This electron-rich ethylene also underwent reaction with methylketene prepared *in situ*. However, the asymmetric substitution pattern in **37** allowed formation of two different 1,4-dipolar intermediates (**38** and **39**). Whereas **38** apparently underwent ring closure to cyclobutanone (**40**), as judged by isolation of 13% of methylmoniliformin (**35**) after the usual acid hydrolysis of crude reaction mixture, the 1,4 dipole **39** stabilized itself by shift of the trimethylsilyl group to give **41**, which was actually the only isolable product of attempted separation of the crude mixture (Scheme 10). The unintended reaction **39**→**41** points at the presumably irrevocable limits of the [2+2] cycloaddition reactions between trimethylsilyloxyethylenes and ketenes [*28*].

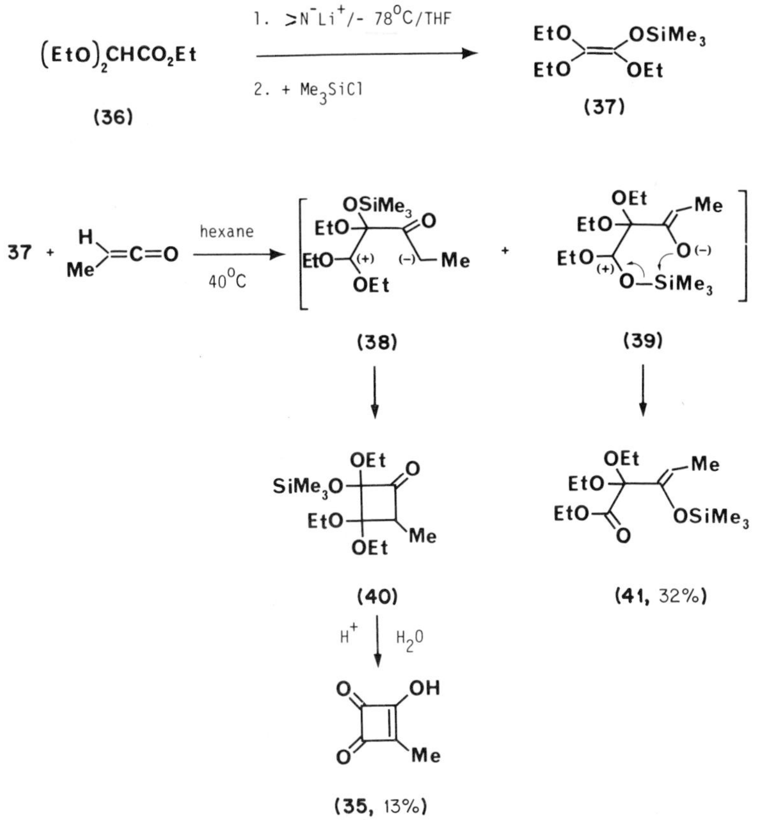

$(EtO)_2CHCO_2Et$ (36) → 1. $>N^-Li^+$/− 78°C/THF; 2. + Me_3SiCl → $(EtO)_2C{=}C(OSiMe_3)(OEt)$ (37)

37 + H(Me)C=C=O → hexane, 40°C → [(38) + (39)]

(38) → (40); (39) → (41, 32%)

(40) → H^+, H_2O → (35, 13%)

Scheme 10

IV. Arylmoniliformins

As mentioned in Section I, the arylmoniliformins have been hitherto available by the ''fluoroolefin'' route, which possesses a number of disadvantages as far as versatility and experimental conditions are concerned. We prepared a number of arylmoniliformins (Table I) employing the same synthetic principle as that for the alkylmoniliformins. However, a characteristic difference in comparison with alkylmoniliformins is noteworthy. For the *in situ* preparation of phenylketenes (**42**) we used the phenylacetyl chlorides, which are available in great variety. The primarily formed cyclobutanones (**43**) eliminated alcohol spontaneously even under mild reaction conditions (room temperature) and formed the ''push–pull'' cyclobutenones (**44**) as the main cyclic products. To achieve yields of 60–80%, at least 2 equivalents of phenylacetyl chloride were necessary: 1 equivalent for

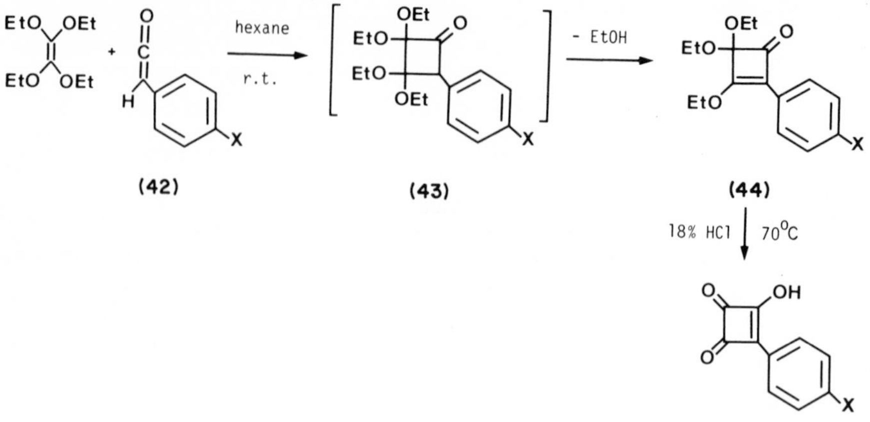

Scheme 11

[2+2] cycloaddition and 1 equivalent for intercepting of alcohol eliminated from **43**. The hydrolysis of **44** with aqueous HCl was very easy (Scheme 11). Although only four parasubstituted phenylmoniliformins (**45**) were prepared by the "ketene" route, the versatility of the described approach to this class of oxocarbons is beyond doubt.

V. Squaric Acid

Squaric acid (**1**) is a deceptively simple organic molecule that has captured the imagination of chemists since its discovery in 1959 [*1,29,30*]. In Section I, attention was drawn to the structural and chemical similarity of moniliformin (**3**) and squaric acid (**1**). Because of the recently discovered postemergent herbicidal activity of the alkyl and thioalkyl squarates [*8*], we looked for a synthesis of squaric acid (**1**) that was more facile than the existing one [*31*]. Again, in this case we stayed with our ketene [2+2] cycloaddition route (Scheme 12).

Squaric acid (**1**) is a cyclobutane derivative with oxidation level 7. Synthesis of hydrolyzable, readily accessible precursors possessing this oxidation level was an obvious target. For this purpose, methoxyketene (**25a**) [*32*], acetoxyketene (**25b**) [*33*], chloroketene (**25c**) [*33*], ethoxyketene (**25d**) [*34*], and methylthioketene (**25e**) [*35*] were produced *in situ* in the presence of tetraalkoxyolefins by dehydrochlorination of the corresponding acid chlorides with Et_3N. First, the [2+2] cycloaddition of **25c** to **18** (R = Me or Et) was tried. According to the NMR spectra the crude, unstable reaction mixtures consisted of 80 ± 5% **46** (X = Cl; R = Me or Et). These crude cyclobutenol esters, the separation and full characterization of which were rendered impossible by their thermal in-

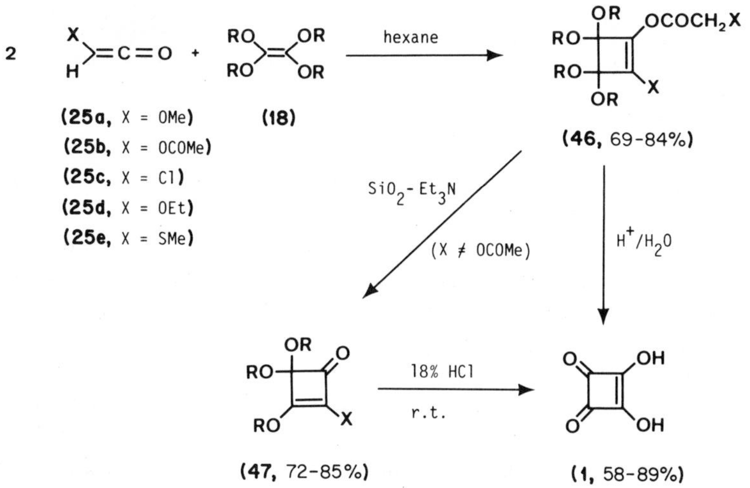

(25a, X = OMe) (25b, X = OCOMe) (25c, X = Cl) (25d, X = OEt) (25e, X = SMe) (18) hexane (46, 69–84%) SiO_2-Et_3N (X ≠ OCOMe) H^+/H_2O (47, 72–85%) 18% HCl r.t. (1, 58–89%)

Scheme 12

stability, could be hydrolyzed to **1**. To achieve 58–64% yields, the hydrolysis had to be performed with concentrated H_2SO_4. However, this small preparative shortcoming could be avoided by an alternative route.

The [2+2] cycloaddition of oxyketenes **25a**, **25b**, and **25d** as well as methylthioketene **25e** led cleanly to the corresponding cyclobutenol esters **46** [*31*]. These could be either smoothly hydrolyzed with 18% HCl at room temperature to squaric acid (**1**) or converted by treatment with the SiO_2–Et_3N system to the hitherto unknown monoorthoester of squaric acid (**47**) (Scheme 12). However, there were two exceptions: (a) Under identical experimental conditions (SiO_2–Et_3N, ether, room temperature), the cyclobutene-1,2-diol diester **46a** underwent twofold displacement-induced fragmentation, initiated by cleavage of an ester group by some external nucleophilic species (Scheme 13). Dimethyl squarate (**48**) was formed, presumably via intermediate cyclobutenone enol ester **47a**. (b) The mixed *O,S*-orthoester **47b** afforded by hydrolysis the monomethyl thio ester

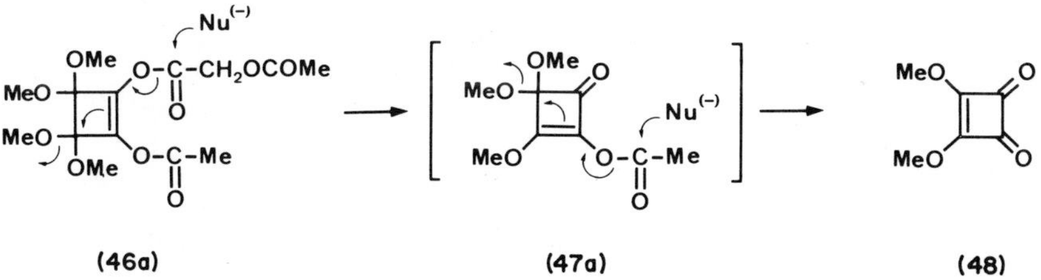

(46a) (47a) (48)

Scheme 13

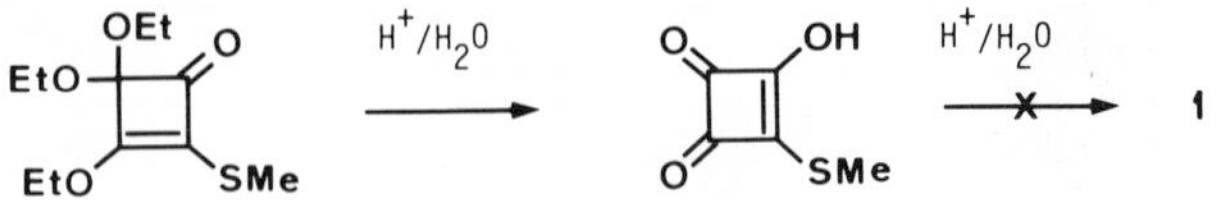

Scheme 14

of squaric acid (**49**), which was surprisingly resistant to further hydrolysis (18% HCl, 100°C) (Scheme 14). It appears, therefore, that the [2+2] cycloaddition of an alkylthio- or perhaps also arylthio-substituted ketene to tetraalkoxyethylenes followed by acid hydrolysis of the resulting products might constitute an easy method for synthesis of the hitherto unknown monothio esters of squaric acid.

The yields of all steps of the versatile synthesis of **1** shown in Scheme 12 are rather good. In fact, this is the first simple synthesis of squaric acid (**1**) that uses readily accessible starting materials and can easily be accomplished in any laboratory [*31,36*].

Because of the similarity of the chemical nature of cycloaddends (i.e., alkoxyketene and nucleophilic acetylene), the elegant synthesis of squaric acid (**1**) performed by Pericás and Serratosa [*30*] must be mentioned here (Scheme 15). Di-*tert*-butoxyethyne (**50**) when heated in refluxing benzene, eliminated isobutylene to form *tert*-butoxyketene (**51**) by a concerted process involving an intramolecular hydrogen shift. The ketene **51** reacted *in situ* with the parent acetylene **50** to afford the cyclobutenone **52** in quantitative yield. Oxidation with *N*-bromosuccinimide (NBS) in CCl_4 led to di-*tert*-butyl squarate (**53**) in 83% yield, from which squaric acid (**1**) could be quantitatively liberated by trifluoroacetic acid.

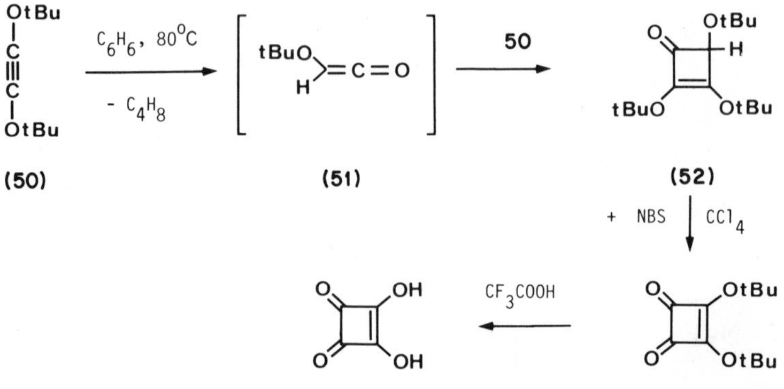

Scheme 15

Serratosa [37] found that dichloroketene reacted with **50** to give low yields (5–10%) of di-*tert*-butyl squarate.

To summarize, the investigation of new synthetic routes to the fungal metabolite moniliformin (**3**) has helped to design and, as shown in this chapter, to realize a general synthesis of four-membered oxocarbons involving the easy and high-yield [2+2] cycloaddition reaction of a great variety of ketenes with tetraalkoxyethylenes as a cyclobutane-forming step.

Acknowledgment

The author is very much indebted to Dr. H.-R. Blattmann, Mr. E. Christen, Dr. H.-P. Fischer, Dr. H. Greuter, Dr. P. Martin, Dr. H. Sauter, and Dr. T. Winkler for their considerable help and valuable discussions during the work on the oxocarbon project and to Dr. B. de Sousa for helpful comments on and Mrs. Ch. Fletcher for excellent technical assistance with the manuscript.

References

1. R. West and J. Niu, *in* "The Chemistry of the Carbonyl Group" (J. Zabicky, ed.), Vol. 2, p. 241. Wiley (Interscience), New York, 1970.
2. S. Cohen, J. R. Lacher, and J. D. Park, *J. Am. Chem. Soc.* **81,** 3480 (1959).
3. E. J. Smutny, M. C. Caserio, and J. D. Roberts, *J. Am. Chem. Soc.* **82,** 1793 (1960).
4. W. Ried and A. H. Schmidt, *Angew. Chem.* **84,** 1048 (1972).
4a. A. H. Schmidt and W. Ried, *Synthesis* p. 1 (1978).
5. D. Belluš, H.-P. Fischer, H. Greuter, and P. Martin, *Helv. Chim. Acta* **61,** 1784 (1978).
6. R. J. Cole, J. W. Kirksey, H. G. Cutler, B. L. Doupnik, and J. C. Peckham, *Science* **179,** 1324 (1973).
7. J. P. Springer, J. Clardy, R. J. Cole, J. W. Kirksey, R. K. Hill, R. M. Carlson, and J. L. Isidor, *J. Am. Chem. Soc.* **96,** 2267 (1974).
8. H.-P. Fischer and D. Belluš (to Ciba-Geigy AG), Ger. Offen. 2,616,756 (1975).
8a. D. Belluš and H.-P. Fischer, *in* "Advances in Pesticide Science" (H. Geissbühler, ed.), Part 2, p. 373. Pergamon, Oxford, 1979.
9. L. Ghosez and M. J. O'Donnell, *Org. Chem.* **35,** Vol. 2, 94 (1977).
10. G. Maahs, *Justus Liebigs Ann. Chem.* **686,** 55 (1965).
11. P. Martin and D. Belluš, unpublished results.
12. P. Martin, H. Greuter, and D. Belluš, *J. Am. Chem. Soc.* **101,** 5853 (1979), footnote 19.
13. N. Morita, T. Asao, and Y. Kitahara, *Chem. Lett.* **1,** 927 (1972).
14. R. W. Hoffmann, U. Bressel, J. Gehlhaus, and H. Häuser, *Chem. Ber.* **104,** 873 (1971).
15. J. W. Scheeren, R. J. F. M. Staps, and R. J. F. Nivard, *Recl. Trav. Chim. Pays-Bas* **92,** 11 (1973).
16. S. Ranganathan, D. Ranganathan, and A. K. Mehrotra, *Synthesis* p. 289 (1977).
17. W. T. Brady and P. L. Ting, *J. Org. Chem.* **40,** 3417 (1975).
18. J. S. Chickos, *J. Am. Chem. Soc.* **92,** 5749 (1970).
19. H. J. Roth and H. Sporleder, *Arch. Pharm. (Weinheim, Ger.)* **303,** 895 (1970).
20. H. J. Roth and H. Sporleder, *Tetrahedron Lett.* p. 6223 (1968); *Arch. Pharm. (Weinheim, Ger.)* **303,** 886 (1970).

21. D. Belluš, *J. Am. Chem. Soc.* **100,** 8026 (1978).
22. D. Belluš, P. Martin, H. Sauter, and T. Winkler, *Helv. Chim. Acta* **63,** 1130 (1980).
23. H. Mayr and R. Huisgen, *Angew. Chem.* **87,** 491 (1975).
24. B. Bak, J. J. Led, L. Nygaard, J. Rastrup-Andersen, and G. O. Sorensen, *J. Mol. Struct.* **3,** 369 (1969); R. Mattes and S. Schroebler, *Chem. Ber.* **105,** 3761 (1972); D. Belluš, H.-C. Mez, and G. Rihs, *J. Chem. Soc., Perkin Trans. 2* p. 884 (1974).
25. Compare, e.g., E. Patton and R. West, *J. Am. Chem. Soc.* **95,** 8703 (1973); A. J. Fatiadi, *ibid.* **100,** 2586 (1978).
26. A. H. Schmidt and W. Ried, *Synthesis* p. 869 (1978).
27. D. Belluš, *Abstr. Pap. 178th Am. Chem. Soc. Natl. Meet. 1979,* Part 2, Abstract ORGN 147 (1979).
28. Compare also A. I. F. Lutsenko, Y. I. Baukov, G. S. Burlachenko, and B. N. Khasapov, *J. Organomet. Chem.* **5,** 20 (1969); L. R. Krepski and A. Hassner, *J. Org. Chem.* **43,** 3173 (1978); W. T. Brady and R. M. Lloyd, *J. Org. Chem.* **44,** 2560 (1979).
29. J. D. Park and S. Cohen, *J. Am. Chem. Soc.* **84,** 2919 (1962); R. West, H. Y. Niu, and M. Ito, *ibid.* **85,** 2584 (1963); G. Maahs, *Angew. Chem.* **75,** 982 (1963); *Justus Liebigs Ann. Chem.* **686,** 55 (1965); H.-D. Scharf and H. Seidler, *Angew. Chem.* **82,** 935 (1970); G. Silvestri, S. Gambino, G. Filardo, M. Guainazzi, and R. Ercoli, *Gazz. Chim. Ital.* **102,** 818 (1972); M. Schroeder and W. Schaefer (to Chem. Werke Hüls), Belgian Patent 855,165 (1976); B. E. Smart and C. G. Krespan, *J. Am. Chem. Soc.* **99,** 1218 (1977).
30. M. A. Pericás and F. Serratosa, *Tetrahedron Lett.* p. 4437 (1977).
31. D. Belluš, *J. Org. Chem.* **44,** 1208 (1979).
32. M. Rey, S. Roberts, A. Dieffenbacher, and A. S. Dreiding, *Helv. Chim. Acta* **53,** 417 (1970).
33. G. Opitz, M. Kleemann, and F. Zimmermann, *Angew. Chem.* **74,** 32 (1962).
34. T. DoMinh and O. P. Strausz, *J. Am. Chem. Soc.* **92,** 1766 (1970).
35. D. Belluš, *Helv. Chim. Acta* **58,** 2509 (1975).
36. D. Belluš (to Ciba-Geigy AG), U.S. Patent 4,159,387 (1976).
37. F. Serratosa, private communication; presented also at the 1st European Symposium on Organic Chemistry, Cologne, August 20–23, 1979.

10

The Chemistry of Squaraines

Arthur H. Schmidt

I. Introduction . . . 185
II. Structure and Nomenclature . . . 186
III. Identification of Squaraines . . . 190
IV. Methods of Preparation . . . 190
A. From Squaric Acid . . . 191
B. From Dialkyl Squarates . . . 192
C. From Substituted Hydroxycyclobutenediones . . . 192
D. From Mercapto- and Hydroselenocyclobutenediones . . . 194
E. From Molecular Rearrangements of Squaramides and Dithiosquaramides . . . 195
F. Miscellaneous Methods . . . 196
V. Table of Squaraines . . . 197
VI. Reactions of Squaraines . . . 197
A. Simple Reactions . . . 197
B. Addition Reactions . . . 211
C. Substitution Reactions . . . 214
D. Squaric Acid Bisamidines and Related Compounds (Exchange of O for N—R and of S for N—R . . . 224
VII. Final Remarks . . . 229
References . . . 230

I. Introduction

Squaric acid (dihydroxycyclobutenedione, **1**) was first synthesized by Cohen, Lacher, and Park in 1959 [*1*]. By the mid-1960's, this compound was being prepared in industry on a technical scale, and samples were given free of charge to research groups [*2*]. Because of this policy, an intensive investigation of squaric acid began.

In 1965 Treibs and Jacob [*3*] examined the action of pyrroles **2** on squaric acid (**1**). These compounds reacted in the molar ratio of 2 : 1 and produced

OXOCARBONS

ISBN 0-12-744580-3

intensely colored condensation products. The attack of the pyrrole molecules took place at opposite carbon atoms of the four-membered ring and not, as might have been expected, at neighboring carbon atoms:

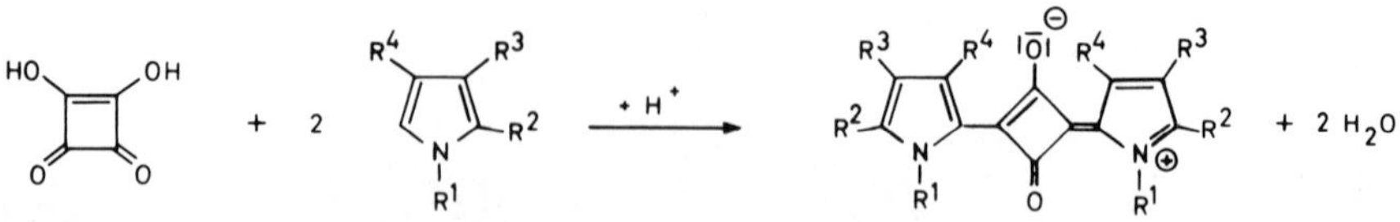

(1) (2) (3)

Thus, for the first time, 1,3-disubstitued derivatives of squaric acid had been synthesized.

A little later two other research groups reported on condensation reactions of squaric acid with azulenes [*4*], tertiary aromatic amines [*5*], and primary as well as secondary amines [*6*]. In each case 1,3-disubstituted derivatives of squaric acid were obtained. For reasons given in the next section, we shall name such compounds, which may be represented by formula **4**, squaraines [*7*].

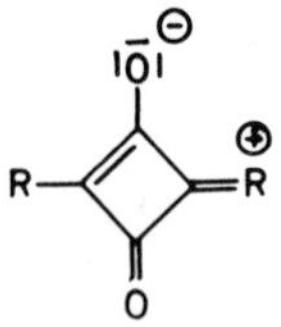

(4)

II. Structure and Nomenclature

Although the squaraines (**3**) with pyrryl substituents were the first examples of this class of compounds, the major part of the work in this field was carried out on squaraines of type **5**. Compounds **5** differ greatly in their physical properties

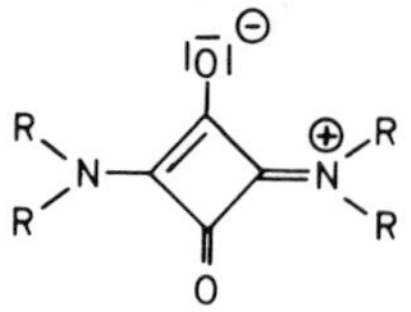

(5) 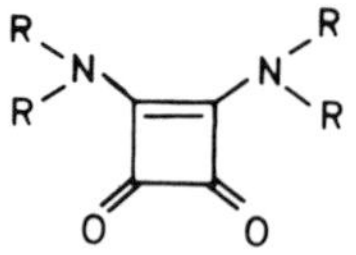 (6)

from the isomeric squaramides **6,** which are usually obtained by the reaction of a dialkyl squarate with a twofold molar amount of an appropriate amine [*8*]. This finding was clear evidence that the amino groups in squaraines of type **5** are attached to opposite carbon atoms of the four-membered ring and not to neighboring ones. Thus, the squaraines were represented by Manecke and

Gauger [*6*], in analogy [*9*] to the squaraines **3**, by the two resonance forms **5A** and **5B**. Other authors accepted this representation of squaraines, but for reasons of simplicity, frequently only one resonance form was used.

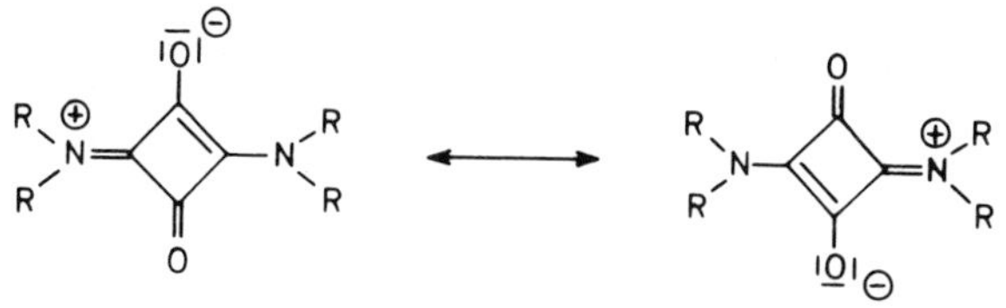

(5A) (5B)

In 1967 Sprenger and Ziegenbein [*10*] used an idealized delocalized formula **4C** for the representation of squaraines. Then, in a review [*11*] that appeared in

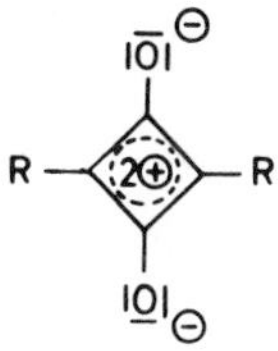

(4C)

1968, they represented the squaraines by the formulas **4** and **4C**, or only by

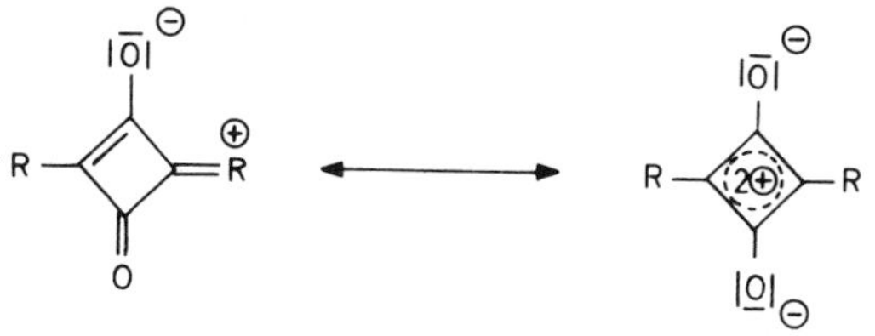

(4) (4C)

formula **4C**. In the following years this representation of squaraines was widely accepted, although not confirmed chemically. Therefore, a structural investigation [*12*] of **7** and **8** by means of UV and photoelectron spectroscopy was of great interest.

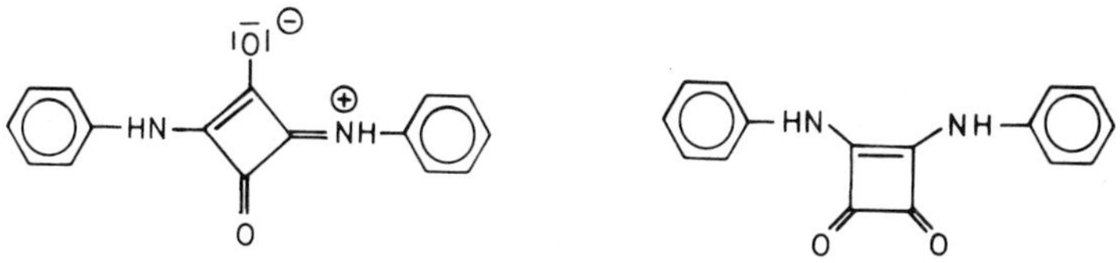

(7) (8)

The investigations proved **7** to be a mesoionic system. Ketoenolate and imminium diolate structures contribute to a large extent to the ground state of **7**. An absorption maximum at long wavelength was ascribed to imminium structures, among them a bis(imminium) ketoenolate structure **4D,** which included the two

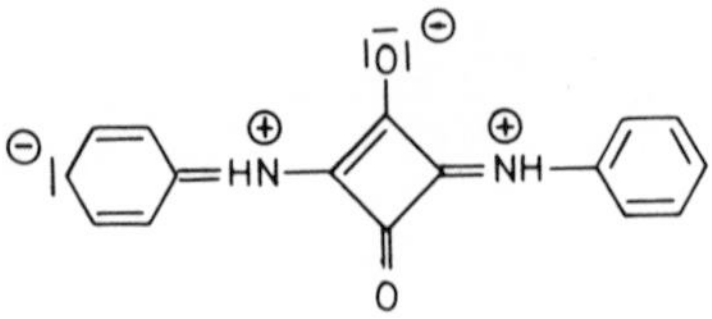

(4D)

phenyl groups in the bond-delocalized system. Interestingly, no evidence was found for an essential contribution of **4C** to the ground state of **7**. For this reason as well as the significance of ketoenolate structures, the squaraines are represented in the remainder of this review by formula **4** exclusively.

For purposes of naming squaraines, we use that resonance structure **4E** in which the positive charge is localized at one carbon atom of the four-membered ring and both substituents R are formally neutral. Since molecules bearing a

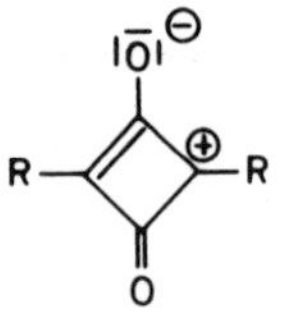

(4E) 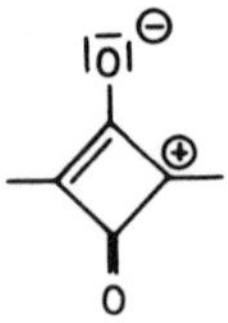 (F)

positive as well as a negative charge are called betaines, we [7] propose naming mesoionic compounds with the structural element **F** squaraines (in German, *Quadraine*). According to this proposal, compounds **7**, **9**, **10**, and **11** are to be named extremely simply, yet in a way that stresses their chemical relationship, in the following way:

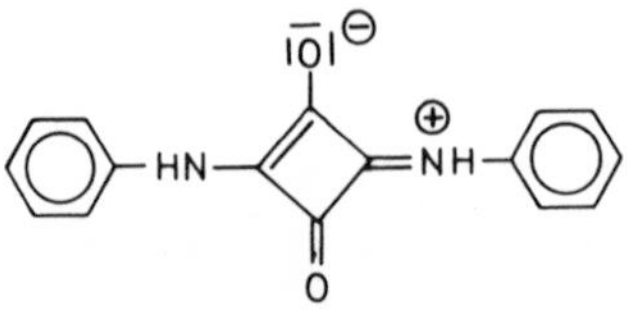

(7)
Bis(anilino)squaraine

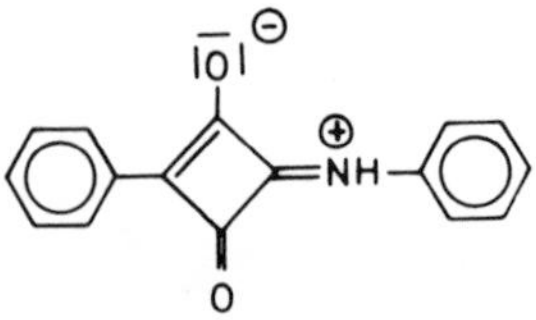

(9)
Phenylanilinosquaraine

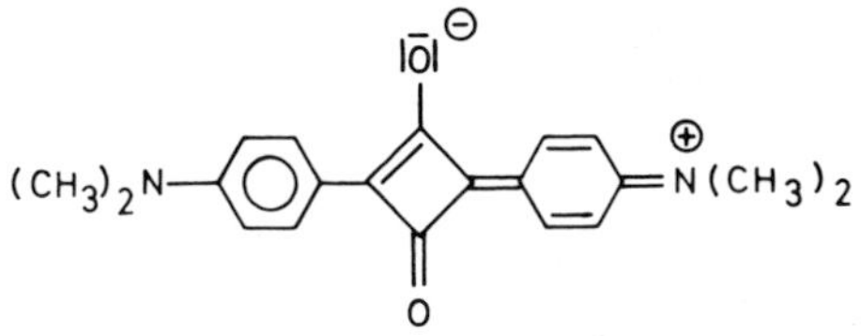

(10)
Bis(4-methylaminophenyl)squaraine

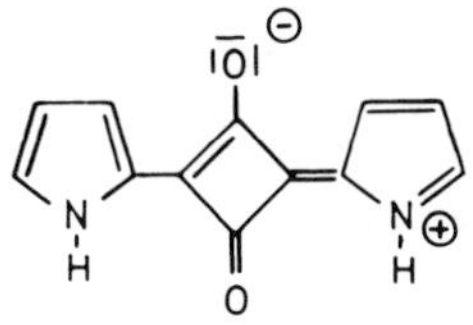

(11)
Bis(pyrryl)squaraine

It is interesting that squaraines may be regarded as derivatives of the hypothetical dihydroxysquaraine (**12**). Since **12** is isomeric with squaric acid (**1**), it may

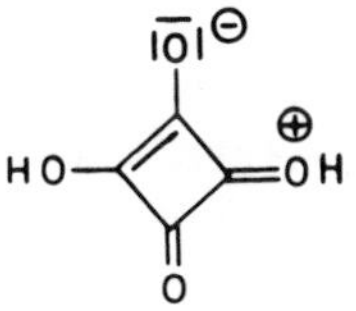

(12)

also be regarded as "isosquaric acid." On this basis, the squaraines **5** are furthermore "isosquaramides."

It will be demonstrated in a later section that the oxygen in squaraines **4** can be replaced by other atoms or functional groups, e.g., S, Se, N—R, thus producing mesoionic compounds with three or four different substituents. These compounds are most conveniently represented by the general formula **13**. For reasons

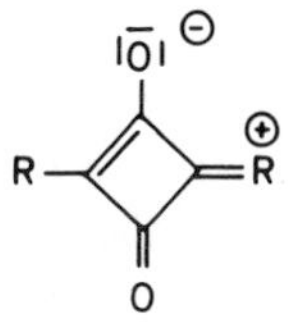

(4) 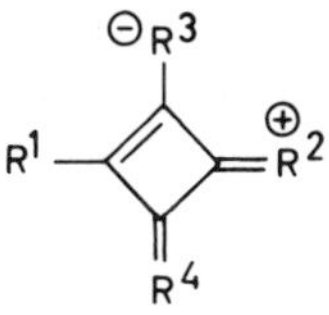 (13)

of standardization we suggest that these compounds also be called squaraines, even if they do not contain any oxygen. This proposal is exemplified by the names of the compounds **14** and **15**.

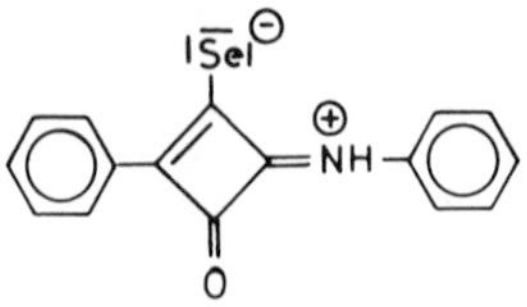

(14)

Phenylanilino monoselenosquaraine

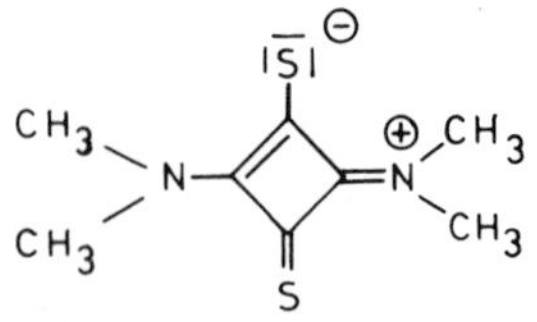

(15)

Bis(dimethylamino)dithiosquaraine

III. Identification of Squaraines

The identification of the squaraines **5** and their distinction from the isomeric squaramides **6** is easily done by means of IR spectroscopy. The C=O groups of cyclobutenediones cause strong absorptions in the region 1800–1740 cm^{-1}; the conjugated C=C bond in a four-membered ring is responsible for an absorption at ca. 1600 cm^{-1}. In agreement with this, the squaramides **6** show strong absorptions in the regions 1840–1770 and 1730–1680 cm^{-1} and at 1620 cm^{-1}. In contrast to this finding, the IR spectra of the squaraines **5** show no bands in the C=O absorption area. This indicates bond delocalization in the four-membered ring. Furthermore, no sharp absorption due to the C=C bond can be seen. Instead, there is a strong and broad absorption band in the area 1650–1500 cm^{-1}, which seems to be characteristic of the bond-delocalized four-membered ring system [*13*].

A comparison of the isomeric compounds **7** and **8** reveals further interesting differences in the physical properties: (a) The squaraine **7** has a distinctly higher melting point than **8** [*14*] [**7**: > 360°C (dec.); **8**: 280°C (dec.]. (b) The squaraine **7** shows in the UV region an absorption at longer wavelength and with a higher extinction coefficient than **8** [*12*].

The IR spectra of the bis(pyrryl)squaraines (**3**) exhibit only weak NH bands. There are no absorption bands in the region above 1720 cm^{-1}. In the IR spectrum of every squaraine **3** there appears a very strong absorption band in the area 1640–1610 cm^{-1}. In this connection it is interesting that the asymmetric stretching vibration of the carboxylate anion appears at 1610–1550 cm^{-1} [9].

The UV spectra of the bis(pyrryl)squaraines **3** show a strong absorption with the maximum in the area 580–540 nm (ϵ = 460,000–144,000) [*9*].

IV. Methods of Preparation

We have reviewed the preparation of squaraines derived from **1** in detail [*7*]. Therefore, we give in the following only a survey of the most important methods of preparation. This survey is followed by a table listing most of the squaraines so far described in the literature.

A. FROM SQUARIC ACID

The vast majority of squaraines has been obtained by the condensation of squaric acid with a twofold molar amount of an appropriate nucleophile. The reaction of **1** with pyrroles [*3,9,15,16*] is normally carried out by heating the compounds over a period of 1–3 hr in ethanol, acetic acid (50%), or a mixture of acetic acid and acetic acid anhydride. The addition of a small amount of a mineral acid catalyzes the condensation reaction. Pyrroles with low reactivity (e.g., 2,4-dimethyl-3,5-diethoxycarbonylpyrrole and its *N*-methyl derivative) can be condensed with **1** only after the addition of concentrated hydrochloric acid.

Bis(amino)squaraines are usually obtained on heating **1** with a primary or a secondary amine in an alcohol, in dimethylformamide, or in dimethyl sulfoxide [*6,17–19*]. The condensation reaction may again be catalyzed by mineral acids (e.g., sulfuric acid) and affords the squaraines **5** in moderate to excellent yield. Good results have also been obtained by the use of a 20% excess of squaric acid (**1**) in methanol as solvent. If the reaction is carried out in dimethylformamide or in dimethyl sulfoxide, it is advantageous to remove the water that is formed in the condensation process continuously.

Numerous squaraines **5** have been obtained by using a mixture of *n*-butanol and benzene as the reaction medium [*20*]. *n*-Butanol has been chosen as the solvent for the condensation of squaric acid with tertiary aromatic amines [*5*].

Squaric acid (**1**) and methyl-substituted quinolinium, benzthiazolium, or benzselenazolium iodide were easily condensed by means of a twofold molar amount of a base (quinoline, sodium ethoxide). In Scheme 1 the details of this process are outlined for the example of *N*-ethyl(4-methylquinolinium) iodide [*10*].

Scheme 1

Under conditions similar to those outlined in this section, several other nucleophiles have been condensed with squaric acid (**1**) to give the corresponding squaraines (see Section V).

B. FROM DIALKYL SQUARATES

The reaction of dialkyl squarates with 2 moles of a primary or secondary amine leads normally to squaramides **6** [*8*]. However, several weakly basic amines react in a different way [*21–23*]. For example, the reaction of diethyl squarate **19** with a substituted *N*-methylaniline (**20**) produces the amido ester **21** and the squaraine **22**:

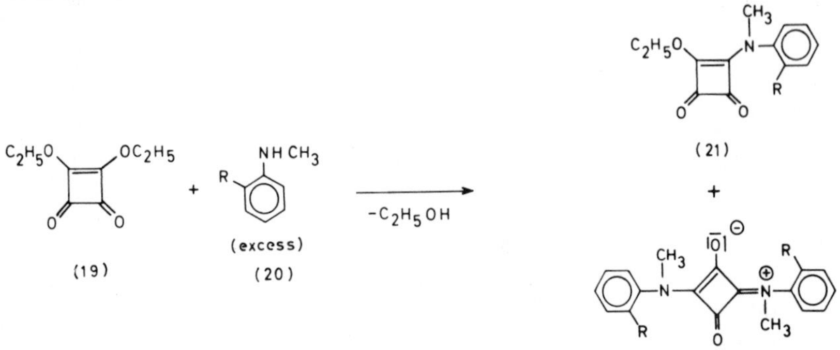

C_2H_5O OC_2H_5 (19) + NH CH_3 R (excess) (20) $\xrightarrow{-C_2H_5OH}$ (21) + (22a, R = H) (22b, R = CH_3)

In several cases the squaraines are formed as the main or only product. Thus, the reaction of dialkyl squarates with an appropriate amine may be used for the preparation of special squaraines.

Several years before this discovery, Treibs and Jacob [*15*] had already reported that squaraines were generated on heating diethyl squarate **19** with indoles or with the "Emil Fischer base" (1,3,3-trimethyl-2-methyleneindoline). The solvent of choice was acetic acid or acetic anhydride, respectively.

C. FROM SUBSTITUTED HYDROXYCYCLOBUTENEDIONES

Mechanistic studies on the reaction of squaric acid (**1**) with pyrroles provided an entry into the series of pyrrylhydroxycyclobutenediones **23** [*15*]. Since these

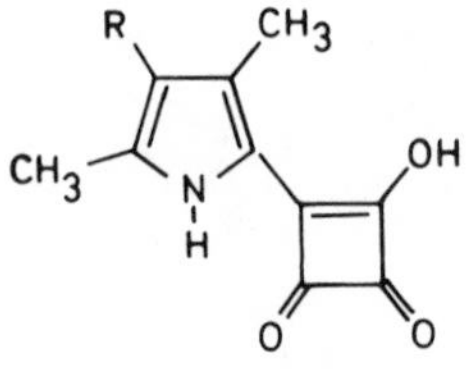

(23)

hydroxycyclobutenediones are closely related to squaric acid, the question arose whether **23** would enter into condensation reactions with nucleophiles. Under essentially the same conditions as were used for the preparation of the bis(pyrryl)squaraines **3**, the hydroxycyclobutenediones **23** reacted with pyrroles, indoles, or the "Emil Fischer base" to give the condensation products **24** [*15*].

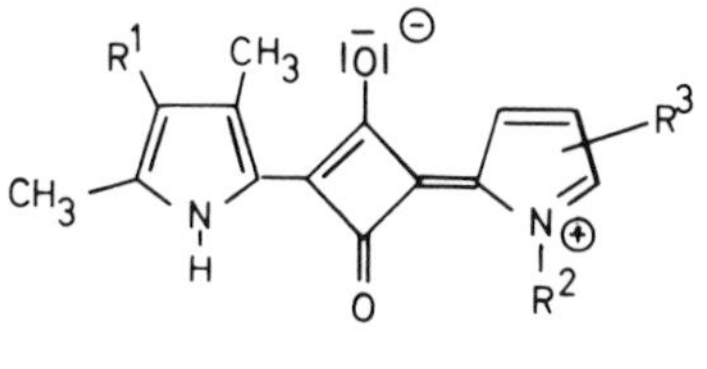

(24)

The IR spectra of these compounds show the same characteristics as the spectra of the bis(pyrryl)squaraines **3**, thus indicating their structural relationship.

The reaction of phenylhydroxycyclobutenedione (**25**) [*24*] with primary and secondary amines proceeded in an analogous fashion. On heating **25** and an

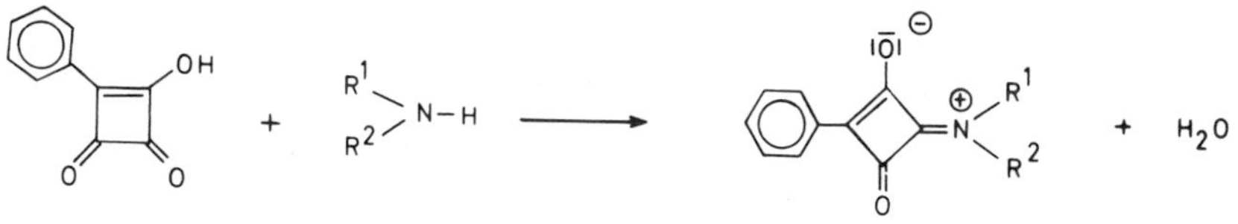

(25)

(26, R^1= aryl ; R^2= H , alkyl)

(27, R^1= alkyl ; R^2= H)

equimolar amount of a primary or secondary amine in *n*-butanol for a period of 4–6 hr, the squaraines **26** and **27** were obtained in moderate to good yield [*25,26*]. With primary aromatic amines the condensation reaction proceeded on heating for only a short period of time in solvents such as ethanol; in tetrahydrofuran the condensation took place even at room temperature.

Compounds **26** form yellow to orange-yellow crystals, which are only slightly soluble in solvents of low polarity but readily soluble in dimethylformamide, dimethyl sulfoxide, and nitrobenzene. Their decomposition points indicate a high polarity. On the other hand, compounds **27** are yellowish-white and readily soluble in nonpolar solvents. The phenyl residue on the nitrogen atom must thus be responsible for the difference in the properties of these squaraines.

The reaction of **25** with amines under mild conditions leads to the ammonium salts **28**, the stability of which increases with increasing basicity of the amine. At higher temperatures, the salts **28** readily lose water and form the squaraines **26** or **27** [*26*], as demonstrated in Scheme 2.

Quite unexpectedly, the reaction of **25** with N-monosubstituted *o*-phenylenediamines did not occur according to Scheme 2. In addition to the

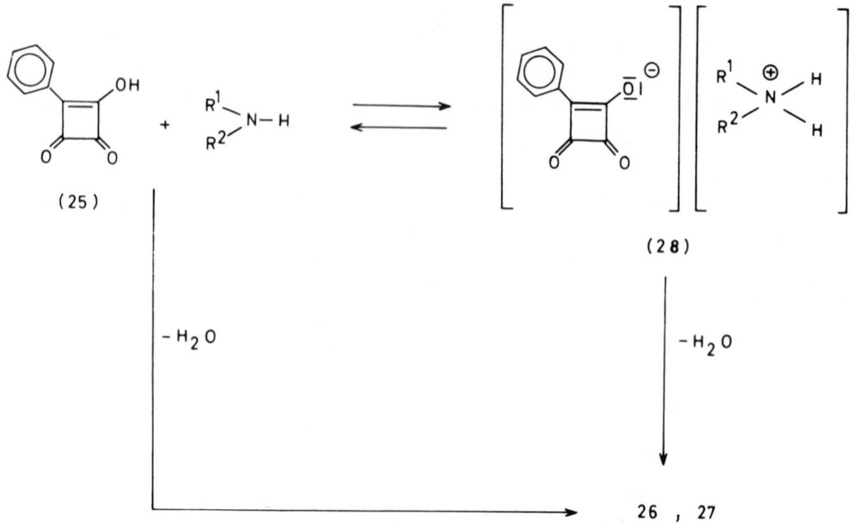

(25) (28) 26, 27 $-H_2O$ $-H_2O$

Scheme 2

condensation, an oxidation process was observed leading to compounds **29** as deep violet crystals:

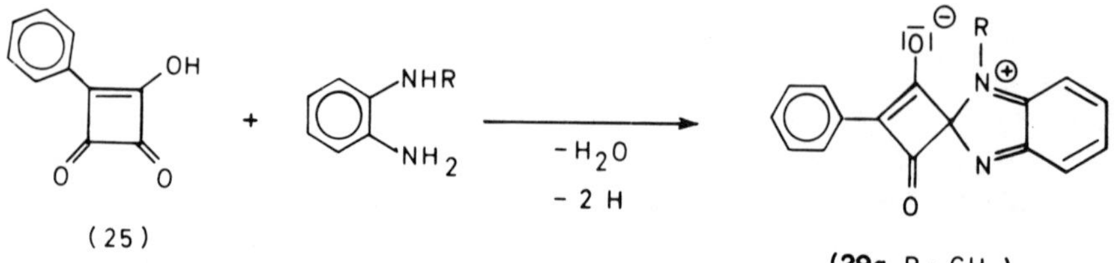

(25) $-H_2O$ $-2H$

(**29a**, R = CH_3)
(**29b**, R = C_6H_5)

Interestingly, the aromatic moiety of the diamine is destroyed during this reaction and appears in the product **29** as a quinoid one. Compounds **29** may be regarded spirosquaraines. They are the first representatives of a hitherto almost unexplored class of squaraines [*27*].

D. FROM MERCAPTO- AND HYDROSELENOCYCLOBUTENEDIONES

In 1972 we [*28*] succeeded in preparing phenylmercaptocyclobutenedione (**30a**), the first sulfur analog of a hydroxycyclobutenedione and one of the first sulfur-substituted cyclobutenediones ever prepared. Compound **30a** reacted with an equimolar amount of a primary aromatic amine in tetrahydrofuran/ethanol, even in the cold, to afford deeply colored condensation products (**31**) [*29,30*].

The same products were obtained by using the pyridinium salt **30b** instead of **30a**. The structure of the thiosquaraines **31** has been confirmed by X-ray analysis [*30*].

(**30a**, $M^{\oplus} = H^{\oplus}$)

(**30b**, $M^{\oplus}$ = pyridinium)

a: $-H_2O$

b: $-H_2O$, − pyridine

(31)

The above-mentioned procedure was extended a little later to the preparation of the selenosquaraines **33**, the only compounds of this type so far known [*30,31*]. Recently, in a similar reaction, the dithiosquaraine **35** was prepared

+ H_2O

(32)

(33)

[*32*]. On heating the dipotassium salt **34** with an excess of morpholine, **35** was generated in almost quantitative yield. Interestingly, in this reaction a sulfur atom migrates from one position of the four-membered ring to a neighboring position. The reaction is therefore related to the molecular rearrangements presented in the next section.

$2 K^{\oplus}$

(excess)

(34)

(35)

E. FROM MOLECULAR REARRANGEMENTS OF SQUARAMIDES AND DITHIOSQUARAMIDES

Several molecular rearrangement reactions have been described so far which lead to squaraines. When the squaramides **36** were heated with a catalytic amount of sulfuric acid in *n*-butanol, the squaraines **37** were obtained in low yield [*6,19*]:

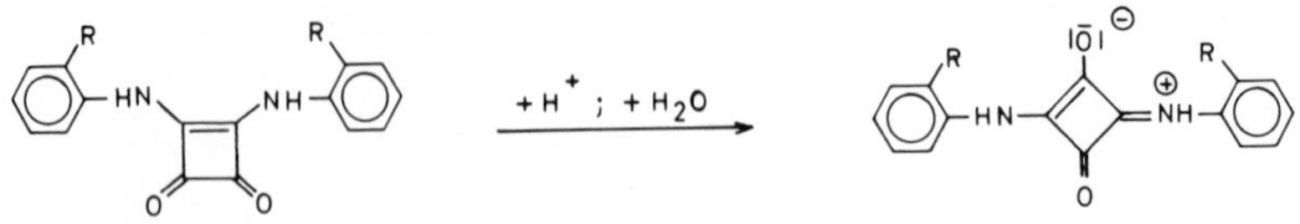

(**36a**, R = OH)
(**36b**, R = H)

(37)

It was possible to extend this method to the preparation of several dithiosquaraines. On heating the dithiosquaramide **38** in an excess of an amine, the dithiosquaraines **39** were generated in high purity and good yield [*33*]:

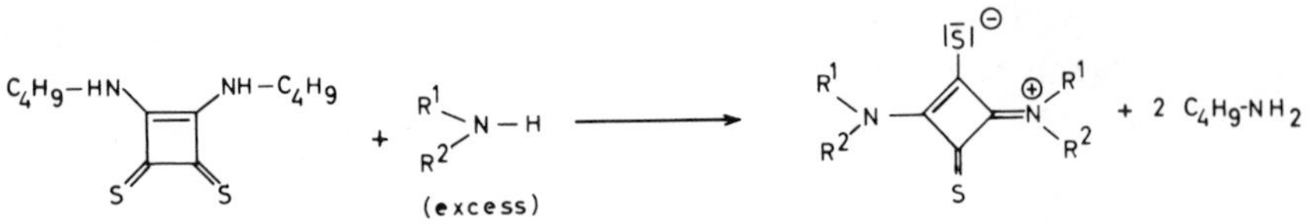

(38)

(excess)

(39)

The preparation of bis(morpholino)dithiosquaraine **35** described in the last section may be seen in the context of this section from a mechanistic viewpoint. It seems probable that **34** and morpholine react first to give bis(morpholino)dithiosquaramide (**40**), which then rearranges in the above-mentioned way to the squaraine **35**:

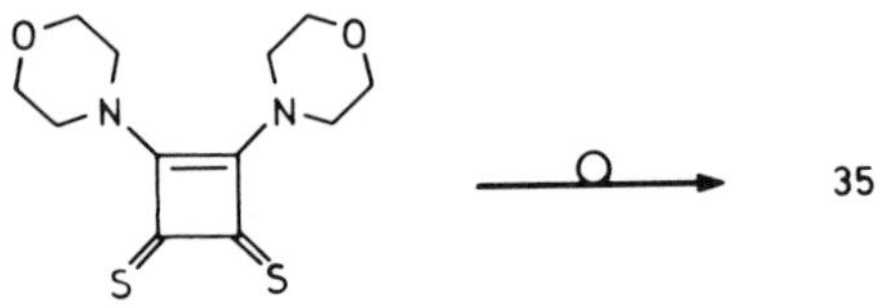

(40)

F. MISCELLANEOUS METHODS

Numerous examples have been reported for the generation of squaraines from other squaraines by means of substitution reactions. These transformations are discussed in the next section. Two special procedures for the generation of squaraines are described in Scheme 3. Treatment of **41** with a mixture of HCl and CH_3OH afforded the squaraine **42** [*34*]. The oxidation of **43** with bromine or mercuric oxide gave the squaraines **44a** and **44b** in 20–50% yield [*35*].

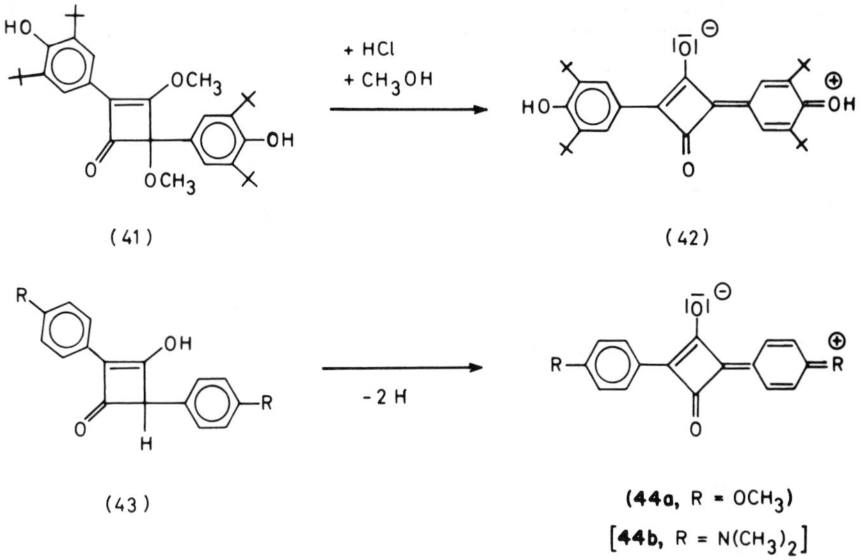

+ HCl
+ CH_3OH

(41) (42)

−2 H

(43)

(44a, R = OCH_3)
[44b, R = $N(CH_3)_2$]

Scheme 3

There have been several reports of reactions in which squaraines have been obtained as by-products [*7*].

V. Table of Squaraines

Table I is a survey of most of the squaraines so far prepared, arranged according to the preparation procedures outlined in Section IV.

VI. Reactions of Squaraines

A. SIMPLE REACTIONS

We first pointed out [*29*] that some squaraines have an appreciable acidity and are weak acids. Thus, phenylarylamino monothiosquaraines are easily dissolved in 10% sodium bicarbonate solution. After careful acidification the thiosquaraines are recovered unchanged from these solutions.

On the addition of dilute sulfuric acid to a solution of equimolar amounts of **45** and cyclohexylamine in tetrahydrofuran, the cyclohexylammonium salt **47** was isolated. This finding is seen as proof for the formation of **46** as an intermediate and for the ease of its hydrolysis. (Text continued on p. 210.)

Table I. Survey of Squaraines

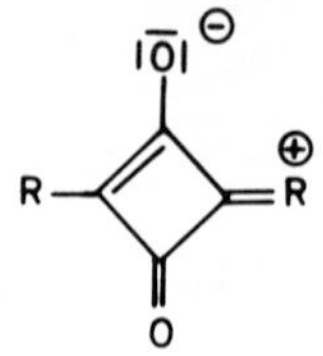

A. Squaraines obtained from the reaction of squaric acid with nucleophiles

R	Yield (%), color	mp (°C)	Ref.
1. Pyrroles			
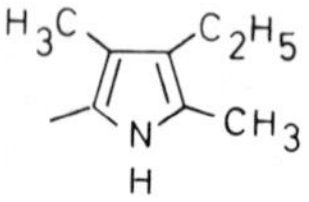 H_3C, C_2H_5, CH_3, N, H	90	265	*9*
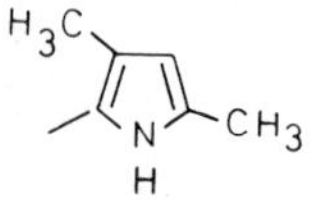 H_3C, CH_3, N, H	75, violet	240–250; 286 (subl.)	*9*
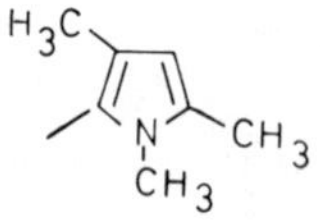 H_3C, CH_3, N, CH_3	79, blue-black	242 (subl.)	*9*
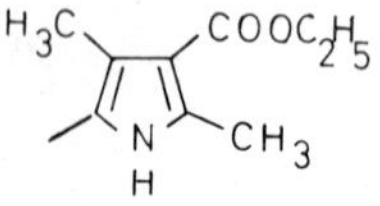 H_3C, $COOC_2H_5$, CH_3, N, H	60, green-violet	296	*9*
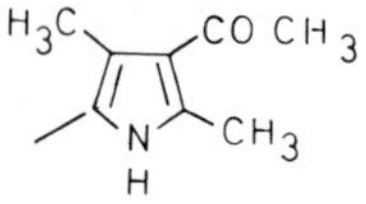 H_3C, $CO\ CH_3$, CH_3, N, H	76	299	*9*
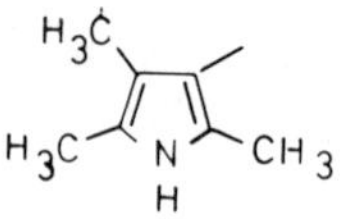 H_3C, H_3C, CH_3, N, H	34, dark green	295 (dec.)	*16*
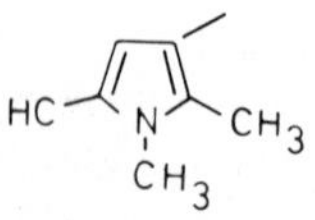 HC, CH_3, N, CH_3	40, dark green	292 (dec.)	*16*
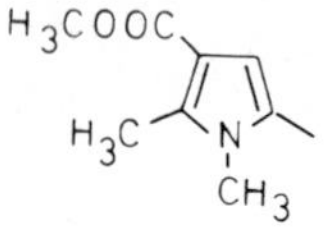 H_3COOC, H_3C, N, CH_3	75, blue-violet	285 (dec.)	*16*

2. Primary and secondary amines

–NH–	74 87	>350 >360	*18* *20*
OH, –NH–	77.5	>350	*18*
HO– –NH–	71; 94.5	>350	*18*
NO_2, –NH–	76	279	*18*
O_2N, –NH–	79	>350	*18*
O_2N– –NH–	82	>350	*18*
$(H_3C)_2N$– –NH–	85.5	>350	*18*
COOH, –NH–	94	339–340	*18*
$H_2C{=}HC$, –NH–	69.5	>350	*18*
CH_3, H_3C– –NH–, CH_3	46	>350	*18*
CH_3, H_3C– –NH–	53	328 (dec.)	*20*
H_3CO, –NH–	81	278 (dec.)	*20*

(*Continued*)

Table I (*Continued*)

$2\text{-}C_2H_5C_6H_4\text{-}NH\text{-}$	80	303 (dec.)	*20*
$3\text{-}H_3C\text{-}C_6H_4\text{-}NH\text{-}$	86	194	*20*
$3\text{-}O_2N\text{-}4\text{-}H_3C\text{-}C_6H_3\text{-}NH\text{-}$	93	>360	*20*
$4\text{-}Cl\text{-}2\text{-}CH_3\text{-}C_6H_3\text{-}NH\text{-}$	84	274 (dec.)	*20*
$3\text{-}HOOC\text{-}C_6H_4\text{-}NH\text{-}$	94	>360	*20*
$4\text{-}C_2H_5O_2C\text{-}C_6H_4\text{-}NH\text{-}$	92	>360	*20*
$2,6\text{-}(CH_3)_2C_6H_3\text{-}NH\text{-}$	25	355 (dec.)	*20*
$4\text{-}H_3C\text{-}OC\text{-}C_6H_4\text{-}NH\text{-}$	94	360	*20*
$2\text{-}C_6H_5\text{-}C_6H_4\text{-}NH\text{-}$	82	297 (dec.)	*20*
$C_6H_5\text{-}N{=}N\text{-}C_6H_4\text{-}NH\text{-}$	95	>360	*20*
1-Naphthyl-NH-	88 85	343–344 340 (dec.)	*18* *20*
2-Naphthyl-NH-	63	313	*20*

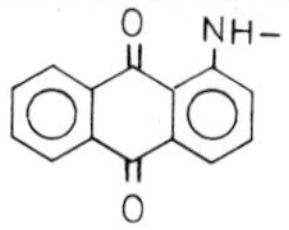 O NH– / O	80	>360	*20*
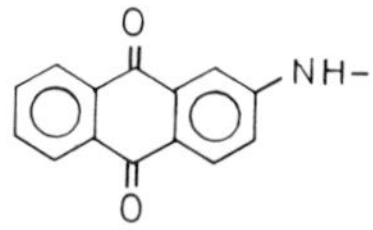 O NH– / O	80	>360	*20*
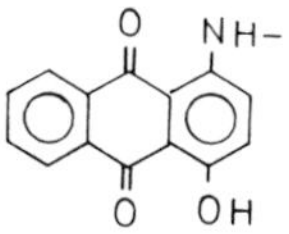 O NH– / O OH	96	353 (dec.)	*20*
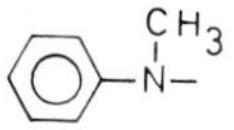 CH_3 –N–	92 48	199 192	*18* *20*
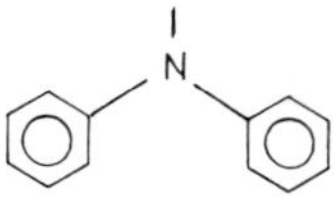 N	64 66	389–390 360	*18* *20*
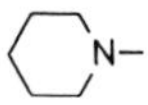 N–	61	281–283	*20*
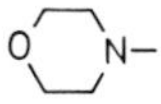 O N–	72	340	*20*
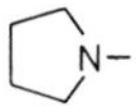 N–	—	287–293 (dec.)	*20*
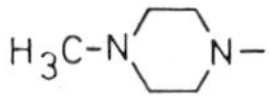 H_3C-N N–	77	233–236	*20*
3. Tertiary aromatic amines			
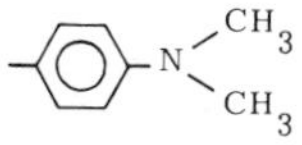 –N(CH_3)(CH_3)	60	276	*5*
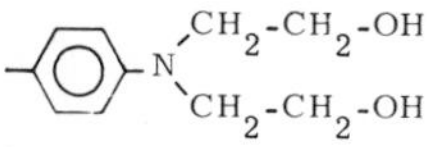 –N(CH_2-CH_2-OH)(CH_2-CH_2-OH)	—	>230 (dec.)	*5*

(*Continued*)

Table I (*Continued*)

	—	274–276 (dec.)	*5*
	—	>250	*5*
4. *N*-Alkylated heterocycles			
	30	320 (dec.)	*20*
	80	300 (dec.)	*10*
	42	286 (dec.)	*10*
	92 88	301 (dec.) 347 (subl.)	*10* *15*
5. Miscellaneous nucleophiles			
$R^1, R^2, R^3, R^4, R^5 = H$	90 green	240 (dec.)	*4*
$R^1, R^2 = CH_3; R^3, R^5 = H; R^4 = CH(CH_3)_2$	—, green	245 (dec.)	*4*
$R^1, R^4 = H; R^2, R^3, R^5 = CH_3$	—, green	Does not melt below 350	*4*
	— violet	—	*8*

R	Yield (%), color	mp (°C)	Ref.
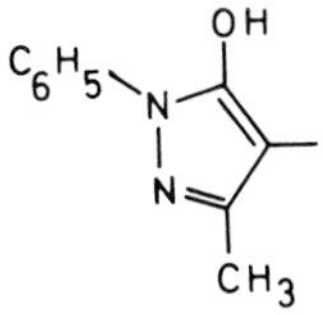 [structure: 1-phenyl-3-methyl-5-hydroxypyrazol-4-yl]	54, yellow-red	170	*11*
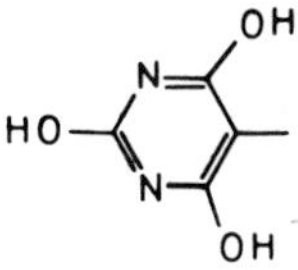 [structure: 2,4,6-trihydroxypyrimidin-5-yl]	72, dark red	330	*11*
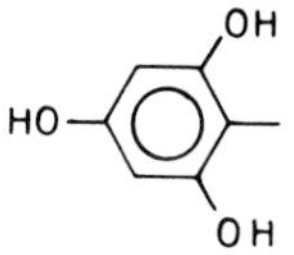 [structure: 2,4,6-trihydroxyphenyl]	67, red-violet	340–345	*9*
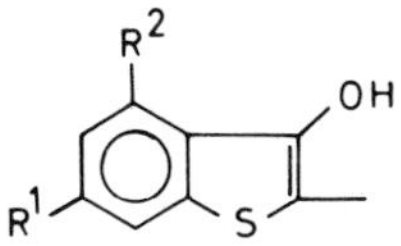 [structure: 3-hydroxybenzo[b]thiophen-2-yl with R^1, R^2]			
$R^1, R^2 = H$	12, violet	290	*11*
$R^1 = Cl, R^2 = CH_3$	35, brownish	326 (dec.)	*11*

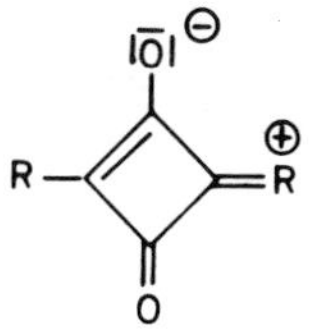

B. Squaraines obtained from the reaction of dialkyl squarates with nucleophiles

R	Yield (%), color	mp (°C)	Ref.
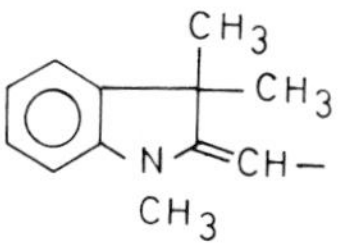 [structure: 1,3,3-trimethylindolin-2-ylidenemethyl]	40, —	346 (subl.)	*9*
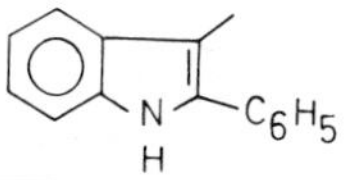 [structure: 2-phenylindol-3-yl]	—	400	*9*

(*Continued*)

Table I (*Continued*)

2-methylindol-3-yl (structure)	—	365	*9*
2,4-dihydroxy-3-methylphenyl (structure)	ca. 10, violet	360	*15*
$C_6H_5N(CH_3)$–	50 84	199 192	*18* *22*
2-$NO_2C_6H_4NH$–	65	279	*21* *22*
4-$O_2NC_6H_4NH$–	Main product	>350	*23*
4-$O_2NC_6H_4N(CH_3)$–	No reaction	—	*22*
C_6H_5NH–	11	360	*23*
piperidino (N–)	5.5	281–283	*23*

C. Squaraines obtained from the reaction of substituted hydroxycyclobutenediones with nucleophiles

R^1	R^2	Yield (%), color	mp (°C)	Ref.
1. Pyrroles and related compounds				
2,4-dimethylpyrrol-3-yl (CH_3, NH, CH_3)	3,5-dimethylpyrrol-2-ylidene (CH_3, NH, CH_3)	—, violet	240–250 (dec.) 286 (subl.)	*15*

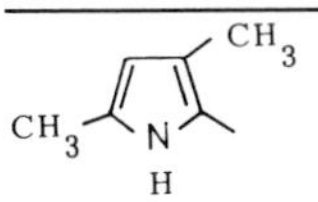 CH_3, CH_3, N, H	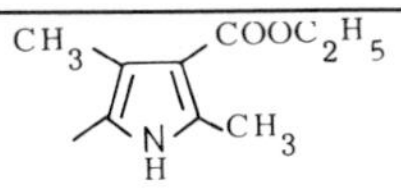 CH_3, $COOC_2H_5$, N, H, CH_3	75, violet	183	*15*
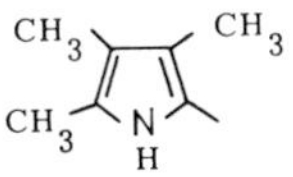 CH_3, CH_3, CH_3, N, H	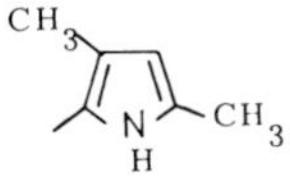 CH_3, N, H, CH_3	65, violet	184	*15*
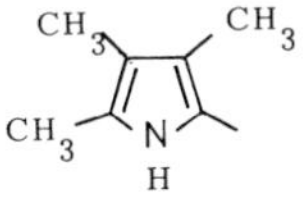 CH_3, CH_3, CH_3, N, H	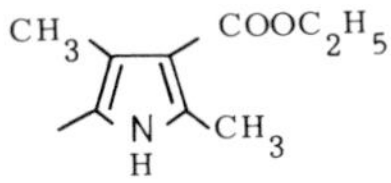 CH_3, $COOC_2H_5$, N, H, CH_3	60, violet	296	*15*
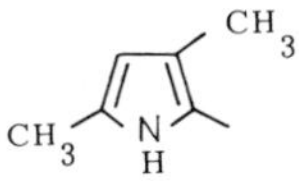 CH_3, CH_3, N, H	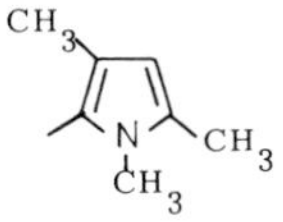 CH_3, N, CH_3, CH_3	75, violet	240	*15*
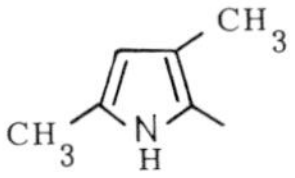 CH_3, CH_3, N, H	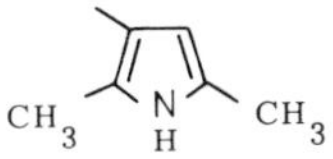 CH_3, N, H, CH_3	35, green	250	*15*
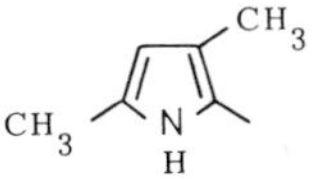 CH_3, CH_3, N, H	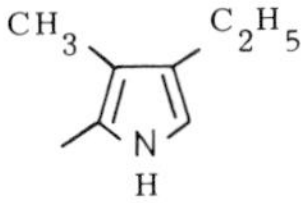 CH_3, C_2H_5, N, H	100, violet	272 (subl.)	*15*
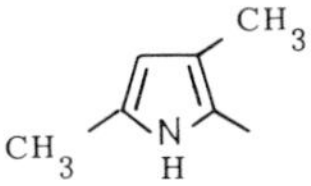 CH_3, CH_3, N, H	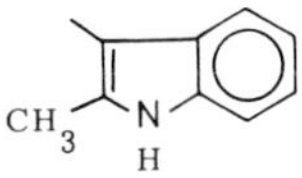 CH_3, N, H	72, green	265	*15*
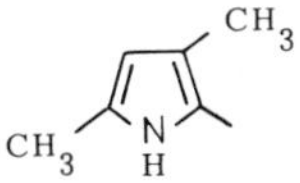 CH_3, CH_3, N, H	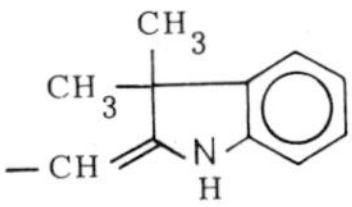 CH_3, CH_3, —CH, N, H	90, gray-blue	270 (subl.)	*15*
2. Primary and secondary amines				
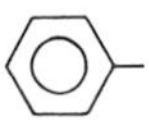	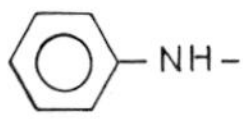 NH–	72, yellow needles	275	*25*
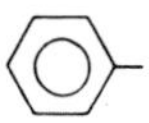	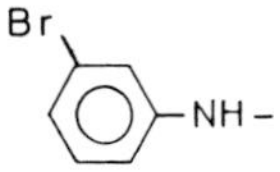 Br, NH–	83, yellow needles	280	*25*
	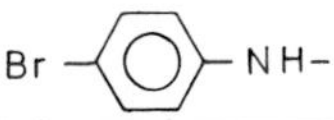 Br, NH–	80, yellow needles	284	*25*

(*Continued*)

Table I (*Continued*)

C_6H_5–	HO–C_6H_4–NH– (4-)	76, yellow powder	306	*25*
C_6H_5–	1-Naphthyl–NH–	80, yellow powder	276	*25*
C_6H_5–	2-Naphthyl–NH–	61, yellow powder	262	*25*
C_6H_5–	O_2N–C_6H_4–NH– (3-)	78, yellow powder	273	*25*
C_6H_5–	C_6H_5–N=N–C_6H_4–NH– (4-)	78, yellow powder	267	*25*
C_6H_5–	H_2N–C_6H_4–NH– (4-)	72, red powder	>320	*25*
C_6H_5–	HOOC–C_6H_4–NH– (4-)	52, yellow crystals	311–312	*26*
C_6H_5–	2-Ph–C_6H_4–NH–	85, yellow crystals	256–257	*26*
C_6H_5–	2,6-(CH_3)$_2$$C_6H_3$–NH–	68, yellow crystals	245 (dec.)	*26*
C_6H_5–	C_6H_5–N(CH_3)–	63, yellow crystals	186	*26*
C_6H_5–	6-Methyl-dibenzo[d,f][1,3]diazepin-5-yl (N–, CH_3)	70, yellow-orange	300 (dec.)	*26*

Phenyl–	n-C_4H_9-NH—	89, light yellow	197–198	*26*
Phenyl–	C_6H_5-CH_2-NH–	64, brownish	239–240	*26*
Phenyl–	Cyclohexyl-NH–	65, light yellow	211–213	*26*
Phenyl–	3-Pyridyl-NH–	73, yellow powder	242 (dec.)	*26*
Phenyl–	2-(NH–)phenyl-$N^{\oplus}$(H)= cyclobutene (O, $O^{\ominus}$, phenyl)	48, yellow crystals	220 (dec.)	*50*

$$R^1—C_4(^{\ominus}R^3)(=R^4)=\overset{\oplus}{R}{}^2$$

D. Squaraines obtained from the reaction of mercapto- and hydroselenocyclobutenediones with amines

R^1	R^2	R^3	R^4	Yield (%) color	mp (°C)	Ref.
Phenyl–	C_6H_5-NH–	S	O	78, red	213–215	*29*
Phenyl–	2-Br-C_6H_4-NH–	S	O	54, dark red	178–180	*29*
Phenyl–	4-Br-C_6H_4-NH–	S	O	51	214–216	*30*

(Continued)

Table I (*Continued*)

C_6H_5–	2-Cl-C_6H_4-NH–	S	O	43	173–175	*30*
C_6H_5–	3,4-Cl_2-C_6H_3-NH–	S	O	47	210–211	*30*
C_6H_5–	2-CH_3-C_6H_4-NH–	S	O	57	174–175	*30*
C_6H_5–	4-H_3C-C_6H_4-NH–	S	O	63, red	206–207	*29*
C_6H_5–	3-H_3CO-C_6H_4-NH–	S	O	39	175–178	*30*
C_6H_5–	4-H_3CO-C_6H_4-NH–	S	O	62, red	200–202	*29*
C_6H_5–	2-OH-C_6H_4-NH–	S	O	49	245–252	*30*
C_6H_5–	4-HO-C_6H_4-NH–	S	O	72	207–211	*30*
C_6H_5–	3-HO-C_6H_4-NH–	S	O	55	216–218	*30*
C_6H_5–	3,5-$(H_3C)_2$-C_6H_3-NH–	S	O	74	225–231	*30*
C_6H_5–	1-naphthyl-NH–	S	O	42	183–191	*30*

C_6H_5–	C_6H_5–NH–	Se	O	38, brown	160 (dec.)	*31*
C_6H_5–	Br–C_6H_4–NH–	Se	O	—	—	*30*

E. Squaraines obtained from the molecular rearrangement of dithiosquaramides

R^1	R^2	R^3	R^4	Yield (%) color	mp (°C)	Ref.
CH_3	CH_3	S	S	88	>287 (dec.)	*33*
—$(CH_2)_4$—		S	S	99	>290 (dec.)	*33*
—$(CH_2)_5$—		S	S	80	>295 (dec.)	*33*
—CH_2—CH_2—O—CH_2—CH_2—		S	S	80	>304 (dec.)	*33*

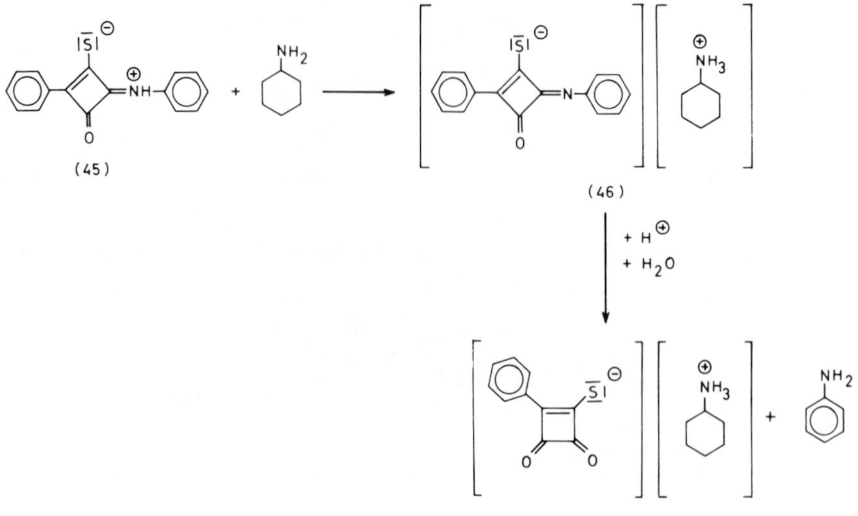

(45) (46) (47)

In agreement with its acidic character the thiosquaraine **45** reacts with an aqueous solution of $HgCl_2$ to give an immediate precipitate of **48**:

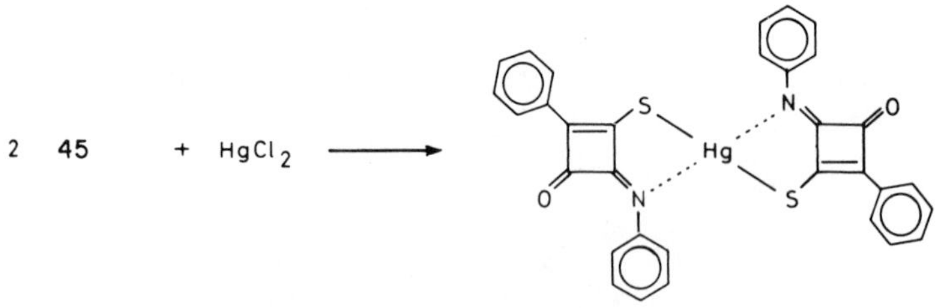

2 **45** + $HgCl_2$ ⟶ (48)

The formation of metal chelates from bis(amino)squaraines and various metal salts has been described in detail by Manecke and Gauger [*6,19*].

When an etheral solution of diazomethane was added to **45**, a clean reaction was observed affording the thiomethyl ether **49**:

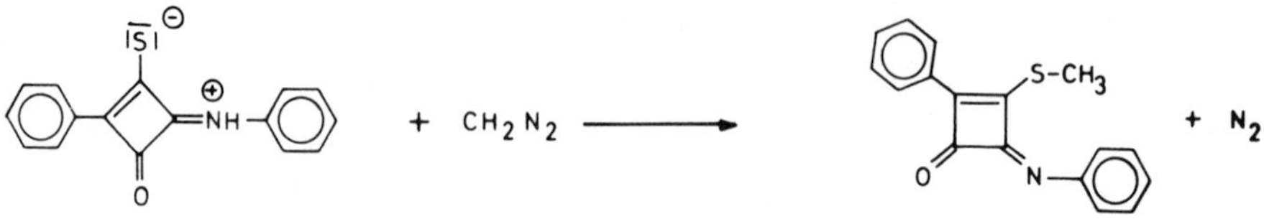

(45) + CH_2N_2 ⟶ (49) + N_2

This rearrangement was the first example of a successful alkylation of a squaraine [*29*]. It still awaits thorough investigation. Hünig *et al.* [*22*] reported the alkylation of squaraines by means of methyl iodide and a strong base such as potassium *tert*-butoxide or ethyldiisopropylamine. Starting from the squaraine **7**, they obtained the dimethylated product **50** in 84% yield:

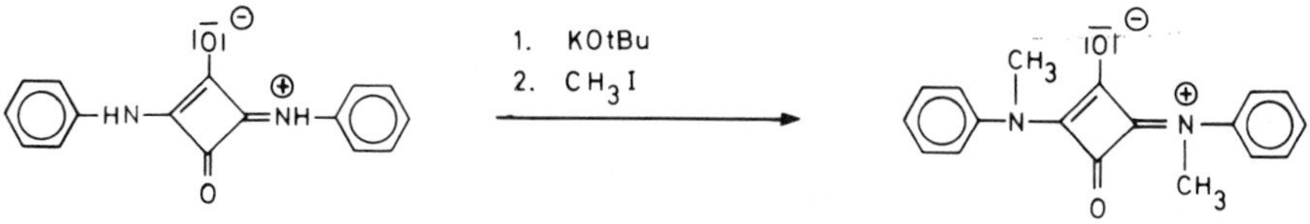

1. KOtBu
2. CH_3I

(7) (50)

The squaraine **51**, however, rearranged under these conditions to afford **52**. This is comparable to the formation of **49** from **45**.

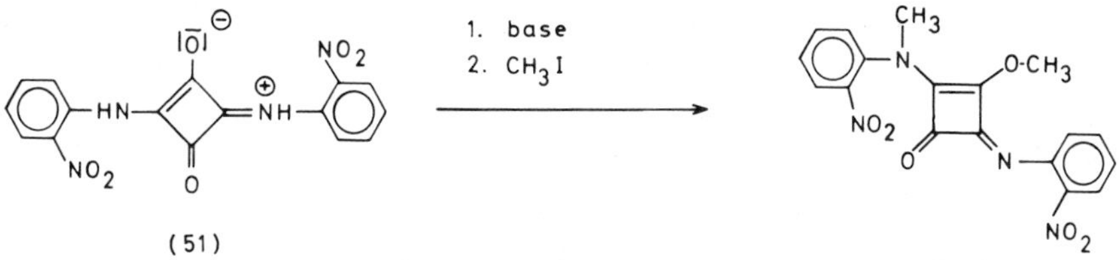

1. base
2. CH_3I

(51) (52)

B. ADDITION REACTIONS

During an investigation of the action of pyrroles on squaric acid, Treibs and Jacob [*15*] discovered several interesting reactions of the bis(pyrryl)squaraines. Their observation that bis(pyrryl)squaraines react with pyrrylhydroxycyclobutenediones to form addition compounds such as **53** and **54** proved to be of great preparative value.

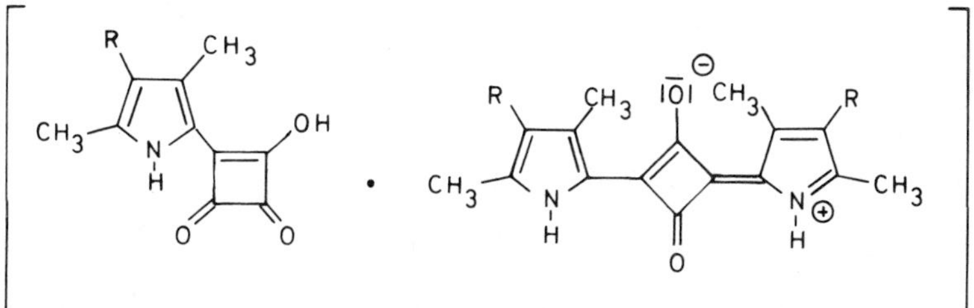

(**53,** R = H)
(**54,** R = $COOC_2H_5$)

When treated with an aqueous solution of potassium cyanide, the addition compounds **53** and **54** decompose to afford the potassium salts **55** and **56**; the latter compound is formally generated by the addition of potassium cyanide to the appropriate substituted bis(pyrryl)squaraine. Treatment of **55** with hydrochloric

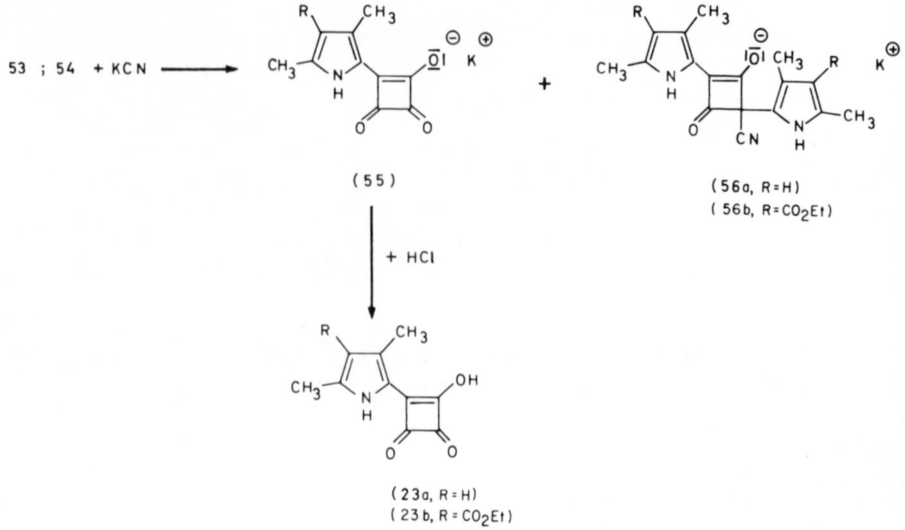

53 ; 54 + KCN → (55) + (56a, R=H) (56b, R=CO_2Et)

(55) + HCl → (23a, R = H) (23b, R = CO_2Et)

Scheme 4

acid afforded the pyrrylhydroxycyclobutenediones **23** (Scheme 4), which were used for the generation of unsymmetric bis(pyrryl)squaraines (see Section IV,C).

A thorough product analysis showed that the reaction of **1** with pyrroles leads not only to bis(pyrryl)squaraines but also to colorless by-products. The elemental analysis of the by-product **59** obtained from the reaction of **1** with 2,4-

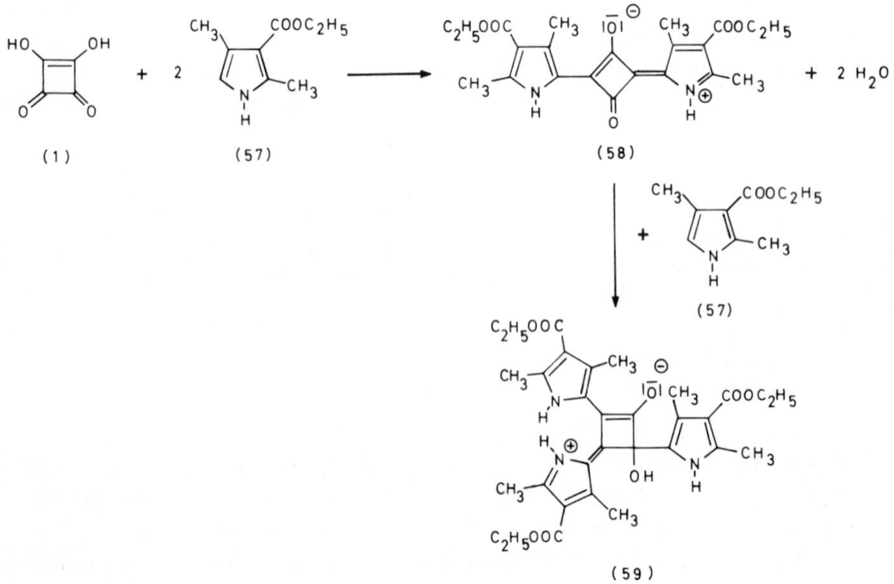

(1) + 2 (57) → (58) + 2 H_2O

(58) + (57) → (59)

Scheme 5

dimethyl-3-ethoxycarbonylpyrrole (**57**) indicated a reaction of **1** with **57** in the molar ratio 1 : 3. A possible pathway for the generation of **59** is outlined in Scheme 5. According to this scheme, **59** is generated by the addition of one molecule of pyrrole **57** to the squaraine **58**. Proof of this assumption was derived from the action of the pyrrole **60** on the squaraine **58**, which led in good yield to the addition product **61**. When **59** was heated with **1**, the squaraine **58** was

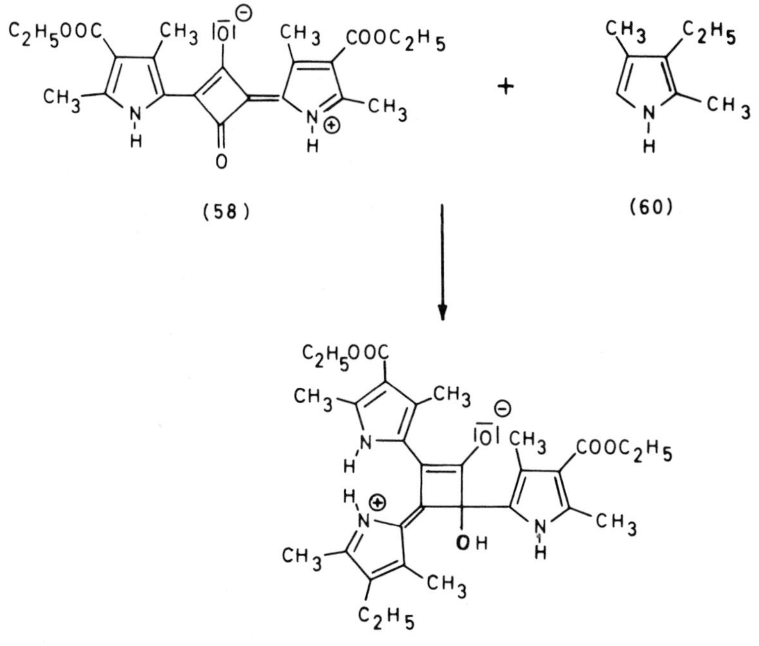

(58) (60)

(61)

formed, indicating that one of the three pyrryl substituents is relatively weakly bonded to the four-membered ring. Formation of **58** was also observed on treatment of **59** with diazonium salts [*9*].

The reaction of 2-methyl-3-ethoxycarbonylpyrrole with squaric acid (**1**) afforded the squaraine **62** only in low yield. Surprisingly, the tripyrryl compound **63** was generated as the main product and could be isolated in 79% yield:

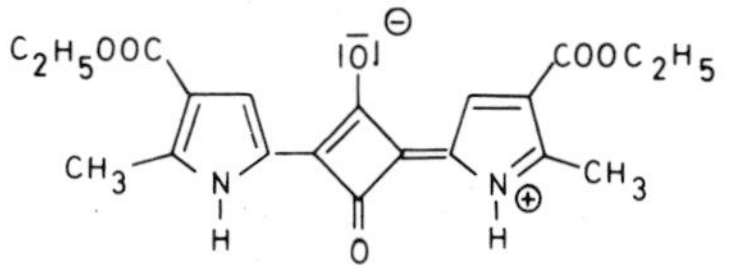

(62) 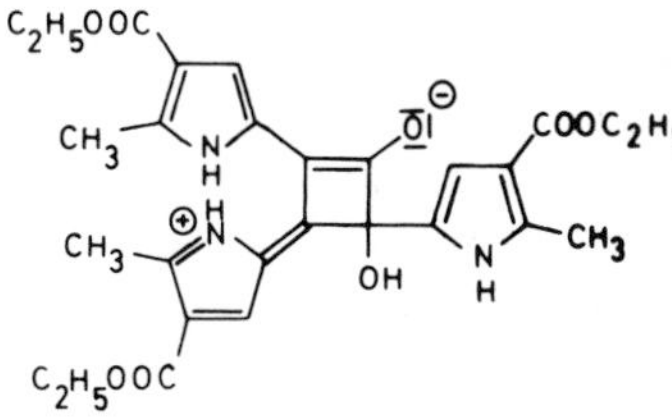 (63)

When resorcinol and squaric acid were melted together, the mixture became deep blue; after a short period of time, the blue color vanished because of the addition of a third molecule of resorcinol to the intermediate squaraine. The structure of the colorless addition product is **64**. The reaction of **1** with phloroglucinol (molar

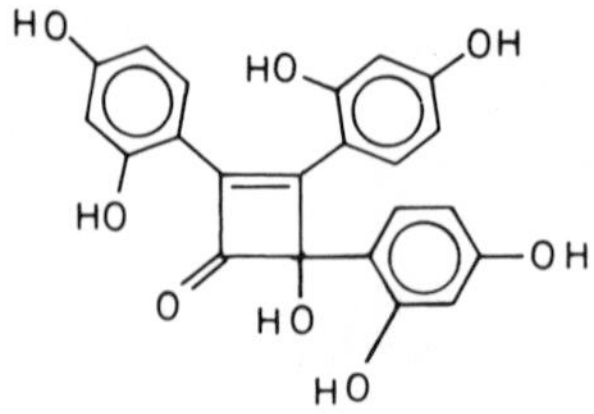

(64)

ratio 1 : 2) in boiling 60% acetic acid led to the red squaraine **65**. When **65** was suspended in pyridine and acetic anhydride was added, yellow crystals of the heptaacetyl derivative **66** were precipitated:

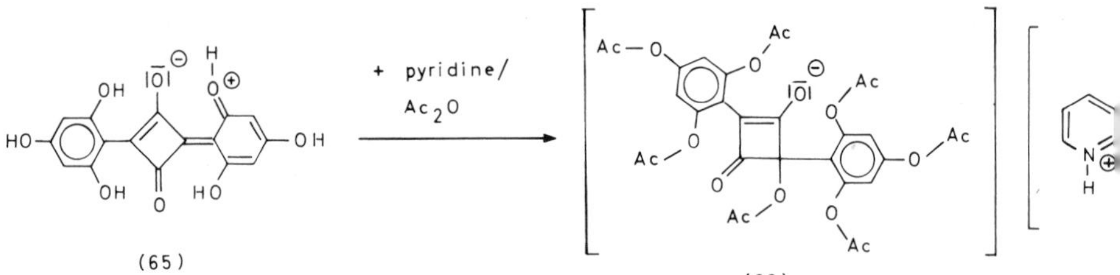

(65) + pyridine/ Ac_2O → (66)

The reduction of a squaraine has also been reported [*35*]. The action of $SnCl_2$ on **67** in a mixture of CF_3COOH and HCl led to the formation of the cyclobutenolone **68**. On ommission of the trifluoroacetic acid no reduction of the squaraine took place; instead, dimerization to **69** was observed, as can be seen from Scheme 6.

C. SUBSTITUTION REACTIONS

1. Reaction with Phosgene and with Thionyl Chloride (Exchange of O for Cl)

In a fundamental investigation, Seitz *et al.* [*36*] found that diaminosquaraines **70** react readily with phosgene and with thionyl chloride. By using chloroform or dichloromethane as a solvent, they obtained the highly hygroscopic salts **71**, which were isolated and characterized as the hexachloroantimonates **72**. In the presence of a catalytic amount of dimethylformamide, the chlorides **71** were obtained in almost quantitative yields [*37*]. The chlorides **71** proved to be highly reactive. Some of the transformations carried out on them are summarized in Scheme 7. The compounds **74** have also been prepared by an alternative route

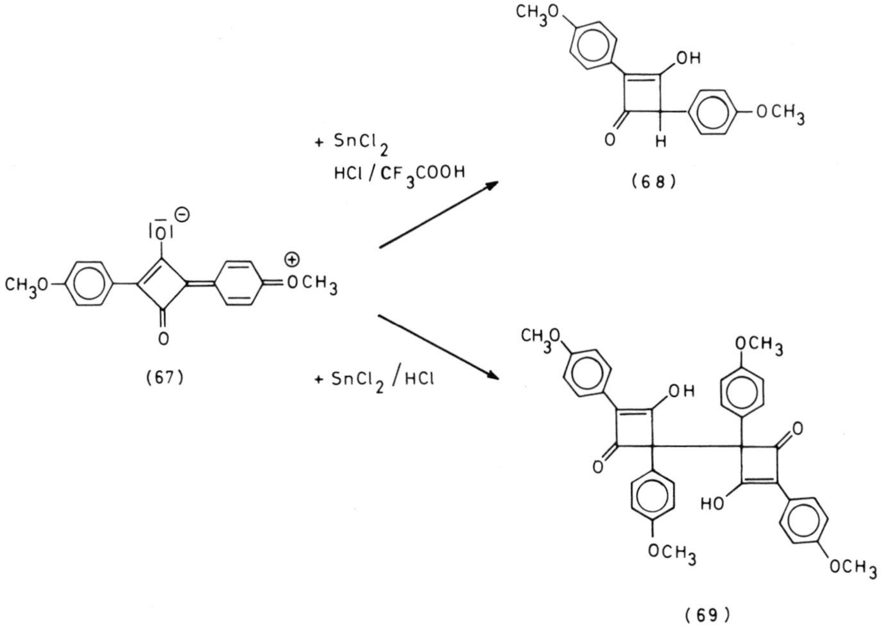

Scheme 6

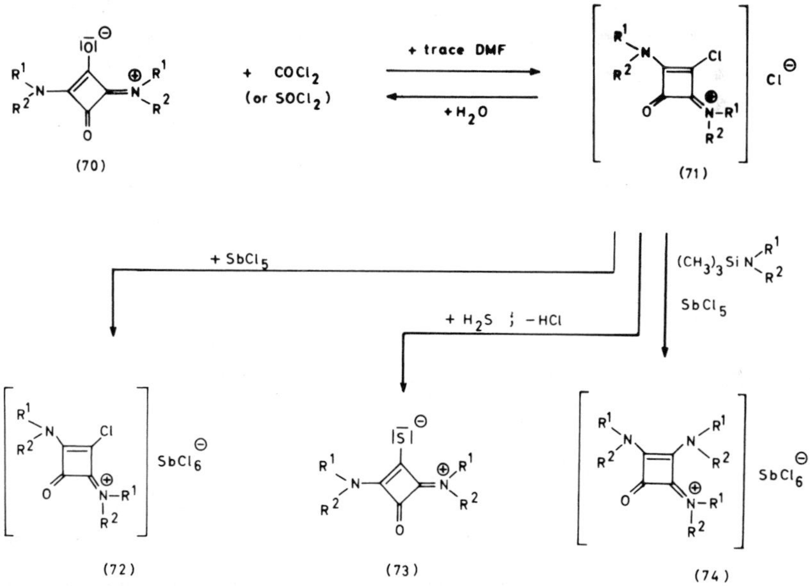

Scheme 7

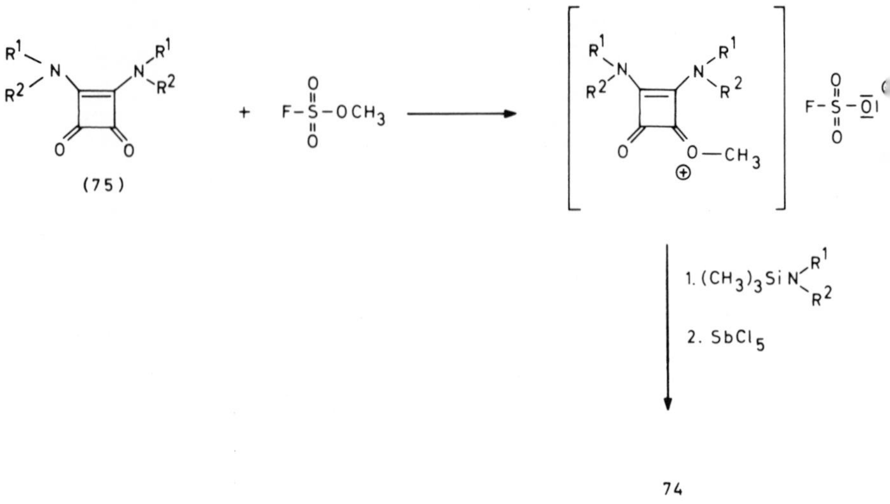

Scheme 8

starting from squaramides. The reaction pathway is outlined in Scheme 8 [*37*]. The compounds **74** obtained according to Scheme 7 are summarized in Table II.

In this connection it should be mentioned that Treibs *et al.* [*16*] had already succeeded much earlier in the preparation of the hydrobromides of bis(pyrryl)iminosquaraines **77**, which are closely related to the hexachloroantimonates **74**. When the aminocyclobutenediones of the type **76** were heated with a substituted pyrrole in a mixture of ethanol and HBr, the hydrobromides **77** were obtained as violet crystals in good yield:

Table II. Compounds **74** Obtained from Bis(amino)squaraines[a]

Compound no.	R^1	R^2	Yield (%)	Recrystallized from	mp (°C)
74a	CH_3	CH_3	57	CH_3NO_2/ether	>175 (dec.)
74b	—$(CH_2)_4$—		30	CH_3OH	>145 (dec.)
74c	—$(CH_2)_5$—		48	C_2H_5OH	>148 (dec.)
74d	—CH_2—CH_2—O—CH_2—CH_2—		31	CH_3OH	>172 (dec.)

[a] From Seitz *et al.* [*37*].

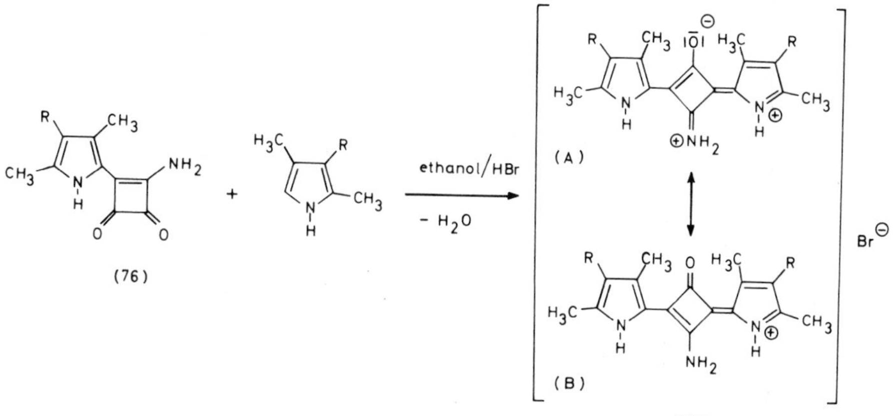

(76)

(77)

The acidity of the compounds **77** was too low to allow the generation of "free" iminosquaraines from them. In agreement with this, the IR spectra of **77** exhibit a strong C=O band at ca. 1750 cm^{-1}, indicating a high contribution of the resonance structure (B) to the ground state of **77**.

2. Exchange of the Pyrryl Substituents (Exchange of C for C')

In 1968 Treibs and Jacob [*15*] demonstrated that the substituents in squaraines can be changed with relative ease. Thus, action of the pyrrole **57** on the bis(indolyl)squaraine **78** afforded **58**. Much later this type of reaction gained considerable importance in the transamidation of bis(amino)squaraines (Section VI,C,3).

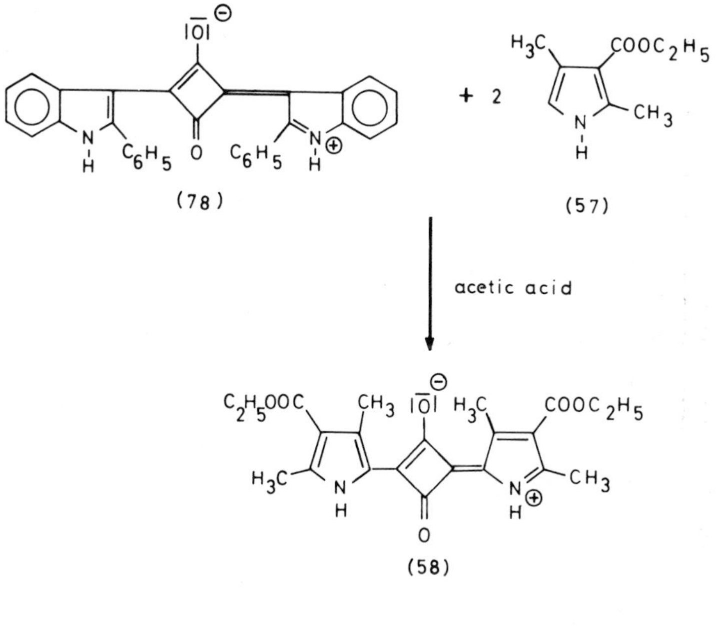

(78)

(57)

(58)

Table III. Transamidation of Squaraines **70** to Give Squaraines **79**[a]

Starting material				Product					
Compound	R^1	R^2	Reaction with	Compound	R^3	R^4	Solvent	Yield (%)	mp (°C)
70a	2-$NO_2C_6H_4$	H	$C_6H_5NH_2$	**79a**	C_6H_5	H	Without	90	>350
70a	2-$NO_2C_6H_4$	H	1-$C_{10}H_7NH_2$	**79b**	1-$C_{10}H_7$	H	DMF	100	344
70a	2-$NO_2C_6H_4$	H	1,2-$C_6H_4(NH_2)_2$	**79c**	2-$NH_2C_6H_4$	H	1,2-Dichlorobenzene	93	243
70a	2-$NO_2C_6H_4$	H	piperidine ($C_5H_{10}NH$)	**79d**	$-(CH_2)_5-$		DMF	69	281–283
70a	2-$NO_2C_6H_4$	H	$C_6H_{11}NH_2$	**79e**	C_6H_{11}	H	DMF	89	340 (dec.)

70a	$2\text{-}NO_2C_6H_4$–	H	C_6H_{11}–CH_2–NH_2	**79f**	C_6H_5–CH_2–	H	DMF	90	350 (dec.)
70b	$4\text{-}O_2N$–C_6H_4–	H	C_6H_{11}–CH_2–NH_2	**79f**	C_6H_5–CH_2–	H	Without	72	350 (dec.)
70c	$2,5\text{-}Cl_2C_6H_3$–	H	piperidine ($C_5H_{10}NH$)	**79g**	–$(CH_2)_5$–		DMF	98	281–283
70c	$2,5\text{-}Cl_2C_6H_3$–	H	C_6H_{11}–CH_2–NH_2	**79f**	C_6H_5–CH_2–	H	Without	80	350 (dec.)

(Continued)

Table III (*Continued*)

Starting material			Reaction with	Product			Solvent	Yield (%)	mp (°C)
Compound	R^1	R^2		Compound	R^3	R^4			
70c	$Cl_2C_6H_3$–	H	$O(CH_2CH_2)_2NH$	**79h**	$-CH_2-CH_2-O-CH_2-CH_2-$		DMF	97	>340
70d	$Cl-C_6H_4$–	H	$C_6H_5-CH_2-NH_2$	**79f**	$C_6H_5-CH_2-$	H	Without	44	>340
70e	C_6H_5–	H	$(CH_2)_5NH$	**79g**	$-(CH_2)_5-$		DMF	84	>340
70e	C_6H_5–	H	$C_6H_5-CH_2-NH_2$	**79f**	$C_6H_5-CH_2-$	H	Without	95	>340
70f	C_6H_5–	CH_3	$C_6H_5-CH_2-NH_2$	**79f**	$C_6H_5-CH_2-$	H	Without	80	>340

[a] From Hünig and Pütter [*21*] and Ehrhardt *et al*. [22].

3. *Transamidation of Squaraines (Exchange of N for N′)*

Hünig *et al.* [*21,22*] discovered that the amino substituents of squaraines **70** are easily exchanged for other amino groups at elevated temperature. Such substitution reactions proceed cleanly provided that an excess of amine R^3R^4NH is used and that this amine is more basic than the amine R^1R^2NH, which is to be eliminated. When the components were heated to 150°–200°C, neat or in a solvent, the reaction was finished within a few minutes; the new squaraines are normally obtained in high purity. The preparative value of this method is that it

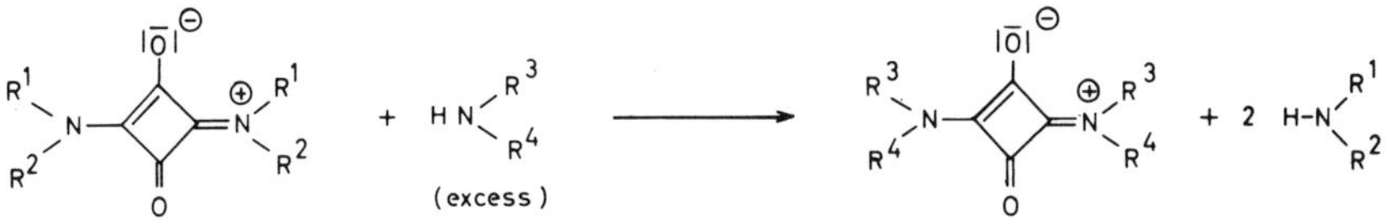

(70) + HNR^3R^4 (excess) ⟶ (79) + 2 H-NR^1R^2

opens a fairly general route to squaraines derived from aliphatic amines. Thus far, only a few squaraines of this type have been prepared by other routes. Table III gives a survey of squaraines that have been obtained by the transamidation procedure.

Moreover, the process of transamidation allows the synthesis of unsymmetrically substituted squaraines **80**, as can be seen from the following reaction scheme:

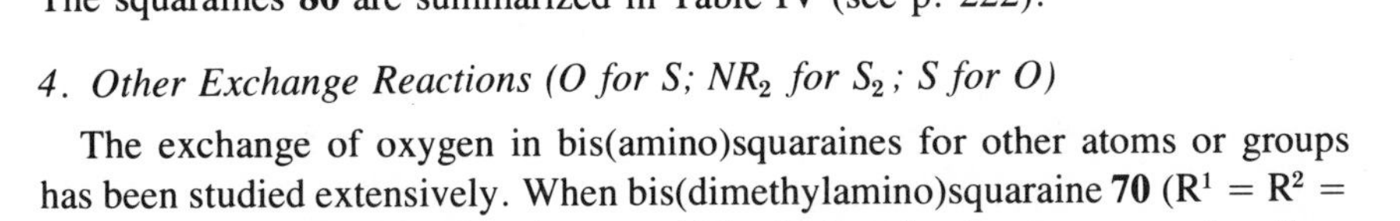

(70) + H-N(CH_3)-Ph ⟶ (80) + H-NR^1R^2

The squaraines **80** are summarized in Table IV (see p. 222).

4. *Other Exchange Reactions (O for S; NR_2 for S_2; S for O)*

The exchange of oxygen in bis(amino)squaraines for other atoms or groups has been studied extensively. When bis(dimethylamino)squaraine **70** ($R^1 = R^2 = CH_3$) was heated with phosphorus sulfide (P_4S_{10}) in the presence of sodium hydrogen carbonate in dimethoxyethane, the dithiosquaraine **83** ($R^1 = R^2 = CH_3$) was isolated in 40% yield [*36*]:

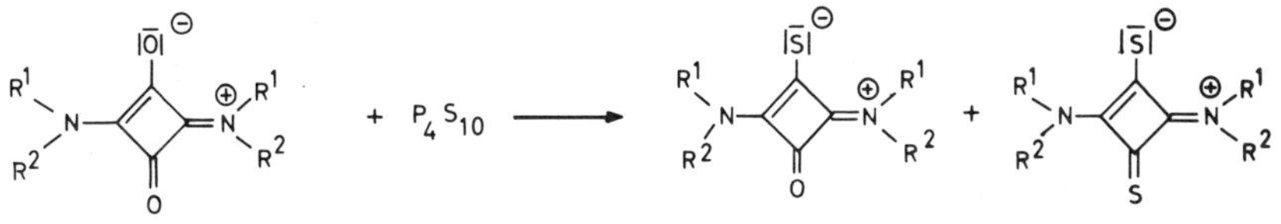

(70) + P_4S_{10} ⟶ (73) + (83)

Table IV. Unsymmetrically Substituted Squaraines **80** Obtained from Transamidation Reactions[a]

Compound	R^3	R^4	R^5	Yield (%), color	mp (°C)
80a	NO_2	H	H	43, red crystals	192
80b	H	NO_2	H	75, yellow	330 (dec.)
80c	Cl	H	Cl	74, yellow	218

[a] From Ehrhardt *et al.* [*22*].

The use of dichloromethane as a solvent afforded mixtures of monothiosquaraine **73** and dithiosquaraine **83**. Separation of these compounds proved extremely difficult [*36*]. The monothiosquaraines obtained according to this procedure are summarized in Table V.

It has been demonstrated that the exchange of sulfur for oxygen in bis(amino)squaraines can be accomplished in high yield by the use of ethoxycarbonyl isothiocyanate [*38*].

Surprisingly, when **81** was treated with potassium hydrogen sulfide, no exchange of oxygen was observed. Instead, the dimethylamino groups were removed, giving rise to the generation of the dipotassium salt **84** [*39*]. By means of

Table V. Monothiosquaraines **73**[a]

Compound	R^1	R^2	Yield (%)	mp (°C)
73a	CH_3	CH_3	11	263
73b	$—CH_2—CH_2—O—CH_2—CH_2—$		53	302
73c	$—(CH_2)_5—$		63	305

[a] From Seitz *et al.* [*36*].

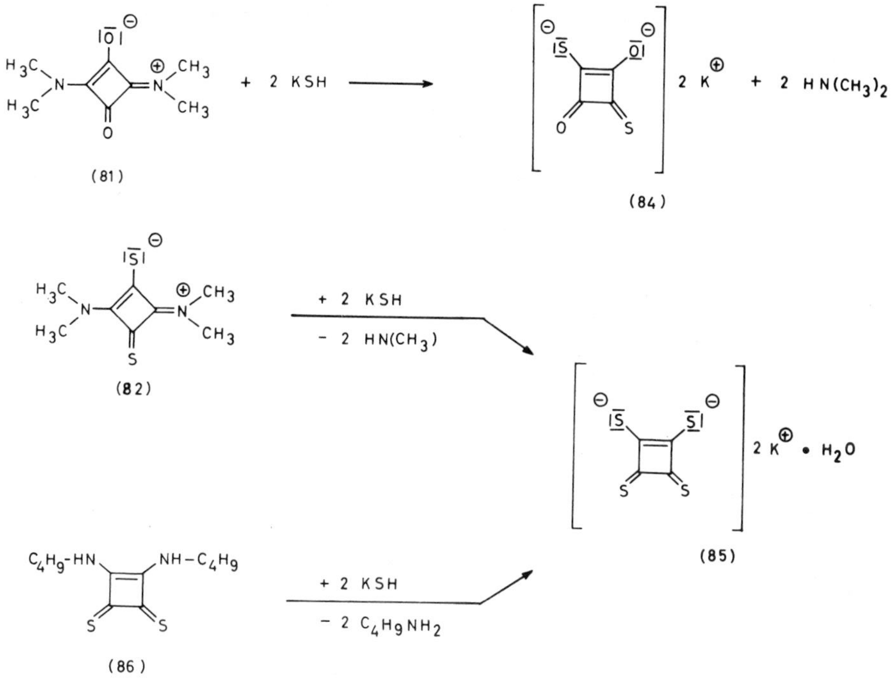

$(81) + 2\ KSH \rightarrow [(84)]\ 2\ K^{\oplus} + 2\ HN(CH_3)_2$

(81) (84)

(82) $\xrightarrow[-\ 2\ HN(CH_3)]{+\ 2\ KSH}$ (85) $2\ K^{\oplus} \cdot H_2O$

C_4H_9-HN NH-C_4H_9 (86) $\xrightarrow[-\ 2\ C_4H_9NH_2]{+\ 2\ KSH}$ (85)

(82) (85) (86)

Scheme 9

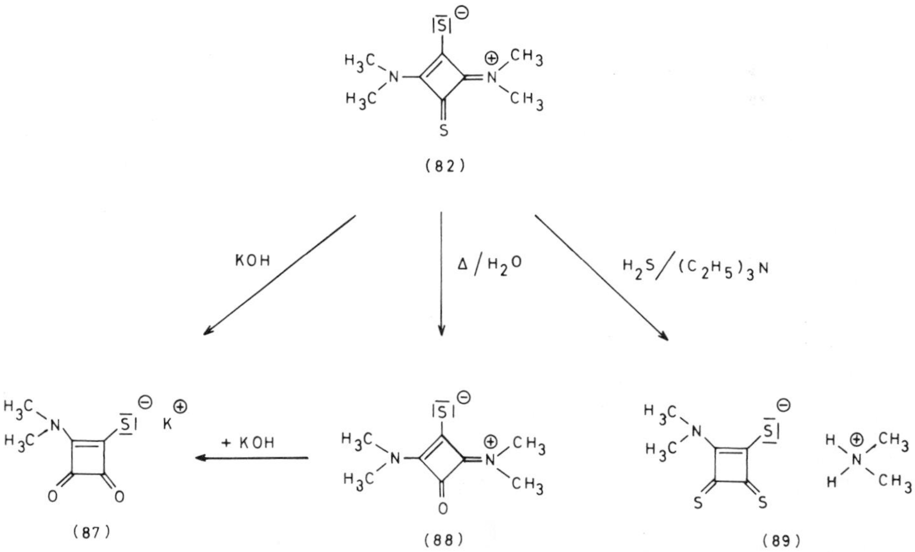

(82)

KOH → (87)

Δ / H_2O → (88)

$H_2S / (C_2H_5)_3N$ → (89)

(88) $\xrightarrow{+\ KOH}$ (87)

(87) (88) (89)

Scheme 10

this reaction the sulfur analog of squaric acid has been synthesized and isolated in the form of its dipotassium salt **85**.

As shown in Scheme 9, treatment of **82** with an equimolar amount of freshly prepared potassium hydrogen sulfide led to the formation of **85**. This compound crystallized as yellow-orange crystals, which were purified by dissolution in ethanol/water and precipitation with ether [*39,40*]. Scheme 9 also contains an alternative method [*40*] for the preparation of **85**.

The preparative value of the previously described substitution reactions of squaraines has been demonstrated by the synthesis of a variety of differently substituted derivatives of squaric acid [*41*]. A few of these transformation reactions are summarized in Scheme 10 (see p. 223).

D. SQUARIC ACID BISAMIDINES AND RELATED COMPOUNDS (EXCHANGE OF O FOR N—R AND S FOR N—R)

When a solution of *p*-tosyl isocyanate (**90**) and the squaraine **70** in toluene was heated at reflux for 6 hr, the product was an iminosquaraine (**91**) that represents a new type of squaraine [*42*]:

(70) + $H_3C-C_6H_4-SO_2-NCO$ (90) ⟶ (91) + CO_2

In contrast to this, the reaction of **70** with **90** in the absence of a solvent led to the formation of squaric acid bisamides **92** (diiminosquaraines). Under much milder conditions, compounds **92** were obtained by the action of *p*-tosyl isocyanate on the highly reactive dithiosquaraines **83**.

Hünig *et al.* [*43*] succeeded in the preparation of several squaric acid bisamidinium salts. Transamidation of the squaraine **93** with *N*-phenyl-1,3-diaminopropane led to the formation of the squaraine **94**. When **94** was heated in polyphosphoric acid to 170°C, ring closure was observed in quan-

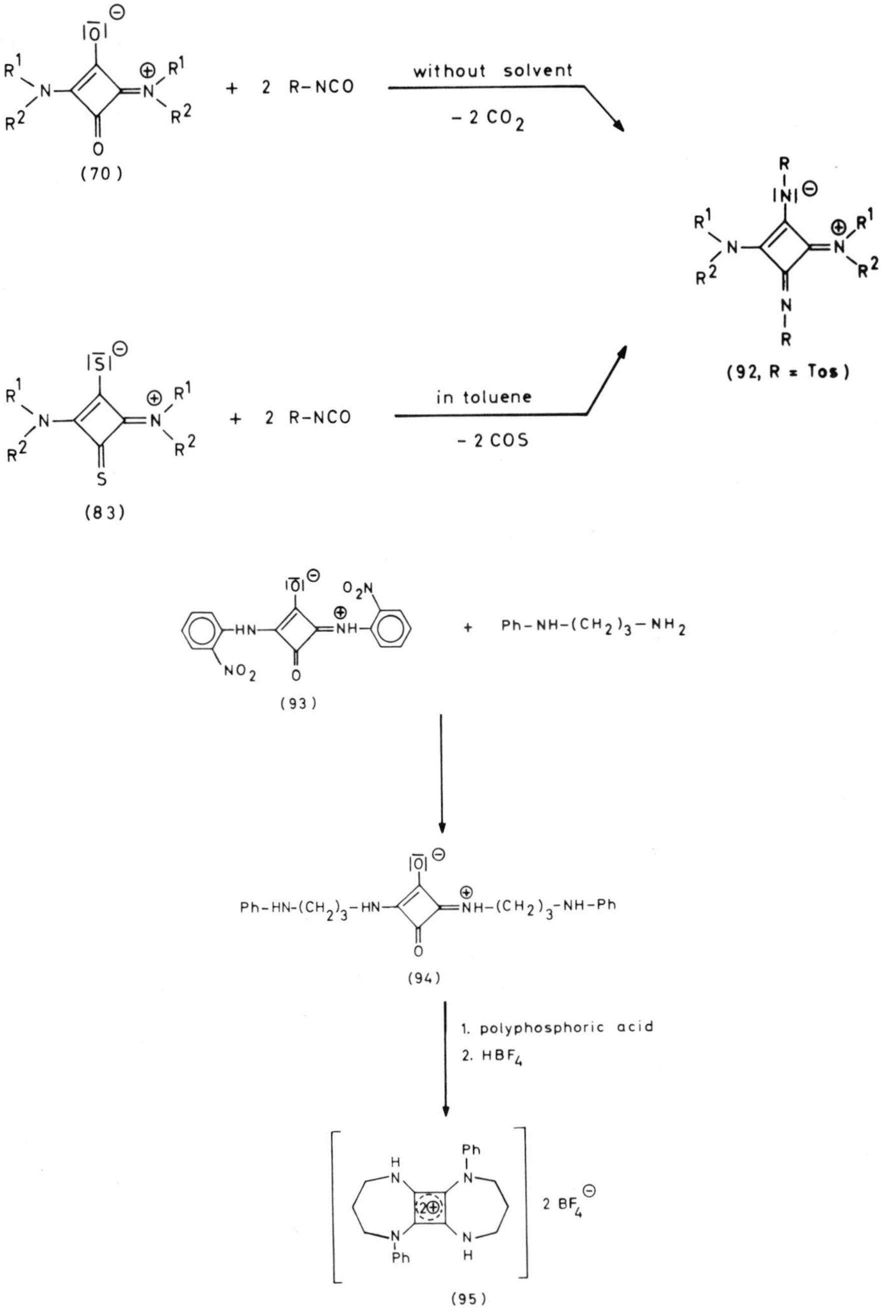

(70) + 2 R–NCO — without solvent, $-2\ CO_2$ → (92, R = Tos)

(83) + 2 R–NCO — in toluene, $-2\ COS$ → (92, R = Tos)

(93) + Ph–NH–$(CH_2)_3$–NH_2 → (94)

(94) — 1. polyphosphoric acid; 2. HBF_4 → (95) $2\ BF_4^{\ominus}$

Scheme 11

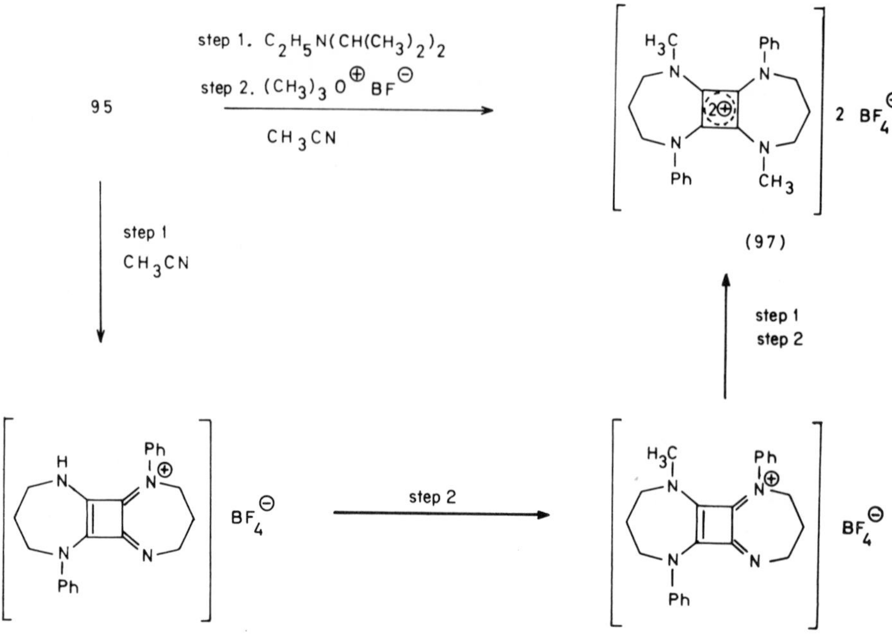

Scheme 12

titative yield. After the addition of HBF_4, the ring-closed product was isolated as the tetrafluoroborate **95,** as can be seen from Scheme 11. One-step and multistep procedures for the methylation of **95** to give **97** are outlined in Scheme 12.

In a similar fashion [*45,46*] it was also possible to prepare the bisamidinium salt **99**. When **99b** was added to a mixture of dimethylformamide and triethylamine and the resulting suspension was stirred for a short period of time, the free bisamidine **100** was formed. Under alkaline conditions the bisamidinium salt **99** is readily oxidized to give the tetraazabinaphthylene (cyclobutaquinoxaline) **101.** Compounds **100** and **101** are easily interconverted by oxidation/reduction (Scheme 13).

Compounds closely related to the tetraazabinaphthylene **101** have been easily prepared by the reaction of **85** with hydrazines [*32*]. This is demonstrated in Scheme 14. As shown in the scheme, the reaction of **85** with hydrazine did not stop at the stage of compound **102**. Instead, the two remaining sulfur groups in

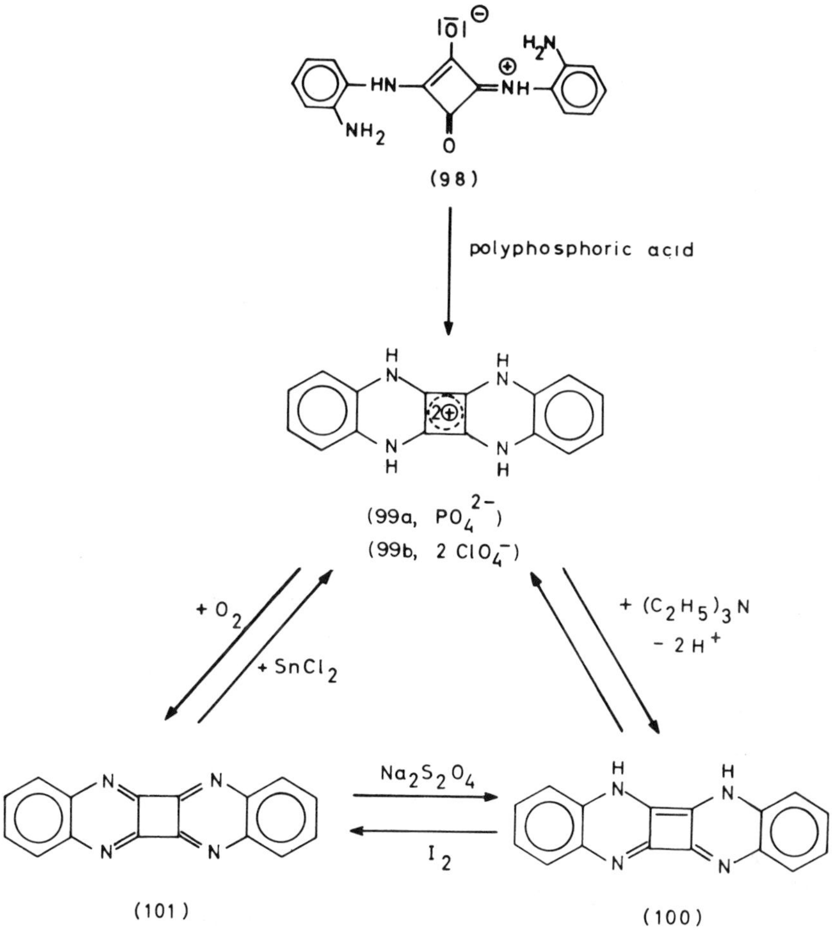

Scheme 13

102 were also exchanged for hydrazine, including an oxidation step. Finally, the tetrahydrazonocyclobutane **103** was generated. Compound **103** forms olive green crystals that are air stable. They are readily recrystallized from a mixture of dimethylformamide and water.

In a similar process, **85** reacted with methylhydrazine as well as with phenylhydrazine to give the N-substituted compounds **104a** and **104b** (wherein the four methyl groups of **104a** have been replaced by phenyl groups), respec-

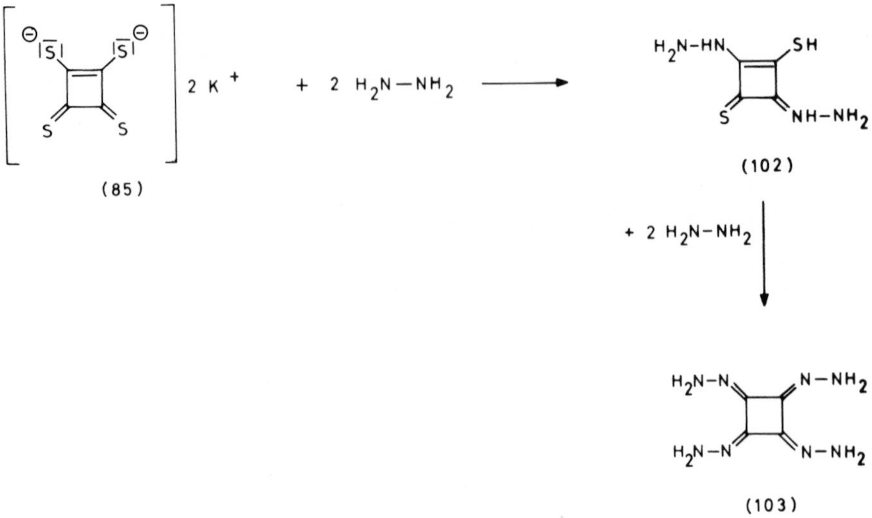

Scheme 14

tively. When **104a** was treated with an aqueous solution of formaldehyde, a ring closure was observed to give the tricyclic system **105**:

The synthesis of the compounds **103**, **104a**, and **104b** has also been accomplished by the reaction of the dithiosquaramide **86** with hydrazines [*32*].

In 1974 the highly interesting compounds **107** were prepared [*47*]. Despite their structural similarity to squaraines, no resonance forms of type **13** can be drawn. However, they may well be represented by an uncharged resonance form (cyclobutane structure). For these reasons, **107**—as well as several oxygen analogs that were prepared by the reaction of triphenylphosphoranylideneketene with carbonyl compounds [*48*]—are to be regarded as derivatives of tetraoxocyclobutane. To complete this survey the synthesis and one aspect of the reactivity of compounds **107** are outlined. As seen in Scheme 15, the reaction of the phosphorane **106** with a carbonyl compound (R^2 = H) in the molar ratio 1 : 1 leads to the formation of tetraoxocyclobutane derivatives **108.** When the reaction is carried out in the molar ratio 2 : 1 it is possible to isolate **107.** The action of 1 mole of any carbonyl compound (R^4 = H) on **107** finally affords the unsymmetrically substituted tetraoxocyclobutane derivatives **109.** Cyclobutanes **108** and **109,** prepared according to Scheme 15, are summarized in Table VI.

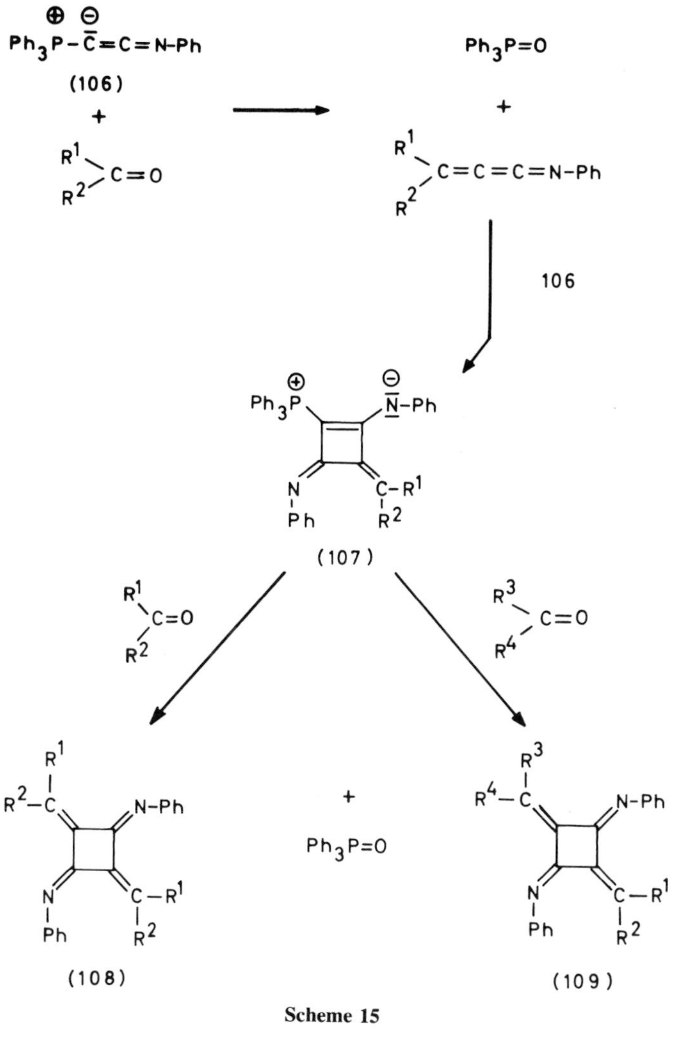

$$Ph_3\overset{\oplus}{P}-\overset{\ominus}{\bar{C}}=C=N\text{-}Ph \; (106) + R^1R^2C=O \longrightarrow Ph_3P=O + R^1R^2C=C=C=N\text{-}Ph$$

106

(107)

(108) + $Ph_3P=O$ (109)

Scheme 15

VII. Final Remarks

We hope to have demonstrated in this review that the chemistry of squaraines is a fascinating chapter of modern organic chemistry. Moreover, the results obtained so far are not only of purely scientific importance. For several years squaraines have been investigated intensively in industry with respect to their practical use. The growing number of patents on squaraines indicates the strong industrial interest in this class of compounds.

Table VI. Compounds **108** and **109** Obtained from the Reaction of Phosphoranes with Carbonyl Compounds[a]

Compound	R^1	R^2	R^3	R^4	Yield (%), color	mp (°C)
108a	$O_2N-C_6H_4-$	H	—	—	65, dark red	245
108b	$H_3C-C_6H_4-$	H	—	—	36, red	239
108c	1-naphthyl	H	—	—	59, red	248
109a	$H_3C-C_6H_4-$	H	$O_2N-C_6H_4-$	H	40, red-violet	221
109b	1-naphthyl	H	$O_2N-C_6H_4-$	H	35, red-violet	218

[a] From Bestmann and Schmid [*47*].

Acknowledgment

I am grateful to Dr. R. Lantzsch, Leverkusen, for stimulating discussions and to Dr. O. Weissbach of the Beilstein Institut, Frankfurt am Main, for his advice concerning the nomenclature of several compounds. Furthermore, I should like to express my thanks to Dr. T. W. Flechtner of Cleveland State University for perusal of this review.

References

1. S. Cohen, J. R. Lacher, and J. D. Park, *J. Am. Chem. Soc.* **81,** 3480 (1959).
2. Chemische Werke Hüls, "Neues aus Hüls," p. 47. Folge, 1964.
3. A. Treibs and K. Jacob, *Angew. Chem., Int. Ed. Engl.* **4,** 694 (1965).
4. W. Ziegenbein and H. E. Sprenger, *Angew. Chem. Int. Ed. Engl.* **5,** 893 (1966).
5. H. E. Sprenger and W. Ziegenbein, *Angew. Chem. Int. Ed. Engl.* **5,** 894 (1966).
6. G. Manecke and J. Gauger, *Tetrahedron Lett.* p. 3509 (1967).
7. A. H. Schmidt, *Synthesis* (accepted for publication).
8. A. H. Schmidt, *Synthesis* (accepted for publication).
9. A. Treibs and K. Jacob, *Justus Liebigs Ann. Chem.* **699,** 153 (1966).
10. H. E. Sprenger and W. Ziegenbein, *Angew. Chem., Int. Ed. Engl.* **6,** 553 (1967).

11. H. E. Sprenger and W. Ziegenbein, *Angew. Chem., Int. Ed. Engl.* **7,** 530 (1968).
12. E. W. Neuse and B. R. Green, *J. Am. Chem. Soc.* **97,** 3987 (1975).
13. J. Gauger and G. Manecke, *Chem. Ber.* **103,** 2696 (1970).
14. E. W. Neuse and B. R. Green, *J. Org. Chem.* **39,** 3881 (1974).
15. A. Treibs and K. Jacob, *Justus Liebigs Ann. Chem.* **712,** 123 (1968).
16. A. Treibs, K. Jacob, and R. Tribollet, *Justus Liebigs Ann. Chem.* **741,** 101 (1970).
17. G. Manecke and J. Gauger, *Tetrahedron Lett.* p. 1339 (1968).
18. J. Gauger and G. Manecke, *Chem. Ber.* **103,** 2696 (1970).
19. J. Gauger and G. Manecke, *Chem. Ber.* **103,** 3553 (1970).
20. H. E. Sprenger, German Patent, D.O.S. 1,618,211 (1971).
21. S. Hünig and H. Pütter, *Angew. Chem., Int. Ed. Engl.* **11,** 431 (1972).
22. H. Ehrhardt, S. Hünig, and G. Pütter, *Chem. Ber.* **110,** 2506 (1977).
23. E. Neuse and B. Green, *Justus Liebigs Ann. Chem* p. 619 (1973).
24. E. J. Smutny, M. C. Caserio, and J. D. Roberts, *J. Am. Chem. Soc.* **82,** 1793 (1960).
25. W. Ried, W. Kunkel, and G. Isenbruck, *Chem. Ber.* **102,** 2688 (1969).
26. W. Ried, A. H. Schmidt, G. Isenbruck, and F. Bätz, *Chem. Ber.* **105,** 325 (1972).
27. W. Ried and G. Isenbruck, *Chem. Ber.* **105,** 337 (1972).
28. A. H. Schmidt, W. Ried, P. Pustoslemsek, and H. Dietschmann, *Angew. Chem., Int. Ed. Engl.* **11,** 142 (1972).
29. A. H. Schmidt, W. Ried, P. Pustoslemsek, and W. Schuckmann, *Angew. Chem., Int. Ed. Engl.* **14,** 823 (1975).
30. P. Pustoslemsek, Dissertation, Universität Frankfurt am Main, 1977.
31. A. H. Schmidt, W. Ried, and P. Pustoslemsek, *Angew. Chem., Int. Ed. Engl.* **15,** 704 (1976).
32. G. Seitz, R. Matusch, and R. Schmiedel, *Chem.-Ztg.* **101,** 557 (1977)
33. G. Seitz, K. Mann, and R. Schmiedel, *Chem.-Ztg.* **99,** 332 (1975).
34. L. A. Wendling, S. K. Koster, J. E. Murray, and R. West, *J. Org. Chem.* **42,** 1126 (1977).
35. D. G. Farnum, B. Webster, and A. D. Wolf, *Tetrahedron Lett.* p. 5003 (1968).
36. G. Seitz, H. Morck, K. Mann, and R. Schmiedel, *Chem.-Ztg.* **98,** 459 (1974).
37. G. Seitz, R. Schmiedel, and R. Sutrisno, *Synthesis* p. 845 (1977).
38. G. Seitz and R. Sutrisno, *Synthesis* p. 831 (1978).
39. G. Seitz, K. Mann, R. Schmiedel, and R. Matusch, *Chem.-Ztg.* **99,** 90 (1975).
40. R. Allmann, T. Debaerdemaeker, K. Mann, R. Matusch, R. Schmiedel, and G. Seitz, *Chem. Ber.* **109,** 2208, (1976).
41. G. Seitz, R. Schmiedel, and K. Mann, *Arch. Pharm. (Weinheim, Ger.)* **310,** 991 (1977).
42. G. Seitz, R. Schmiedel, and K. Mann, *Arch. Pharm. (Weinheim, Ger.)* **310,** 549 (1977).
43. H. Ehrhardt and S. Hünig, *Tetrahedron Lett.* p. 3515 (1976).
44. S. Hünig and H. Pütter, *Angew. Chem. Int. Ed. Engl.* **11,** 433 (1972).
45. S. Hünig and H. Pütter, *Angew. Chem., Int. Ed. Engl.* **12,** 149 (1973).
46. S. Hünig and H. Pütter, *Chem. Ber.* **110,** 2532 (1977).
47. H. J. Bestmann and G. Schmid, *Angew. Chem., Int. Ed. Engl.* **13,** 473 (1974).
48. G. H. Birum and G. N. Matthews, *J. Am. Chem. Soc.* **90,** 3842 (1968); C. N. Matthews and G. H. Birum, *Acc. Chem. Res.* **2,** 373 (1969); H. J. Bestmann, *Angew. Chem. Int. Ed. Engl.* **16,** 349 (1977).

Index

Aqueous chemistry of oxocarbons, 43–58
  complexes, 56–57
  pK_a values, 43–46
  structures in solution, 49–56
    croconic acid, 52–53
    deltic acid, 49–50
    rhodizonic acid, 53–56
    squaric acid, 50–52
  thermodynamics, 46–49

Croconate and croconic acid,
  absorption spectra, 10–12, 51, 95–99, 127–128
  anions and anion-radicals, 8–10
  aqueous chemistry, 52–53
  bond lengths, 4
  croconate violet, 62–65
  croconic acid blue, 68–72
  diamagnetic anisotropy and aromaticity, 6–8
  dicyanomethylidene derivatives, 59–77
  2-(dicyanomethylidene)croconate salts, 65–68
  electrical conductivities, 73–76
  Jahn–Teller effects, 133–139
  MCD spectra, 95–99
  normal vibrations, 131
  oxidation and reduction products, 8–10
  π-electron energy levels, 11
  pK_a values, 45–46
  Raman spectra, 130
  reactions of croconate with malononitrile, 62–73
  semioxocarbon derivative, 115
  synthesis and history, 1–3
  theoretical calculations, 10–12, 88–95, 128
  thermodynamic data, 48, 50, 51
  thiooxocarbon derivatives, 38–39

Deltate and deltic acid, 1–2, 5–6
  anions and anion-radicals, 10
  aqueous structure, 49–50
  electronic spectra, 11
  force constants, 5–6
  IR and Raman spectra, 5–6
  pK_a values, 45
  semioxocarbon analog, 115–116
  theoretical calculations, 11
  thermodynamic data (ionization), 48
  thiodeltic acid derivatives, 16, 36–38
Dithiosquaric acid diamides (DTSD), 22–38
  1,2-DTSD, syntheses and properties, 23–25
  1,3-DTSD, syntheses and properties, 25–28
  reactions with electrophiles, 33–36
  reactions with nucleophiles, 28–33

Electrical conductivities, 73–76
Excited states, 79–100
  electronic spectra, 95–99
  MCD spectra, 95–99
  perimeter model for annulenes, 80–88
    electronic states and spectra, 85–88
    molecular orbitals, 80–85
  perimeter model for oxocarbon dianions, 88–95
    electronic states and spectra, 91–95
    molecular orbitals, 88–91

Jahn–Teller effects, 121–140
  deformations of oxocarbon ions, 135–138

excitation profile of $C_6O_6^{2-}$, 138–139
Raman intensities, 133–134
theory, 123–127

Ketenes, 170–183
alkylmoniliformin precursors, 175–179
arylmoniliformin precursors, 179–180
moniliformin precursors, 171–175
squaric acid precursors, 170–171, 175, 180–183

Moniliformins, 101–119, 169–180
acidities, 111–112
alkylmoniliformins, 175–179
amides and ammonium salts, 113
arylmoniliformins, 179–180
biological and toxicological properties, 108
^{13}C NMR, 106
crystals, 102–103
derivatives, 112
emission spectra, excited states, 105–108
IR spectra, 104
mass spectra, 105
oxidation of, 112
semioxocarbon series, 114–115
structure and physical properties, 105
syntheses, 109–111, 169–180
thio analogs, 113
UV spectra, 106–107

Phenylogs, naturally occurring, 116–117

Raman spectra, 121–140
Jahn–Teller effects, 133–139
measured spectra, 128–133
quantum mechanical theory, 123–127
Rhodizonate and rhodizonic acid,
absorption spectra, 10–12, 51, 95–98, 127–128
acid ionization, thermodynamics, 48, 50
anions and anion-radicals, 8–10
aqueous solution structure, 53–56
^{13}C NMR, 54
conductivity of dipotassium salts, 74
excitation profile, 133
history, 1–3
Jahn–Teller effects, 134–140
malononitrile, reaction with, 72–73
MCD spectrum, 95–98
normal vibrations, 132
oxidation and reduction, 8–9
π-electron energy levels, 11
pK_a values, 45
Raman spectrum, 130
semioxocarbon analog, 115
synthesis, 1–3
theoretical calculations, 10–12, 88–95, 128

Squaraines, 185–231
acidity, 197
additions, 211–214
alkylation and rearrangement, 210–211
hydrolysis, 197, 210
IR spectroscopy, 190
melting points, 190
metal chelates, 210
molecular rearrangements, 195–196, 209
pyrolic, 185–186, 192–193, 198, 211–213, 217–218
reactions of, 197, 210–229
squaric acid bisamidines, 223–230
ring closures, 224–228
reaction with p-tosyl isocyanate, 223–224
structure and nomenclature, 186–190
substitution reactions, 214–223
with acid chlorides, 214–217
exchange of pyrryl substituents, 217
transamidation, 218–222
thiosquaraines, 218, 222–224
syntheses, 191–197
from squaric acids, 191–192, 198–203
from dialkylsquarates, 192, 203–204
from substituted hydroxycyclobutenediones, 192–194, 204–207
from substituted hydroxycyclobutenediones, 192–194, 204–207
from mercapto- and hydroselenocyclobutenediones, 194–195, 207–209
table of squaraines, 198–209
tetranitrogen derivatives of squaric acid, 224–228
transformations, one squaraine to another, 196–197
UV spectra, 190

Squarates and squaric acid, 1-4
absorption spectra, 10-12, 51, 95-99, 127-128
anions and anion-radicals, 8-10
aqueous complexes, 56-57
aqueous structure, 50-52
birefringence, 152-154
bond lengths, 144
crystal structure, 142-147
dicyanomethylidene derivatives, 60-61
conductivities, 74
dielectric properties, 154-160
elastic properties, 159-160
hydrogen bonds, 142-147
Jahn-Teller effects, 133-138
MCD spectra, 95-99
NMR (^{13}C and ^{1}H), 149-152
normal vibrations, 131
oxidation and reduction products, 8-10
oxothioxocarbon dianions, 19-22
syntheses, 19-22
ester reactions with amines, 21-22
phase transitions, 141-142, 160-165
physical properties, 141
pK_a values, 45-46, 141
Raman spectra, 129, 147-149
squaraine precursor, 191-192, 198-203
syntheses, 3-4, 169-171, 180-183
tetranitrogen derivatives, 12-13, 224-228
theoretical calculations, 10-12, 88-95
on phase transitions, 160-165
thermodyanamic data, 48, 50
thiooxocarbon derivatives, 16-19
hydrazine reactions, 227-228
spectroscopic data, 18-19
syntheses, 16-18, 222-223
vibrational spectra, 121, 147-149

Tetraalkoxyethylenes, 172-173, 176-181
Thiooxocarbons, 15-41
$C_4S_4^{2-}$
reactions, 16-18
spectroscopy, 18
synthesis, 16-18, 223
mixed C_4 oxo-thiooxocarbons, 19-22
mixed C_5 oxo-thiooxocarbons, 38-39
thiodeltic acid derivatives, 36-37